Contaminated Forests

NATO Science Series

A Series presenting the results of activities sponsored by the NATO Science Committee. The Series is published by IOS Press and Kluwer Academic Publishers, in conjunction with the NATO Scientific Affairs Division.

A. **Life Sciences**	IOS Press
B. **Physics**	Kluwer Academic Publishers
C. **Mathematical and Physical Sciences**	Kluwer Academic Publishers
D. **Behavioural and Social Sciences**	Kluwer Academic Publishers
E. **Applied Sciences**	Kluwer Academic Publishers
F. **Computer and Systems Sciences**	IOS Press
1. **Disarmament Technologies**	Kluwer Academic Publishers
2. **Environmental Security**	Kluwer Academic Publishers
3. **High Technology**	Kluwer Academic Publishers
4. **Science and Technology Policy**	IOS Press
5. **Computer Networking**	IOS Press

NATO-PCO-DATA BASE

The NATO Science Series continues the series of books published formerly in the NATO ASI Series. An electronic index to the NATO ASI Series provides full bibliographical references (with keywords and/or abstracts) to more than 50000 contributions from international scientists published in all sections of the NATO ASI Series.
Access to the NATO-PCO-DATA BASE is possible via CD-ROM "NATO-PCO-DATA BASE" with user-friendly retrieval software in English, French and German (© WTV GmbH and DATAWARE Technologies Inc. 1989).

The CD-ROM of the NATO ASI Series can be ordered from: PCO, Overijse, Belgium.

Contaminated Forests

Recent Developments in Risk Identification and Future Perspectives

edited by

Igor Linkov
Harvard University and
Menzie-Cura and Associates, Inc., U.S.A.

and

William R. Schell
University of Pittsburgh, U.S.A.

Kluwer Academic Publishers

Dordrecht / Boston / London

Published in cooperation with NATO Scientific Affairs Division

Proceedings of the NATO Advanced Research Workshop on
Contaminated Forests
Kiev, Ukraine
27–31 May 1998

A C.I.P. Catalogue record for this book is available from the Library of Congress.

ISBN 0-7923-5738-8 (HB)
ISBN 0-7923-5739-6 (PB)

Published by Kluwer Academic Publishers,
P.O. Box 17, 3300 AA Dordrecht, The Netherlands.

Sold and distributed in North, Central and South America
by Kluwer Academic Publishers,
101 Philip Drive, Norwell, MA 02061, U.S.A.

In all other countries, sold and distributed
by Kluwer Academic Publishers,
P.O. Box 322, 3300 AH Dordrecht, The Netherlands.

Printed on acid-free paper

Printed in the Netherlands

TABLE OF CONTENTS

viii

PREFACE

Concentrations of pollutants in the atmosphere have increased dramatically over the last century and many of these changes are attributable to anthropogenic activities. Influence of acid rain on forest development has been studied but that of other pollutants such as toxic chemicals, heavy metals and radionuclides has not been explored extensively. Natural ecosystems, especially forests, tend to accumulate many of these pollutants which subsequently can affect ecosystem health. These contaminants may be very damaging to the environments in Eastern Europe where the rapid disappearance of forests results not only from contamination, but also from poor forest management practices.

The NATO ARW in Kiev was a further important step in the development of forest radioecology. It has summarized what has been done since the previous topical meeting held in Stockholm in 1992 when the field was just developing and assimilating information collected after the Chernobyl NPP accident. A shift from empirical data collection to modeling and understanding of the fate and transport processes of radionuclides was clearly demonstrated in the workshop presentations and discussions. Although 22 countries were represented the number of participants was relatively small and allowed fruitful discussions in the working groups and between participants.

The workshop agenda as a whole was designed to reduce uncertainty in our current knowledge on forest radioecology. This was done through group consensus and individual expert judgment elicitation. Three working groups were provided during the workshop namely: a) Modeling, b) Measurements and Data, c) Countermeasures and Risk Assessment. The objectives of these groups were to review achievements and gaps in the current knowledge as well as to establish priorities for the future research. The groups were run by two Group Co-Chairs and Rapporteurs. The Individual Expert Elicitation was conducted according to the protocol developed prior to the workshop for the group of 8 experts nominated by participants. The Individual Expert Elicitation was directed towards the gaps in modeling.

According to many participants, the workshop was unusual because it provided space for informal discussions rather than just regular plenary presentations. The meeting gave the CIS participants new insights and contacts and many formal and informal collaborations were established.

Igor Linkov and William R. Schell
January, 1999.

ACKNOWLEDGEMENTS

The editors would like to thank Dr. Yu. Kutlahmetov and his colleagues of the Kiev Institute of Radioecology for their help in the workshop organization and hospitality. We also thank workshop participants for their contribution to the book and peer review of manuscripts. Special thanks to Jason Givens-Doyle and Alla Burmistrov for editorial and technical assistance. The workshop agenda was prepared in collaboration with the International Union for Radioecology. Financial support for the workshop organization was provided in the main parts by NATO.

PRIORITIES IN FOREST RADIOECOLOGY: MODEL DEVELOPMENTS FOR INTEGRATED ASSESSMENT OF TOXIC ELEMENT TRANSPORT

W.R. SCHELL
Graduate School of Public Health, University of Pittsburgh, PA, USA

I. LINKOV
Department of Physics, Harvard University, Cambridge, MA, USA

Abstract

An integrated conceptual model describing the fate and transport of toxicant metals and radionuclides from terrestrial to aquatic ecosystems has been initiated. The model should provide information on the long time transport and predict toxicant distribution from atmospheric sources to geochemical sinks. Bio-geo-chemical process control the concentration of toxicants in temporary and permanent reservoirs. It is the concentration process which can cause excess chemical and radionuclide dose to the environment and ultimately to man. Forest models are being developed to describe the radionuclide distribution in plants and animals residing in the terrestrial environment. We believe that the next step in descriptive modeling is the upper soil layers of organic matter which provide an intermediate reservoir for radionuclides, but the hydrological effects of overland flow and infiltration have not yet been included nor demonstrated. The approach proposed here is a linked three sector model consisting of: 1) Terrestrial ecosystem-catchment basin, 2) Interface ecosystem including soil-water, the litter layer, peatlands, wetlands, and 3) Aquatic ecosystem including rivers, reservoirs, estuaries. The goals of the present study are to develop, test and validate integrated models for use in prediction and in establishing policies to limit excess chemical and radionuclide dose to the exposed population. Multimedia modeling using Geographical Information System and the latest computer software can provide a technical foundation and lead to predictions in assessing global industrial effluents, their effects on the ecosystem and on man.

1. Introduction

The environmental transport of toxic elements deposited on catchment basins leading to lakes and estuarine zones is time and moisture dependent. Two process must be considered in modeling, namely, transport by overland flow through the upper soil layers and infiltration where moisture provides the driving force for metals transport. The net transport of each chemical element depends upon the resistance of moisture transport and the resistance caused by element soil-matrix bonding.

1

I. Linkov and W.R. Schell (eds.), Contaminated Forests, 1–14.
© 1999 *Kluwer Academic Publishers. Printed in the Netherlands.*

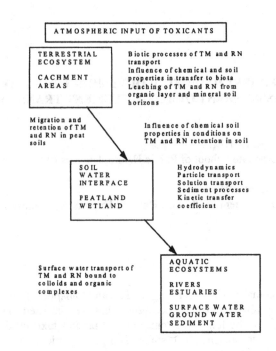

ATMOSPHERIC INPUT OF TOXICANTS

TERRESTRIAL
ECOSYSTEM

CACHMENT
AREAS

Biotic processes of TM and RN
transport
Influence of chemical and soil
properties in transfer to biota
Leaching of TM and RN from
organic layer and mineral soil
horizons

Migration and
retention of TM
and RN in peat
soils

Influence of chemical soil
properties in conditions on
TM and RN retention in soil

SOIL
WATER
INTERFACE

PEATLAND
WETLAND

Hydrodynamics
Particle transport
Solution transport
Sediment processes
Kinetic transfer
coefficient

Surface water transport of
TM and RN bound to
colloids and organic
complexes

AQUATIC
ECOSYSTEMS

RIVERS
ESTUARIES

SURFACE WATER
GROUND WATER
SEDIMENT

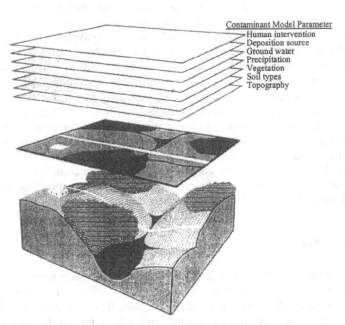

Contaminant Model Parameter
Human intervention
Deposition source
Ground water
Precipitation
Vegetation
Soil types
Topography

Fig. 1. Integrated model development showing a) conceptual Terrestrial, Interface and Aquatic ecosystems and b) Geographical Information System contaminant modeling for fate and transport in a catchment basin

Thus, to describe the transport of toxic compounds to the ocean we must first consider three dimensional flow of material deposited on land from atmospheric sources. At any location the driving forces are due to local hydrological conditions which can mobilize terrestrial material. The time required for such mobilization is important to the health of biological organisms in repositories of rivers, lakes, estuaries and oceans. To describe and quantitate such processes, we must revert to multi-component models which can utilize the ever expanding data base for the testing and validation process. An integrated system model is required to focus on planning and on predicting possible accident and health effects from the consumption of contaminated water and aquatic organisms. Such a model must encompass terrestrial, interface and aquatic components and the material flow between them. Conceptually, the multi-media model can be described as shown in Fig. 1.

2. Materials and Methods

2.1. HYDROLOGICAL PROCESS IN OVERLAND FLOW AND INFILTRATION

Overland flow of precipitation can be defined in terms of cell dimensions, force vectors, flow-line concentrations and inflow combinations [1]. If each cell, for example, is constructed having the dimensions shown in Figure 2 then the resultant force vector for overland flow of precipitation can be established. Each cell is connected to its nearest neighbor, i.e., illustrating the water flow from the previous cell to the next cell, etc., see Figure 2.

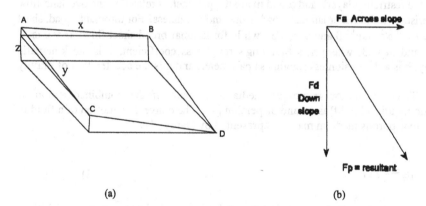

(a)   (b)

Figure 2. Overland flow separated into down slope gradient cells (a) and the resulting vector analysis (b) of the flow direction.

The resultant set of flow rules for the cells, based on, for example elevation, provide the physical meaning of overland flow of precipitation in a catchment basin and finally into a stream. The cells, having once defined the explicit rules, can be treated by partial differential equations and/or by Cellular Automata methods, [2]. What must be modeled is the flow of precipitation on a slope which proceeds down gradient (from A to D in Figure 2) to ponds or to a stream, which also is influenced by vegetation and roots, and where the radionuclides or trace metals are carried into the river, reservoir or sea. To cause overland flow, the rainfall intensity must exceed the infiltration rate through the organic layer and soil. In addition, saturation in overland flow results when the storage capacity of the soil is completely full so that all subsequent additions of water at the surface, irrespective of the rate of application, is forced onto the surface for ponding or for run-off when there is an elevation difference. In forests, the litter layer provides a large porous or capilliary zone where precipitation can accumulate and on a slope, can be transported by lateral flow in this organic rich zone. If cementation or crusting of the surface soil is present, no infiltration occurs and the flow hydrograph of the catchment basin would reflect this condition by more rapid peaking after a precipitation event. Such cementation is present in certain regions such as the Chernobyl Exclusion Zone.

The kinematic forms of the momentum equations for subsurface and overland flow respectively, can be expressed as:

$$Q = \frac{KA}{\gamma} \sin \beta \text{ for } 0 \leq A < \omega \gamma D \tag{1}$$

$$Q = \alpha A^m \tag{2}$$

where Q is the discharge (m³/sec), A is the flow cross-sectional area (m²), K is the effective hydraulic conductivity of the soil, w is the width of the element orthogonal to the streamlines, γ is the effective porosity (total porosity - field capacity soil-water content), β is the local slope (in degrees), D is the thickness of the hydrologically active soil profile (above the restricting layer), and α and m are the parameters related to the overland flow characteristics (turbulent or laminar flow), slope and roughness. For turbulent broad, sheet flow: $\alpha = n^{-1} \omega^{-2/3} \tan^{1/2} \beta$ and m = 5/3 ; while for laminar broad sheet flow: $\alpha = g \omega^{-2/3} \tan\beta/kv$ and m = 3; where n is Manning's roughness coefficient, v is the kinematic viscocity, k is a dimensionless roughness parameter, and g is the acceleration of gravity [3].

Tracer migration in porous media is a result of the combined action of convection, molecular diffusion and dispersion [4]. The convective motion of a fluid in a continuous porous medium may be represented by Darcy's law:

$$v_a = -Ki \tag{3}$$

where v_a is the apparent velocity, or filtration velocity v_f, K is the tensor of permeabilities and the hydraulic head. The mass flow of tracer, ϕ_c, which traverse a surface of representative elementary volume V is:

$$\Phi_c = - \int_v (vC)dt \tag{4}$$

where C is the concentration of the tracer which moves at velocity v [4]. For molecular diffusion, the particles of a tracer involves dilution which occurs identically in the case of both running and still water and may be described by Ficks law which says that substance transport through molecular diffusion is proportional to the concentration gradient, the two dimensional mass balance equation is:

$$\frac{dC}{dt}+v_x\frac{dC}{dx}=D_o\,d^2\frac{C}{dx^2}+D_o\,d^2\frac{C}{dy^2} \qquad (5)$$

where C is the concentration in the point; x, y and t are position and time coordinates; the component of velocity in the x direction and D_o is the coefficient of molecular diffusion or molecular diffusivity and depends on the nature of the substance but, in free water is of the order of 10^{-5} cm/sec [4].

2.2. METALS-SOIL-WATER INTERACTION

The partitioning of a radionuclide or metal between liquid and solid phase is defined by the distribution coefficient, K_d as:

$$K_d=\frac{S}{C} \qquad (6)$$

The retardation factor, which accounts for the slower movement, due to matrix sorption, of the radionuclide or element than the water (precipitation) which flows unimpeded, is:

$$R=1+\frac{\rho}{\theta}K_d \qquad (7)$$

The one dimensional solute transport equation can then be written by:

$$\frac{dC}{dt}=\frac{D}{R}d^2\frac{C}{dx^2}-\frac{U}{R}\frac{dC}{dx}-\lambda C \qquad (8)$$

where
λ = decay constant (sec $^{-1}$)
C = concentration in solution (pCi/ml, or μg/ml)
S = concentration sorbed on solid matrix (pCi/g, or μg/g)
ρ = density in gm/cm^3 (dry) of the bulk soil
θ = moisture content in ml water per cm^3 (wet) of the bulk soil
U = advection velocity (cm/sec)
D = diffusivity (cm^2/sec)

Therefore, the rate of change with time of the concentration of metals in solution is equal to the rate of change due to dispersion minus the rates of change due to advection, geochemical reactions (such as adsorption desorption and precipitation) and radioactive decay. The total adsorption (S) is the amount of a radionuclide sorbed or precipitated per unit mass of the porous medium. The partitioning of a radionuclide between the liquid and solid phase is experimentally determined and is defined as the distribution coefficient, K_d. This equilibrium sorption isotherm is a mathematical relationship which describes the radionuclide distribution between the liquid and solid phase. The effect of the K_d value on resistance for radionuclide or element migration is shown by Equation 6. It is apparent that as K_d increases, the radionuclide migration rate decreases. Thus, the migration of the radionuclides is retarded with respect to the migration of the water. The effect of K_d is taken into account by the retardation factor, R Equation 7. Equation 8 shows that the effective diffusivity and effective advection velocity are reduced by the factor I/R.

3. Results and Discussion

3.1. MEASUREMENTS OF TRACE METALS AND RADIONUCLIDES IN CONTAMINATED SOIL PROFILES.

In a sandy soil near the ASARCO copper smelter, Puget Sound, WN, the surface moisture is 24% and diminishes to 11% at about 12 cm, Table 1. Horizontal migration of elements would occur mostly in the upper 8 cm in the sandy soil. Below this level migration would be by vertical transport and would be slow because of the low water amount which would cause dissolution of elements. This dissolution or resistance to transport has an additional friction term represented by the distribution coefficient, K_d.

The Tacoma Smelter began operation as a Pb smelter in the 1890's and changed to Cu smelter in 1914. This information provides a starting point in time for the input of heavy metals impacting the large estuarine zone of Puget Sound. The high sand concentration in soils of the region provides a matrix which has a much lower resistance for transport of metals with precipitation over the past 100 years. To obtain data on the time dependent migration of the various elements and radionuclides we need a fixed time of pollutant input to develop and test the model under development. The above data provides information on four elements measured in soil by Crecellius et al [5].

TABLE 1. Data on Soil Core-1 near ASARCO, Smelter, Tacoma, Washington [5]

Depth (cm)	Moisture %	pH	TOC %	Sand %	Silt %	Clay %	As ug/g	Cd ug/g	Pb ug/g	Cu ug/g
2.5	24	4.1	6.46	84	7	2	418	9.3	924	1146
8	18	4.0	3.22	82	7	1	225	1.3	34	83
13	13	4.5	1.4	87	4	1	137	0.95	48	82
28	11	4.8	0.83	91	4	1	9.4	0.1	8.4	18
64	11	5.1	0.57	91	4	1	3.4	0.05	4.6	13
Equation $Y=kx^b$							power	power	power	Power
Curve fit R							0.92	099	0.90	0.99

In sand dominated soil, the resultant of overland flow and infiltration can be expressed as a power function of the transport in the vertical dimension as :

$$y = k \, x^b \tag{9}$$

where y is the depth of the soil core slice, x is the concentration in μg/g dry weight, k is the coefficient for the element, and b is a coefficient for migration which depends upon the type of soil. For the sandy soil measured here, the coefficients b for each element are: As-1.64, Cd-1.70, Pb-1.60, Cu-1.40. These elements seem to be migrating vertically at about the same rate as indicated by the average value for b (SD of 1.58 [13]). The Pb power curve is not as good as for the other elements but still within an acceptable error. Since this is the vertical migration, we must determine the overland flow and dissolution of the elements near the surface.

The precipitation of the area is about 100 cm/yr, often distributed equally. Since the infiltration is limited as indicated by the low porosity at the 28 cm layer-11%, most of the precipitation at the sample site would be through overland flow in the upper 12 cm where the organic matter is found. If vegetation was high and roots were present, such as in forests, more unconsolidated soil would provide overland flow which could extend to lower depths. To estimate the resistance of element migration by overland flow, the K_d values measured in organic rich ecosystems such as in peat bogs might be used to approximate the K_d values, as a first estimate.

The velocity for the frontal migration in the soil column of the four elements is about to the 27 cm level. This would correspond to about 70 years after the beginning of copper refining or a vertical migration velocity of 0.386 cm/yr.

Soil samples from deciduous and coniferous forests were taken 800 m south of the Shippingport Atomic Power Station, SAPS, and BeaverValley Nuclear Power station, BVNP, in the winter of 1987. The goals of the project were to determine if measureable amounts of radioactive effluent had been released since the plants began operation [6]. Soil core sections were taken inside decidious and conifer stands to a depth of 30 cm comprising the litter and fermentation layer, the A and the B layers which were recognizable by color. Significant compaction occurred in driving the core through the thick litter layer which was frozen at sampling time of year and the original profile depths were approximated. The sample location was at a table top topography with a regional precipitation of abut 92 cm/yr.

The results of the measurements of soil profiles for ^{137}Cs and ^{210}Pb shows that the atmospheric deposition from nuclear weapons test fallout since the maximum in 1962 on these forest is typical for this latitude and precipitation amount; for deciduous forests 12.1(3.5) and for Conifer 8.4 (2.9) pCi/cm^2. No increase of ^{137}Cs has occurred at the sampling site from the nearby nuclear reactors. The ^{137}Cs is likely fixed to the clay soil and will not be transported vertically. Overland flow is not important for ^{137}Cs and ^{210}Pb transport at the forest sampling sites measured in Western Pennsylvania by Breslin, [6]. The porosity of the forest soil provides for vertical transport of the moisture and solubility properties, K_d, of the elements would dominate in their transport.

TABLE 2. Profiles of ^{137}Cs and ^{210}Pb in sediment cores collected near nuclear reactors in Western Pennsylvania [6].

Depth	Moisture %	pH	TOC	Clay %	^{137}Cs pCi/cm^2(SD}	^{210}Pb pCi/cm^2(SD)
Deciduous						
0-15	28.4	4.0	NM		7.9(2.8)	5.9(2.3
15-25	27.4			24	3.0(1.7)	1.6(1.3)
25-30	23.3			24	1.1(1.1)	1.0(1.0)
Coniferous						
0-15	25.3	4.0	NM		5.0(2.3	15.6(4.0)
15-25	24.1			24	1.6(1.3)	1.0(1.0)
25-30	24.3			24	1.6(1.3	0.7(0.8)

In soil profiles taken from high rainfall regions of Japan, namely the national forests of Oguroi and Mt Kanmuri, the fallout Pu and Am decreased exponentially with depth [7]. As much as 60-70% of the activity is associated with organic materials. In sequential leaching studies, the 0.1N sodium citrate was very effective in extracting the Pu in organic materials present. The plutonium must be associated with the organic humic and fulvic acids in the upper litter and soil layers. Americium is shown to be more mobile than plutonium. The integrated fallout deposition of plutonium of 0.7mCi/km^2, is much lower than the value of 1.2 mCi/km^2 measured by Miyake et al, [8] at this lattitude. The authors attribute the lower value to run off by rain or to resuspension.

Microbial activity effects sorption processes in upper loamy sand soil layers. Iodide is more strongly sorbed in non sterile soil than in sterile soil, i.e., the K_d value continues to increase with time over at least 30 hours, while in sterile soil, the K_d value is low and almost constant with time [9]. This indicates that the iodate is reduced by microbial activity. "It appears that the large portion of the dissolved organically bound iodine becomes bound to solid organic soil material by biochemical reactions"[9]. Brauer and Strebin, [10] measured ^{129}I in water and soil samples near a nuclear fuel reprocessing site in New York and near the Hanford reactor complex and found that forest communities are efficient in accumulating ^{129}I from the atmosphere. The exposed plant surfaces to the atmosphere and the litter layer are much higher than in the soil below.

At a mountain top bog in Western Pennsylvania the regional deposition of trace metals and radionuclides is recorded in the sediments where the precipitation is 135 cm/yr. Clear cutting in 1894 caused a rise in the water table which produced a swamp and bog environment in the small undrained basin. Later as reforestation took place, evapotranspiration decreased the water table. The porosity profile is from 79% at the top to 18% at 32 cm depth and the carbon content decreased from 42% to 1.3 %. This is a case of no overland flow, and a repository for atmospheric chemical deposition [11]. The alkali metals Na, K, Rb, Cs concentration profile correlate with the ash weigh. The ^{137}Cs profile reaches a maximum in 1963 (correlates with the maximum weapons fallout} and is found at the 1935 level indicating frontal migration rate of 0.26 cm/yr

3.2. STORAGE RESERVOIRS

A general model for uptake has been developed to approximate the process occurring in forests - FORESTPATH [12]. This model provides an initial description of a contaminant ^{137}Cs and utilizes software for describing partial differential equations relevant to source distribution in compartments and loss from temporary storage, Figure 3. However, the mathematical formalism does not take into consideration chemical speciation of elements interacting with the plant, soil-root transfer processes, presence of ectomycorrhizas, effect of redox potential of elements in the soil horizons, effects and rates of element migration, infiltration to deeper layers and transport by overland flow of elements to the terrestrial - aquatic interface. Even with these limitations, the FORESTPATH model provides a workable unit. Additions to the forest model include overland flow and infiltration components.

Figure 4 shows the several zones required to model the transport of water and toxic constituents in the terrestrial soil ecosystem. Once deposited on land, the toxic elements can be taken up by plants and animals and dispersed by natural hydrological processes.

By assuming that in the interface zone, the important reservoirs are the upper organic rich zone which has a much shorter residence time for many elements than the underlying sediments. The lower sediment horizons provide different residence times for kinetic processes leading to the aquifer. The major short term reservoir for many elements occurs in the litter/organic humus layers which are affected by overland flow. The desorption K_d is important as well as organic material decomposition into particles and colloids which adsorb toxic elements and are subsequently transported over land into rivers, lakes and estuaries. In addition, solution of deposited toxic elements by precipitation takes place and transport to the estuarine zones occurs. After entering into the streams, much shorter residence times of elements are encountered and seasonal processes dominate. It is in the aquatic zone that major uptake of these toxic compounds by biota occur. The chemical and radionuclide dose to man through the food chain is encountered.

3.3. COEFFICIENTS REQUIRED IN MODEL BUILDING FOR THE INTERFACE ZONE

The development of this segment of the model explores the transition zone between the terrestrial, i.e., forest, and the river or reservoir, identified here as the Interface Ecosystem. This transition zone, in concept and definition, could be a wetland marsh area which is transversed by a stream, river, reservoir, surrounding a lake or estuary, and could include rice paddy fields. It also would include the river/reservoir bank habitat (riparian) where aquatic mammals often reside. The watershed is defined as the land surrounding an area from the highest elevation which drains into the major stream, river or lake. A small catchment basin is the land area where precipitation falls and drains into the small streams or rivers which, in turn, drain into major rivers or lakes. Surface run-off is

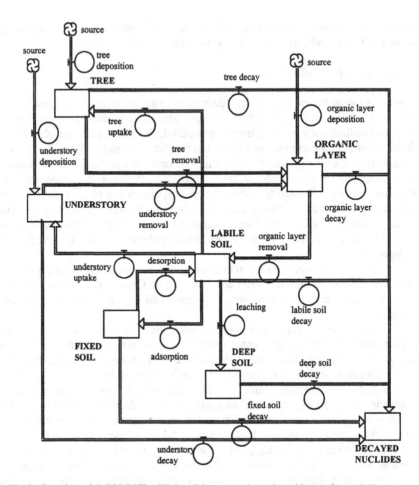

Fig. 3. Generic model, FORESTPATH describing contaminant deposition on forests [12].

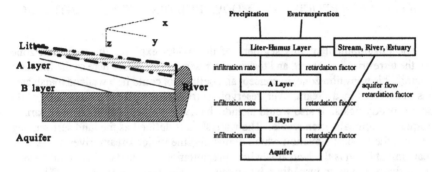

Fig. 4. Reservoirs where the flux of water and toxic elements reside

influenced by landscape conditions of the watershed and is best described and modeled using geographical information system (GIS)-(ARC-INFO) software. Adding another layer to these layer-defining systems can be accomplished for contaminant model parameters by new methods where such analytical information can be linked to GIS. Overland transport and infiltration of deposited pollutants can be modeled where both concentration and dispersion process occur, depending on the time scale. As shown in Figure 1b, concentration of radionuclides and other contaminants occurs in preferred zones based on the ecosystem topography. Such concentration of contaminants can help in reducing clean-up costs for remediation.

The objective of the BOGPATH - Interface - segment of the model development is to provide a conceptual generic model which evaluates the transport and fate of toxic elements in the Interface Ecosystem and provides information on the chemical and radiation dose to man and animals which nest and reproduce in this near-water habitat. Thus, the hydrological features of overland flow of water, flooding of marsh and wetlands by rivers, seasonal effects of ice dams and the effects of changes in river flow must be considered. The specific approaches are to:

a) Identify a database on chemical and radionuclide retentive parameters of bogs and riparian ecosystems.

b) Develop a kinetic model for generic processes occurring in such organic environments encountered in element transport through the interface zone, from forest to river.

c) Use literature values for parameters to test and determine parameter sensitivity for the generic BOGPATH model.

Information on waste disposal in humid zones has shown that the organic rich soil, i.e., peat, has strong adsorption sites for several radionuclides and toxic chemicals and can be used to purify water [13]. These natural bog systems limit the leaching of elements and yet allow water to infiltrate. In fact, studies have shown that such natural marshlands retain certain radionuclides for times of 10^2 - 10^4 years [14, 15]. Recent information has indicated that certain types of peat bogs may satisfy the requirements for isolating radionuclides from low level radioactive wastes.

The distribution coefficient, K_d, values depend upon the organic and mineral content of the soil. In modeling the transport of elements the desorption distribution coefficient becomes the dominant resistance to movement. This desorption K_d is often 5-10 times greater than the adsorption value. Organic rich bogs have a much lower adsorption value for ^{137}Cs than minerotrophic bogs indicating a much higher mobility in these natural organic systems. Because of the clay content of the Spruce flats bog the mobility is much less and the weapons test fallout is distributed near the date of maximum fallout-1963. In organic bogs such as Lefgren, the ^{137}Cs is mobilized and transported up and down the profile by moisture changes over the Seasons and the maximum is found at the top layers. For strontium and cobalt, complexing would occur with the mobility being similiar in mineral and organic layers. The plutoniuim and americium desorption K_d values measured in natural freshwater and salt a water sediment systems is has a range of values of (11 - 47 x 10^4). Measurements of several soil matrices in fresh water showed that the sorption and desorption values are not reversible in sediments, are independent of salinity and have a range of values of 10^5-10^6 ml/g [16, 17, 18]. Bunzl et al [19], measured the ratio of desorption to sorption K_d values of 1-5 times for several elements (Cs, Sr, Ce, Ru, I, Tc) and found that equilibrium is not reached for some elements during 200 days in the upper 0-31cm-Ap Horizon of Parabraunerde soil. The soil type used is dark gray brown, moderate humus in fine sandy loam.

TABLE 3. Distribution coefficient measurements for ^{137}Cs, ^{85}Sr, ^{57}Co, ^{106}Ru and ^{241}Am in bog sediment cores.

Depth (cm)	Moisture %	Carbon %	^{137}Cs K_d 10^4	^{85}Sr K_d 10^3	^{57}Co K_d 10^3	^{106}Ru K_d 10^4	^{241}Am K_d 10^5
	Spruce fl						
7	69	21.40	4.2	2.1	1.1	3.9	1.5
21	35	3.28	1.3	3.6	1.3	5.4	1.2
	Lefgren						
7	89	50.2	0.13	4.53	3.3	8.6	4.5
21	89	52.8	0.14	2.08	4.8	8.7	4.5

3.4. COEFFICIENTS REQUIRED IN MODEL BUILDING FOR THE AQUATIC ZONE

In humid zones, leaching of deposited toxic materials occurs through infiltration by rainfall, surface drainage - overland flow- and infiltration. Once deposited on land and forests, the major problem is leaching of precipitation by run off and infiltration into the ground water. A water balance of precipitation in any area is necessary. In Pennsylvania, for example, the annual water balance is illustrated by: Precipitation-41 inches, evapotranspiration 17-23 inches, infiltration-12-15 inches, run off-6-9 inches. The streams and rivers transport deposited material by precipitation from run off and base flow to temporary storage reservoirs, lakes and finally into the estuarine zone at the sea. During run off and base flow, soil is dissolved and/or particles are transported by the rivers into the estuary. Thus, to model the bio-geo-chemical processes effecting toxic elements, transient and permanent reservoirs must be defined. Figure 5 provides the source input of the TM and RN. In the AQUAPATH model, storage reservoirs are: 1) river, estuary, 2) surface water, 3) deep water, 4) sediment, 5) upper sediment layer, and 6) deep sediment layer. The aquatic biota which reside in these compartments, incorporate toxic elements and provide the hazard potential defined by risk management for man and other mammals through food and water ingestion.

4. Conclusions

A generic model has been formulated which relates hydrological process of overland flow of a contaminated forest ecosystem through an interface ecosystem and into an aquatic ecosystem. These segments provide the framework for process modeling of the total affected ecosystem into the three components. Hydrologic process of infiltration and especially overland flow of precipitation in a catchment basin causes the mobilization of deposited contaminants through the litter and upper organic soil layers and into streams,

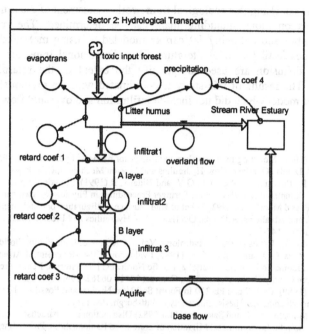

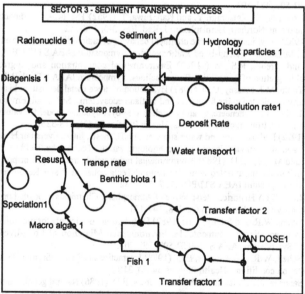

Figure 5. Generic models for describing deposition and hydrological transport of contaminants in (a) soil to river compartments, BOGPATH, and (b) river to reservoir and estuary, AQUAPATH.

14

rivers and estuaries. Using this ecological framework, assessment of the time and effects of contaminants on biota, including man, can be determined. The retardation of contaminant metals and radionuclides can be modeled by using measured distribution coefficient values available inthe literature. Two dimensional transport equations of advection and diffusion are used to quantify the extent of the retardation of the contaminants. The results of this study is coordinated with the previously developed FORESTPATH model which did not include infiltration and overland flow processes.

5. References

1. Scoging, H. (1992) Modeling overland flow hydrology for dynamic hydraulics. In: Parsons, A.J. and Abrahams, A.D. eds., Overland Flow-Hydraulics and Erosion Mechanics, UCL Press. London, 89-103.
2. Manneville, P., Boccara, N., Vichniak, G.Y. and Bidaux, R.(1991). in. (eds.) Cellular Automata and Modeling of Complex Physical Systems, Springer Proceedings in Physics 46, Springer Verlag, Berlin.
3. Grayson, R.B. and Moore, I.D. 1992, Effect of land –surface configuration on catchement hydrology, In: Parsons, A.J. and Abrahams, A.D. eds. Overland Flow-Hydraulics and Erosion Mechanics, UCL Press Limited, London, ,147-175.
4. Gaspar, E., Modern Trends in Tracer Hydrology, Vol. 1, CRC Press, Boca Raton, Florida , 1987, 145 p.
5. Crecelius,E., Apts, C.O., and Cotter, O.A. (1985) Evaluation of the Movement of Arsenic, Cadmium and Lead in Tacoma Soil Profiles. Battelle, Pacific Northwest Laboratories, Report prepared for the Tacoma-Pierce County Health Department, Contract 23112 06362, 19p.
6. Breslin, J.P. (1989) Cs-137 and Pb-210 In Forest Soil Near Nuclear and Fossil Fuel Power Plants in Western Pennsylvania, MS Thesis, University of Pittsburgh, PA, 116p.
7. Yamamoto, M., Komura, K. and Sakanoue, M. (1982) Distribution and charactistics of plutoniumn and americium in soil. In: Environmental Migration of Long-lived Radionuclides. Proceedings of Symposium IAEA STI/PUB/597, 481-494.
8. Miyake, Y., Saruhashi, K., Katsuragi, Y. and Kanazawa, T. (1962) The peak in radioactive fallout in the temperate zone on the Northern Hemisphere. J. Rad. Res. 3-3, 148-152.
9. Behrens, H. (1982) New insights into the chemical behavior of radioiodine in aquatic envirnonments. In: Environmental Migration of Long-lived Radionuclides. Symposium IAEA STI/PUB/597, 27-40.
10. Brauer, F.P. and Strebin, R.S. Jr. (1982) Environmental concentration and migration of ^{129}I., In: Environmental Migration of Long-lived Radionuclides, Symposium IAEA STI/PUB/597, 465-480.
11. Schell, W.R. Tobin, M.J, Massey, C.D. (1987) Evaluation of trace metal deposition history and potential element mobility in selected cores from peat and wetland ecosystems, Sci. Total Environ. 87/88, 19-42.
12. Schell, W.R., Linkov, I., Myttenaere, C. and Morel, B. (1996) A dynamic model for valuating radionuclide distribution in forests from nuclear accidents, Health Physics 70 (3), 318-335.
13. Lettenga, G. (1972) Radioactivity and water supplies , Ph.D. thesis, Interuniversitair Hochschool Reactor Institute and Laboratorium van Chemische Technologie van de Technische, Delft, Netherlands.
14. Schell , W.R. and Massey, C.D. (1982) Environmental radionuclide tracers of transport and diffusion in organic rich soil in the unsaturated zone. In: Isotope Technoques in Water Resources Development,. Proceedings of Symposium IAEA STI/PUB/757, 781-782.
15. Schell,W.R. (1987) A Historical Perspective of Atmospheric Chemicals deposited on a Mountain Top Peat Bog in Pennsylvania, Int. J. Coal Geol. 8, 147-171.
16. Sanchez, A., Schell, W.R. and Sibley, T.H. (1982) Distribution coefficients for plutonium and americum on particulates in aquatic environments. In: Environmental Migration of Long-lived Radionuclides. Proceedings of Symposium IAEA STI/PUB/597, 188-203.
17. Sanchez, A., Schell, W.R. and Thomas, E.D. (1988) Interactions of ^{57}Co, ^{85}Sr, and ^{137}Cs with peat under acidic precipitation conditions. Health Physics 54 (3), 318-335.
18. Schell W.R., Sanchez, A., Underhill, D.W.,and Thomas, E.D. (1986) Natural geochemical repositories for wastes: Field and laboratory meaasurements of diffusion and distribution coefficents in age dated peata bogs. In: T.H. Sibley and C. Myttenaere, Application of Distribution coefficients to radiological assessment models. ElsevierApplied Science Publ., 207-223.
19. Bunzl, K.,Bachuber, H.,and Schimack, W. (1984) Distribution coefficients of ^{137}Cs, ^{85}Sr, ^{141}Ce, ^{103}Ru, ^{131}I, and ^{95m}Tc in various horizons of of cultivated soils in Germany. In: P. Udluft, B. Merkel and K.H. Prosl, eds. Proceedings of the International Symposium, Technical University, Munich, 567-577.

Part 1

Contaminated Forests: Processes, Measurements and Methods

DYNAMICS OF RADIONUCLIDES IN SEMI-NATURAL ENVIRONMENTS

M. BELLI
ANPA - National Environmental Protection Agency - Roma ITALY

K. BUNZL
*GSF - Forschungszentrum fur Umwelt und Gesundheit - Neuherberg
GERMANY*

B. DELVAUX
UCL - Universitè Catholique de Louvain - Louvain la Neuve BELGIUM

M. GERZABECK
ARCS - Austrian Research Centre Seibersdorf - Seibersdorf AUSTRIA

B. RAFFERTY
*RPII - The Radiological Protection Institute of Ireland - Dublin
IRELAND*

T. RIESEN
PSI - Paul Scherrer Institute of Switzerland - Villigen SWITZERLAND

G. SHAW
*ICTSM - Imperial College of Science Technology and Medicine - Ascot
UK*

E. WIRTH
BfS - Bundesamt fur Strahlenschutz - Neuherberg GERMANY

1. Introduction

This paper reports a review of the implementation and achievements of the SEMINAT
Project [1] after the first two years of activities (1996-1997).

SEMINAT (Long-Term Dynamics of Radionuclides in Semi-Natural
Environments: Derivation of Parameters and Modelling) is a project funded by the
European Commission (Research Contract n°FI4P-CT95-0022) in the frame of the
Nuclear Fission Safety Research and Technological Development Programme (1994-
1998).

I. Linkov and W.R. Schell (eds.), Contaminated Forests, 17–21.
© *1999 Kluwer Academic Publishers. Printed in the Netherlands.*

The main aim of SEMINAT activities with respect to semi-natural systems (meadows and forests) is to deliver working computer models of radionuclide behaviour and fate in semi-natural ecosystems typical of the majority of the European Union countries.

Meadows and forests are typical semi-natural ecosystems. Meadows are used extensively in many countries as pastures for cattle, sheep and goats while forests are important to man since they provide wood, paper, wildberries, mushrooms, game and recreational areas. Foodstuffs and other products from both of these types of ecosystem have exhibited persistently high contamination levels following the Chernobyl accident. It is important a) to understand the reasons for this persistence and b) to estimate radiation doses incurred by human usage of these areas.

During the last years our understanding of radionuclide behaviour in semi-natural ecosystems has been improved significantly, especially for boreal forests and middle European meadow systems which have been extensively investigated. Data sets have been obtained which describe the distribution and the cycling of radionuclides (especially ^{137}Cs and ^{90}Sr) within these systems. However, predictive modelling has largely been restricted to aggregated transfer factors which provide good contamination estimates but only for the sites from which data have been obtained directly. For radiation protection purposes advanced models are essential which are applicable to a broad variety of ecosystems. They are needed for dose estimation, countermeasure implementation and environmental management. They should give reliable estimates of the behaviour of radionuclides in semi-natural systems and of external and internal radiation exposure to man. In order to develop such models it is necessary to understand the basic mechanisms of transfer and migration of radionuclides in meadows and forests. The SEMINAT project will therefore identify critical transfer parameters and quantify their influence on the persistence and migration of strontium and caesium in semi-natural systems. At the same time predictive models will be developed on a modular basis which will provide an integrated predictive capability for both forest and meadow ecosystems. Stand-alone modules will be coupled to provide an ecosystem-level model which will be tested and calibrated against site-specific data collected by groups from different countries.

2. SEMINAT Main Achievement (1996-1997)

SEMINAT project is mainly addressed at the development of dynamic models describing the radionuclide behaviour and fate in semi-natural ecosystems, typical of the majority of the EU countries. To achieve the final aim, the activities were organised in 4 work packages (WP) as follows:
- WP1 - Definition of scenarios
- WP2 - Derivation of fluxes and transfer parameters
- WP3 - Modelling
- WP4 - Results analysis

The main objectives for 1996 and 1997 were:
a) to identify the main parameters driving the radionuclides dynamics in semi-natural ecosystems (WP1);
b) to identify the different scenarios and the experimental protocols for the field and laboratory activities (WP1);
c) to determine representative values for the main parameters, quantifying the spatial variabily at large and small scale (WP2);
d) to develop a working version of the forest model (WP3).

2.1 DEFINITION OF SCENARIOS

In the frame of the WP1, forest and meadow sites were chosen. In 7 of forest sites (Lady wood, Buttersteep wood, Ballistone wood, Tarvisio, Weinsberger, Kobernausser and Novaggio forests) studies on the fluxes of caesium are currently taking place within and between various ecosystem components. In the three forest sites in Germany (Hochstadt, Siegenburg and Garching/Alz), the rooting depth of understorey vegetation and mushrooms are under study, comparing the $^{137}Cs/^{134}Cs$ ratio in the different vegetation/mushroom species and in the different soil horizons. The influence of spatial variability of soil parameters on caesium migration and uptake by trees and understorey vegetation is investigated in the Clogheen, Sharahan, Weinsberger, Kobernausser, Tarvisio and Venzone forests. Six study sites have been selected in semi-natural meadow ecosystems, to identify critical transfer parameters and quantify their influence on the persistence of radiocaesium in these ecosystems. In the Cavan site radionuclides fluxes between different soil layers are under investigation. In the other sites the factors influencing the radionuclides fluxes from soil to vegetation (plant species, seasonality and soil erosion processes) are under investigation. All the methods and the experimental protocols used in the investigations were agreed and compared between the participants at the project.

2.2 DERIVATION OF FLUXES AND TRANSFER PARAMETERS

The following sets of data are being collected in the different experimental sites: soil parameters, biomasses in the different compartments, radionuclide content in the different compartments, radionuclide fluxes into and out the compartments. Spatial and temporal variability of these parameters are under investigation in order to estimate reliable rate coefficients to calibrate dynamic models. In all semi-natural ecosystems investigated, an high spatial variability in the radiocaesium content in the environmental compartments has been observed. In almost all sites, the coefficient of variation associated with the total caesium activity in the soil is higher than 50%. This high spatial variability makes it necessary to analyse many samples in order to obtain reliable data for model calibration and validation.

   The observed depth profiles of radiocaesium in soils of natural ecosystems were used to assess the residence half-times of Chernobyl radiocaesium. The first results show that the residence half-times are in the order of few years per cm layer and they vary spatialy and also with depth. This variability has to be considered in the prediction of the transfer of radiocaesium from soil to vegetation.

Radiocaesium inventory in trees has been investigated where the preliminary data indicate that tree canopy, in spite of the little biomass (17% of the tree weight), plays an important role in retention of radionuclides. The canopy stores 41% of total ^{137}Cs contained in the whole tree. In the different sampling sites, the amount of radiocaesium found in the tree biomasses ranges from 1 to 12% of soil inventory. Generally, mushrooms and understorey vegetation account for much less than 1% of the radiocaesium in the soil compartment. Aggregated transfer factors for mushrooms and understorey vegetation show a very high variability with coefficient of variation higher than 100%. The analysis of an extensive data set for a sampling site in Germany revealed that fungal mycelia and fine roots of understorey vegetation occupy distinct layers of forest soil. The temporal changes of radiocaesium activities in mushrooms and understorey vegetation reflect the time-dependent activity concentration of that soil layer, from which radiocaesium is predominantly taken up.

Aggregated transfer factors for meadow vegetation, investigated in Germany along a slope of an alpine pasture, range from 0.2 to 6.6 10^{-3} m^2 kg^{-1} with a coefficient of variation higher than 60%, the high variability could be attributable to the changing plant community along the slope. Aggregated transfer factors assessed in Ireland for a meadow on peaty soils show values ranging from 5 10^{-3} to 2.6 10^{-1} m^2 kg^{-1}. This variability is attributable to the different plant species considered. The highest values were found for the shoots of *Colluna vulgaris*.

To identify the critical parameters and quantify their influence on the persistence and migration of caesium in semi-natural systems, laboratory experiments are in progress. The fate of radiocaesium in the weathering model is validated using a soil weathering sequence derived from sandstone. The soil sequence is: *acid brown soil -> ochreous brown soil -> ochreous podzolic soil -> podzol*. Clay contents are invariably low but the nature of clay minerals markedly differs according to soil type. The associated mineralogical sequence corresponds to the classical transformation process of micas in acid conditions: mica -> vermiculite and HIV -> smectite. The transformation process is not only identified in the various pedons, but also in the various horizons of each pedon, but at different extents. The weathering model is validated using a soil evolution sequence as described above, made of sandy soils with low clay contents (<8% clay), previously fully characterized. The first results show (i) a decrease in the net ^{137}Cs retention value of the Ah layer from the acid brown soil to the podzol, (ii) a "depth-dependance" of the maximum ^{137}Cs net retention value for each profile. The decrease in the net ^{137}Cs retention value in the Ah layers is associated with the transformation process from mica -> vermiculite (acid brown soil) -> from smectite (podzol). In each profile, the maximum value of the ^{137}Cs net retention is associated with the horizon in wich (i) vermiculite is dominant and (ii) Al-interlayering does not take place (complexing organic acids are present). The minimum value of the ^{137}Cs retention is invariably measured in Bw horizons in which Al-interlayering takes place (HIV minerals). The magnitude in ^{137}Cs net retention is directly related to the soil vermiculite content (r = 0.948) estimated by the rubidium assessment methodology.

These results are currently being validated with the soil samples collected in the various sites of the SEMINAT network.

2.3 MODELLING

As a first step in SEMINAT's development of dynamic models appropriate to forest ecosystems a screening model (RIFEQ) was developed which allowed preliminary data from each of the research groups' forest sites to be used in a probabilistic uncertainty analysis of radiocaesium distributions in major components of each ecosystem. The screening model developed uses a combination of aggregated transfer coefficients (T_{agg} values) and biomass estimates in a mass balance calculation. The results of such calculations allow an instantaneous 'snapshot' of the expected ranges of radiocaesium activity concentrations and distributions within forests and provide important insights into the relative importance of individual compartments with respect to the development of dynamic models (site-specific calculations of radiocaesium distributions using RIFEQ are intended to assist with the calibration of dynamic models).

While the RIFEQ screening model is useful in assisting with calibration of dynamic forest models, it does not itself provide a tool for interpreting and forecasting radionuclide behaviour in forests on a temporal basis. For this reason SEMINAT is developing a dynamic modelling capability using data from forest sites within five EU countries. At the outset of the project it was considered desirable to provide users with the ability to undertake model calculations either at a relatively simple or a more complex level, as appropriate to their needs. For this reason, the RIFE1 model, originally conceived and partially developed by the ECP5 [2] project, has been both retained and further developed by SEMINAT. Additionally, RIFE1 has been used as the basis for developing a more complex dynamic model, RIFE2. RIFE1 development into a fully probabilistic code is currently under way.

3. References

1.   Belli, M Editors 1998, SEMINAT: Long-Term Dynamics of Radionuclides in Semi-Natural Environments: Derivation of Parameters and Modelling, Mid-Term Report 1996-997, European Commission, Nuclear Fission Safety Programme, Research Contract N° FI4P-CT95-0022, ISBN 88-448-0295-3, 46p.
2.   Belli, M. & Tikhomirov F.A. Editors; 1996, Behaviour of Radionuclides in Natural and Semi-Natural Environments; Final Report of Experimental Collaboration Project N°5 (1991-1995), European Commission, Belarus, The Russian Federation, Ukraine. International Scientific Collaboration on The Consequences of the Chernobyl Accident (1991-1995), EUR 16531 EN, 1996, ISBN 92-827-5197-X, 147 P.

DYNAMICS OF RADIONUCLIDE REDISTRIBUTION AND PATHWAYS IN FOREST ENVIRONMENTS: LONG-TERM FIELD RESEARCH IN DIFFERENT LANDSCAPES

A.I. SHCHEGLOV

Radioecology Laboratory, Faculty of Soil Science, Moscow State University, 199899 Moscow, RUSSIA

Abstract

The paper analyses the long-term dynamics of radionuclide distribution among the basic compartments of forest ecosystems under conditions of single-time aerial contamination. Forests are shown to be long-term sinks for radionuclides and serve as effective biogeochemical barriers to radionuclide migration. Radionuclide redistribution in the "soil-plant" system is characterised by three main stages of increasing duration and with different dominant processes. The dynamics of the biological availability of radionuclides is primarily determined by landscape conditions and the physico-chemical forms of the initial fallout. Biota dependent migration is the largest contributor to the geochemical cycle of radionuclides in forest landscapes, whereas water migration is practically insignificant.

1. Introduction

Release of technogenetic radionuclides to the environment from nuclear weapon testing, routine and accidental operation of nuclear installations, and other sources promote further development of the global ecological crisis. The present interest in the behavior of radionuclides in contaminated forests is due to the significant contribution of these ecosystems to the global ecological equilibrium. Forests differ from other terrestrial ecosystems in the following:

- High typological diversity;
- Extremely complex structure of trophic chains;
- Multi-floor structure of vegetation cover;
- Permanent accumulation of chemical elements in the tree phytomass;
- Intensive radionuclide leaching in the forest litter layer
- Manifested heterogeneity of the soil profile;
- High radio-sensitivity of some tree species and high dose loads to the local population.

I. Linkov and W.R. Schell (eds.), Contaminated Forests, 23–39.
© 1999 *Kluwer Academic Publishers. Printed in the Netherlands.*

These factors contribute largely to the differences in individual views and priorities of researchers engaged in forest radioecology. Nevertheless, the following issues have always been top priority:

- The effect of forest ecosystems on the initial distribution of radioactive fallout over the contaminated territory;
- The temporal and spatial distribution of radionuclides among the compartments of forest ecosystems in different cenoses, landscapes, and natural zones;
- Basic factors governing radionuclide migration in the "soil-plant" system in different territories ;
- Qualitative and quantitative composition and dynamics of mobile and available forms of radionuclides in soil;
- Parameters and dynamics of radionuclide fluxes between ecosystem compartments;
- Modeling and forecasting of radionuclide migration; and
- Development of effective countermeasures in order to provide for industrial activities in the contaminated forests.

We believe that the integrated methodological approach to the above-mentioned issues can be realized within the framework of a landscape-geochemical concept by considering the investigated territory as a system of elementary landscapes and ecosystems joined to each other by the biogeochemical fluxes of chemical elements. In terms of forest radioecology, this means that radionuclide migration should be studied simultaneously in all of the ecosystem's compartments.

The accident at Chernobyl NPP (ChNPP) has confirmed the need for this type of approach to radioecological research.

2. Materials and Methods

2.1. GENERAL APPROACH AND INVESTIGATED TERRITORY

Our study is aimed at determining basic radionuclide fluxes to and from different compartments of the forest ecosystems in different natural zones and landscapes (Table 1). This approach is based primarily on comprehensive long-term monitoring (Fig 1).

The key sites are located in the most contaminated regions of the central East-European Plane (Russia and Ukraine) and cover basic types of forest, as well as bog ecosystems of the forest-steppe, south taiga, and Ukrainian Poles'e natural zones.

The monitoring network was established in 1986-87 at distances of 5-30, 100-200, and 250-500 km from the Chernobyl NPP. The key sites were where chosen on the basis of geomorphology, soil science, species composition, and forest taxonomy.

The sites have served as a basis for long-term (10 years) detailed radioecological monitoring with annual collection of forest litter, soils, soil solutions, tree and herbaceous vegetation, fungi, forest products, etc. using a standard sampling scheme (Fig 1).

TABLE 1. Basic soil properties over the investigated territory

| Horizon | PH | | Organic | Total | Exchangeable bases | | | Saturation |
	Water	KCl	Matter %	Acidity (mg-equiv./100g of soil)	Ca^{2+}	Mg^{2+}	Al^{3+}	with bases* %
Forest-steppe zone (Tula region, Russia)								
Loamy forest chernozem								
A	6.2	5.3	5.2	2.4	15.7	3.1	No data.	89
AB	6.1	4.9	2.4	3.9	16.3	2.5	-	83
B1	6.4	5	1.5	2.7	15.4	2.4	-	87
B2	6.6	5	0.8	1.8	14.3	2.4	-	90
BC	7.5	6.1	0.5	0.8	14.4	2.5	-	95
South Taiga sub-zone (Bryansk and Kaluga regions, Russia)								
Podzolic sandy soils								
A/A2	4.0-5.2	3.2-4.0	0.4-1.4	1.6-6.5	0.9-2.3	0.1-0.5	0.1-1.0	13-47
B1	4.4-5.6	3.9-4.5	0.1-0.8	0.9-3.6	0.4-1.3	0.1-0.3	0.1-0.9	12-51
B2	4.6-5.7	4.3-4.8	0.1-0.4	0.8-2.9	0.3-0.8	0.1	0.1-0.7	19-47
BC	4.8-6.1	4.4-5.0	0.0-0.1	0.7-1.9	0.5-1.0	0.1-0.6	0.1-0.6	24-58
C	5.1-6.1	3.6-4.9	0.0-0.5	1.3-7.9	1.4-6.4	0.3-2.2	0.1-1.0	41-75
Boggy soils								
O (l+f+h)	3.6	3.1	90.6**	2.3	2.4	0.3	0.5	56
T1 (peat)	4.1-4.9	3.3-4.3	27-85**	2.4-14.2	2.5-11.7	0.3-0.6	0.4-0.6	47-55
T2 (peat)	4.4-5.4	3.3-4.3	19-90**	2.4-6.2	2.2-10.8	0.3-0.4	0.3-0.6	52-65
T3 (peat)	4.1-5.5	3.2-4.5	9-24**	7.3-16.5	3.4-11.4	0.3-0.8	0.3-1.0	36-64
G (gleyic)	4.6-5.9	3.7-4.5	2-7**	5.6-6.0	1.3-16.3	0.2-1.8	0.1-0.9	20-77
CG (gley)	5.1-5.7	4.1-4.6	-	1.2-2.7	2.1-3.7	0.3-0.5	0.1-0.6	48-78
Ukrainian Poles'e (ChNPP Zone of exclusion, Ukraine))								
Podzolic sandy soils								
A/A2	4.4-5.1	3.4-4.3	1.3-3.6	2.8-7.5	0.4-1.7	0.2-0.4	0.3-0.9	14-23
B1	4.8-5.4	4.3-4.5	0.8-0.9	2.2-2.6	0.4-1.0	0.1-0.2	0.3-0.5	19-35
B2	4.8-5.2	4.3-4.6	0.7-0.8	1.9-2.3	0.3-0.5	0.1	0.2-0.4	15-24
BC	4.8-5.2	4.6-4.7	0.4-0.8	0.8-1.5	0.1-0.4	0.1	0.1-0.2	22-33
C	4.7-5.8	4.9-5.2	0.3-0.5	0.5-1.0	0.2-0.6	0.1-0.2	0.0-0.2	23-62
Boggy soils								
T (peat)	4.9-5.6	4.0-5.0	9.3-15.2	9.4-15.5	7.8-16.8	0.6-3.0	0.2-0.4	41-65
A (gleyed)	4.6-6.1	3.9-5.8	3.0-12.0	2.2-16.4	2.8-14.2	0.2-1.3	0.1-0.6	44-79
B (gleyed)	5.3-6.2	4.2-6.0	0.9-2.0	0.5-3.9	1.9-9.8	0.1-2.3	0.0-0.2	53-81
G (gley)	5.2-6.2	4.5-6.1	0.5-1.4	0.5-1.0	1.2-3.1	0.1-0.8	0.0	65-80

* H⁺, % of the total exchangeable cations;   **Loss for ignition at 500 ^{0}C

26

2.2. BASIC FEATURES OF THE SOIL COVER

The investigated area includes several natural zones with different soils:
1. Loamy podzolized chernozem (forest steppe zone, 400-500 km from ChNPP);
2. Podzolic sandy and swamped peaty-gley soils (mixed forests of South taiga zone, 100-300 km from ChNPP);
3. Podzolic sandy and swamped peaty-gley soils (Ukrainian Poles'e including the so-called exclusion zone 5-30 km from ChNPP) (Table 1).

All forest soils (excluding chernozems and grey forest soils) are characterized as acidic, non-saturated with bases, low-humus, and well drained (sand fraction > 95%). These

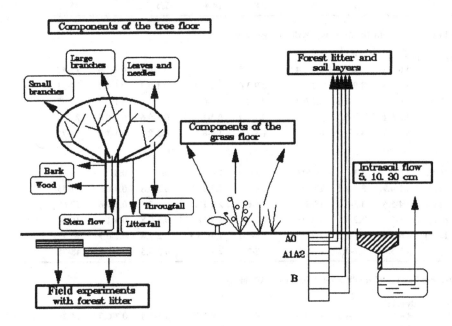

Fig 1. Basic design of radioecological monitoring at a single key plot

soil properties prevent retention of a large amount of radionuclides and promote intensive vertical and horizontal migration. Forest litter is the only horizon of these soils capable of radionuclide retention, since clay minerals are found in the thin sub-layer located directly under the O horizon (forest litter).

Much higher clay content is characteristic for chernozem, resulting in the high ability of this soil for reversible and irreversible Cs absorption. A thick peat layer in the soils of swamp areas provides an ability for a large degree of exchangeability and a low capacity for irreversible absorption. This results in relatively low mobility of ^{90}Sr and high mobility of ^{137}Cs in these soils compared to podzolic soil [1].

3. Results and Discussion

3.1. THE IMPACT OF FOREST ECOSYSTEMS ON THE INITIAL DISTRIBUTION OF RADIOACTIVE FALLOUT

It is believed that forest ecosystems have a significant effect on the large-scale or small-scale distribution of fallout over the contaminated area. We agree with this standpoint and suppose that the ability of forest ecosystems to intercept contaminants causes them to be effective accumulators of fallout. According to our data, deposition in forests are 20-30% higher than in neighboring territories without forests [1] (Table 2).

TABLE 2. The distribution of radioactive fallout among forest and meadow ecosystems

Distance from ChNPP	DEPOSITION (kBq/m2)	
	Meadow	Forest
6 km to the South	14210	17200
28 km to the South	1070	1300

This phenomenon is due not only to direct interception of radioactive particles but also to specific atmospheric turbulence and the local climate over forested areas [2].

Model experiments conducted after the Kyshtym accident [3] showed that from 40 to 90% of the fallout (depending on a range of biological and meteorological factors) could be intercepted by tree crowns. The findings from these experiments are confirmed by data obtained after the Chernobyl accident: the initial interception of fallout by forests can be as high as 80% depending on the type of ecosystem and particle size (Fig. 2).

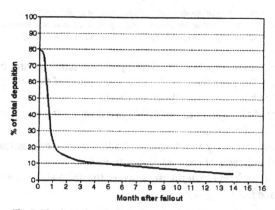

Fig. 2. The dynamics of self-decontamination of the tree layer in the initial period after "Chernobyl" fallout

3.2. LONG-TERM BEHAVIOUR OF RADIONUCLIDES IN THE FOREST VEGETATION

The dynamics of radionuclide behavior in a forest ecosystem is usually described as two sequential stages: (1) fast mechanical self-decontamination of the tree level and (2) decrease in self-decontamination due to root uptake followed by gradual approximation of the system to steady state [4]. However, according to our data these characteristics apply only to forest biogeocenoses in automorphic (dry) landscapes.

Moreover, quantitative studies of the radionuclide content in individual tree components suggest that self-decontamination processes often dominate during the second stage even in the automorphic landscapes (Fig. 3). For hydromorphic forest landscapes, the long-term dynamics of radionuclides in the tree level is very different from that of automorphic landscapes.

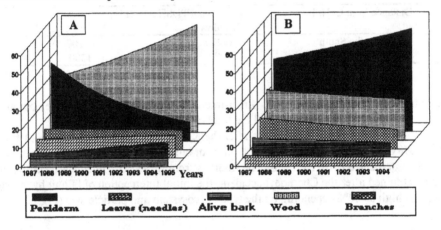

Fig. 3. Exponential approximation of the dynamics of contribution of different organs to total contamination of the pine forest: A - on peaty soils of accumulative landscapes; B- on podzolic sandy soils of hydromorphic landscapes (% of total radionuclide content in the plant)

Therefore, the common practice of dividing the dynamics into two stages is not completely correct. We pose that an adequate description of the fallout dynamics in the tree level should be based on the following criteria:
1. The intensity of changes in radionuclide inventory in the tree floor;
2. The trend of the dynamics;
3. The processes governing the dynamics (radionuclide composition can be used as an additional indicator of radionuclide pathway).

Using the above-mentioned criteria we suggest the following stages of radionuclide behavior in the tree floor (Fig. 4):

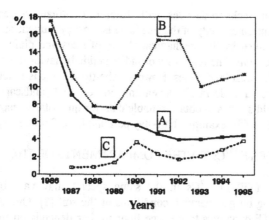

*Fig. 4. The dynamics of ^{137}Cs content in the above-ground
phytomass of forest ecosystems (% of total deposition).*
(A) Mixed forest, alluvial landscape;
(B) Mixed forest, accumulative landscape;
(C) Mixed forest, alluvial landscape in the vicinity of ChNPP (6 km)

1. Intensive mechanical self-decontamination (the first 2-3 months after the fallout).
   During this period, the specific activity of the vegetation is determined by surface
   contamination only, and the radionuclide composition corresponds fully to that of
   the initial fallout (decay corrected). The dynamics of the radionuclide content is
   primarily determined by the rate of mechanical self-decontamination of the plant
   surface. The half-stay period (effective half-life in the stand) at this stage varies
   from 3 weeks to 1 month
2. Increase in radionuclide root uptake (2-3 years after the fallout). The intensity of
   mechanical self-decontamination during this period decreases sharply as this
   process takes place against the background of growing radionuclide concentration
   in the internal plant tissues due to root uptake. This is accompanied by a gradual
   enrichment of the vegetation with ^{137}Cs and ^{90}Sr. The half-stay period at this stage
   is 1.5-2 years.
3. Root pathway dominance (more than 3-4 years since the fallout). The "soil-plant"
   system gradually approaches a quasi-steady state. The dynamics of radionuclide
   content in the plants depends primarily on soil and landscape conditions, the
   physico-chemical properties of initial fallout, and climatic conditions. The
   radionuclide composition is almost completely represented by biologically
   significant, long-lived radionuclides. The half-stay period at this stage (beginning
   from 1991 for the Chernobyl fallout) is 10-15 years for automorphic and even
   longer for the hydromorphic landscapes.

Thus, the general dynamics of radionuclide inventory in the tree floor of both auto-and
hydromorphic landscapes can be depicted as 3 main stages, with half-stay periods

varying from several weeks to decades, with different dominant processes occurring at each stage. The retention capacity of the tree level is most prominent during the first two stages and is determined by the mechanical ability of trees and plants for radionuclide interception and retention. The regularity of radionuclide behavior during these stages is rather similar for all forest landscapes. However, the third stage is much more complex. Parameters vary over a wide range depending on soil and landscape characteristics, requiring further field studies to obtain reliable data for quantitative parameters used for radionuclide migration (for example, half-stay periods) in different ecosystems.

3.3. DYNAMICS OF TF ^{137}Cs TO THE COMPARTMENTS OF THE TREE LEVEL

The transfer factor (TF) of ^{137}Cs in the same species may vary by 2-3 orders of magnitude depending on geochemical conditions of the soil [1]. Our studies confirmed that the long-term TF dynamics to the tree floor further depends on the landscape and physico-chemical properties of the fallout.

Generally, variation in the dynamics of TF ^{137}Cs in forest vegetation can be classified into three basic categories (Fig. 5):

1. A pronounced decreasing trend in the TF (Fig 5-A). This is characteristic of ecosystems in automorphic landscapes and results from decreasing radionuclide availability due to their irreversible fixation by the soil.

2. Pronounced increasing trend of TF. This is characteristic of ecosystems in hydromorphic soils of accumulative landscapes, and in ecosystems located very close to local depressions and floodplains (Fig. 5-B). Similar dynamics is also found in the vicinity of ChNPP (0-5 km zone, Fig. 5-C). The long-term increase in TF is determined by the dominance of the root uptake of ^{137}Cs over the irreversible absorption of ^{137}Cs in soil. In accumulative landscapes the high contribution of the root pathway is mainly due to weak fixation of ^{137}Cs by peats and intensive migration down the soil profile. In the vicinity of ChNPP (0-5 km zone), the increase in TF can be explained by the gradual release of ^{137}Cs from the low soluble fallout particles to the soil solution.

3. No manifested trend of TF against the background of its high annual variation (Fig. 5-D). This type of dynamics occurs in the forests of transitional (semi-hydromorphic) landscapes, and in the middle part of the Chernobyl exclusion zone (about 10-15 km from the NPP). The absence of any visible trend is likely to be caused by mutual compensation of the above-mentioned factors and processes.

Therefore, ten years after the accident, automorphic forest ecosystems are the only systems that can be correctly considered as approaching steady state in terms of the "soil-plant" system and that can be described using an approximately stable half-stay period. Hydromorphic forests or forests located in the vicinity of the accident are not yet in steady state (in terms of radioecology), and any forecasts for these territories are not very reliable. The same is true for radionuclide fluxes and half-stay periods. We would like to emphasize that continued field observations (monitoring) are necessary to get the actual information on changes in radioecological situation in these landscapes.

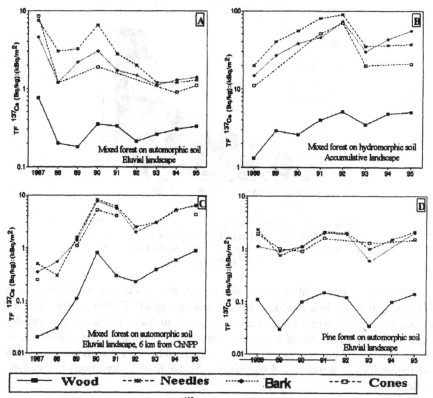

Fig. 5. The long-term dynamics of TF ^{137}Cs to the structural components of Pine depending on soil-geochemical conditions, TF = [Bq/kg (vegetation)]: [kBq/m^2 (soil)]

3.4. RADIONUCLIDE ACCUMULATION IN BIOTA COMPARTMENTS OF FOREST ECOSYSTEMS

The tree level is believed to be the largest sink for radionuclides because of its huge biomass. However, this is only true for the first year after the fallout. Thereafter, the contribution of individual compartments to the total contamination of the ecosystem is correlated with their ability to accumulate radionuclides as well as their biomass (Fig 6). The specific accumulative capacity (irrespective of the biomass) of the different biota compartments for ^{137}Cs varies by 3 orders of magnitude and can be shown in the following descending order: mycobiota (fungi) > mosses and liches > herbaceous plants and shrubs > tree level. The question arises as to what factor is of more significance: the biomass of a compartment or its capacity for radionuclide accumulation?

32

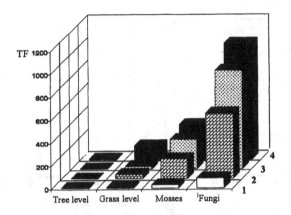

Fig. 6. Accumulative capacity of the forest compartments for 137Cs:
1-   Mixed forest; 2- Pine forest; 3 - Mixed forest, 5 km from ChNPP;
4 - Adler forest on peaty soil. Average-weighted for 5 years,
$$TF_{137Cs} = (Bq/kg)/(kBq/m^2)$$

Our data suggest that the latter is more significant, which is convincingly demonstrated by depicting the data in a flux diagram (see below, Fig.11). Unfortunately data on the individual contribution of the herbal layer (grass-shrub floor), tree level, and mycobiota to the total contamination of the ecosystem are not abundant, and sometimes it is difficult to compare our results with the data obtained by other researchers. We think it is of importance to broaden and standardize (to facilitate comparison) the studies in this field.

3.5 RADIONUCLIDE MIGRATION IN FOREST SOILS

As mentioned previously, the profile of forest soils is very heterogenic and is usually represented by two principally different layers: organic (forest litter) and mineral (soil in itself). These two layers determine the differences in the processes governing radionuclide behavior in forest soils and therefore it seems natural to consider radionuclide migration in the forest soils separately for each layer.

Generally, forest litter is believed to be a long-term sink for radionuclides and serves as a biogeochemical barrier preventing further migration down the soil profile [5]. However, the thickness and inner structure of forest litter varies widely, depending on the natural zone and the characteristics of the ecosystem. This results in a large variation of radionuclide distribution in the litter. According to our recent data, the proportion of radionuclides contained in the forest litter varies from 9 to 90% depending on the natural zone, ecosystem and other factors. For example, in the forest-steppe zone (Tula region) this layer lost more than half of the radionuclide inventory in as little as 2 years after the Chernobyl accident. At the same time, in some ecosystems in the exclusion zone more than 80% of the deposition is still contained in the litter layer.

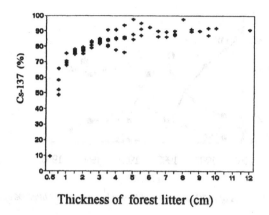

Thickness of forest litter (cm)

Fig. 7. Averaged radionuclide content in the forest litter depending on its thickness
(% of total deposition, 1995)

The high retention potential of the forest litter deserves special consideration. The litter layer is characterized by extremely aggressive leaching conditions, maximum content of soluble organic compounds, high microbiological activity, etc. Therefore, one expects the radionuclide mobility to be high in this layer. Nevertheless, some forest litter still hold up to almost 90% of the deposition (data of 1996).

Our field studies show that the retention capacity of the forest litter is correlated with its thickness, composition, and structure (Fig.7). Moss cover also enhances the retention capacity of forest litter. All these, and perhaps some other factors, have to be taken into account when predicting radionuclide migration in the soil profile. To avoid an increase in the number of parameters, it is necessary to determine the mechanisms governing the retention capacity. In our opinion, the latter are determined by radionuclide interception by plant roots and fungus mycelium which are most abundant in the forest litter. The activities of soil invertebrates are likely to be of significance as well. Unfortunately, we cannot estimate the individual contribution of each process quantitatively. Additional data on retention mechanisms in the forest litter should be obtained to calculate the radionuclide half-stay in this layer as a function of a limited number of parameters (thickness, structure, type of ecosystem, etc.).

As an example we consider ^{137}Cs dynamics in the forest litter of a typical pine ecosystem in the territory of the exclusion zone. This is presented in Fig 8 (field data). The highest rate of self-decontamination and the most prominent trend of downward movement are found to be in the upper sub-layer of the forest litter (Ol). The half-stay period in this layer is about 2 years. Approximately 4-5 years after the fallout, the ^{137}Cs content in this layer stabilized at about 1% of total deposition. The dominant process in this sub-layer is the annual covering of the contaminated surface by relatively clean litterfall (the annual radionuclide infiltration from this layer is low) [6].

The content of ^{137}Cs in the fermentation (Of) and humus (Oh) sub-layers of the forest litter is characterized by a gradual increase followed by some decrease and

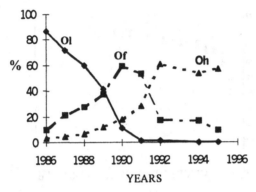

Fig. 8. The dynamics of 137Cs content in different layers of forest litter (% of total deposition)

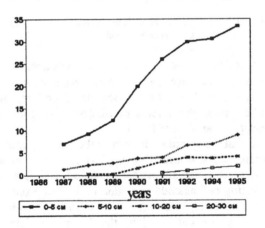

Fig. 9. The dynamics of ^{137}Cs content in different layers of mineral soils of pine forests in the exclusion zone (% of total deposition)

stabilization on certain levels depending on the individual sub-layer and type of ecosystem. At present, the ^{137}Cs content in the Of sub-layer has stabilized at a level of 10-20% of the total content in the soil (except maybe the ecosystems in the vicinity of ChNPP). The quasi steady state in the Oh sub-layer has not yet occurred. Therefore, the calculation of half-stay period in these sub-layers should be made taking into account the present non-equilibrium state.

The behavior of radionuclides in the mineral soil layers is characterized by less complicated dynamics, and is generally determined by physico-chemical migration: diffusion and convection against the background of exchangeable and irreversible sorption. Our field data suggest that the long-term dynamics are

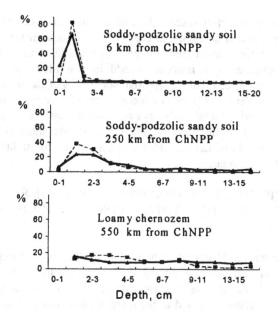

*Fig 10. Distribution of ⁹⁰Sr and ¹³⁷Cs down the profile of forest
soils (% of total deposition, 1995):*
——— ^{90}Sr   ········· ^{137}Cs

characterized by a steady increase in the radionuclide content down the soil profile
(Fig. 9).

The annual increase in radionuclide inventory in the upper 5 cm of the
mineral layer of forest soils varies from 1.5 to 4%. It varies from 0.1 to 1% at a depth
of 5-10 cm in the mineral layer. The maximum and minimum rates correspond to
hydromorphic and automorphic soils, respectively. A special case is the extremely fast
migration of ^{137}Cs in the steppe forests on chernozems (>10% in the layer 0-5 cm layer
and >4% in the 5-10 cm layer). These extreme rates are likely due to the activity of the
soil invertebrates, which are especially abundant in these soils (Fig. 10).

The annual increase of the radionuclide inventory in the soils of different
forest ecosystems depends on the following basic parameters: (1) layer, (2) soil and
ecosystem type, and (3) distance from the accidental unit (fallout properties).

It should be mentioned that the contribution of infiltration to the
redistribution of radionuclides is very low even on sandy soils and is estimated to be
n*0.1-0.01% of total deposition per year [7]. Vertical redistribution of radionuclides
among the layers of the soil profile is determined by different processes (and different
dynamics) in each layer. In the forest litter the processes are mechanical substitution,
decomposition, and blending by soil fauna. In the mineral layers the processes are

mainly physico-chemical factors (except the case of forest chernozem). To determine quantitative parameters it is necessary, in our opinion, to consider the forest litter and mineral soil as two self-sufficient compartments. In the soils of the forest-steppe and steppe zone it is also necessary to introduce an additional soil fauna compartment.

According to our estimations, approximately 50% of the deposition is likely to be transferred from the forest litter to mineral layers of "average" automorphic soil by 2005 [1]. Accumulative forest landscapes are expected by this time to approach a quasi steady state in the system "forest vegetation-forest litter-mineral layers." At the same time, up to 60-70% of radionuclides will stay in the forest litter in the close vicinity of ChNPP (5-km zone).

3.6. THE CONTRIBUTION OF INDIVIDUAL COMPARTMENTS AND SUB-COMPARTMENTS TO THE TOTAL CONTAMINATION OF FOREST ECOSYSTEMS

Based on our long-term field studies, it is possible to develop a conceptual model of ^{137}Cs migration and redistribution among the compartments of contrasting forest ecosystems (Fig 11 A, B), as represented by typical automorphic (A) and hydromorphic (B) ecosystems located in the marginal southern part of the Chernobyl exclusion zone (27 km from ChNPP).

At present (1996), biota accumulates from 6.5 to 44% of the total deposition depending on both landscape and geochemical conditions. This suggests that the general contribution of biota to ^{137}Cs accumulation does not decrease with time. What does take place is a redistribution of ^{137}Cs among the biota sub-compartments: the contribution of the tree floor decreases in spite of its high biomass, whereas the contribution of the herbal (grass and small shrubs) layer and the underground phytomass (including mycelium) increases. The contribution of moss cover and mycobiota (fungi including mycelium) is especially high, with the highest contribution from mycobiota (up to 24%). This exceeds the ^{137}Cs inventory in the tree biomass. We are therefore of the opinion that these new data are of special importance for the modeling of radionuclide behavior in forest ecosystems.

Generally, the contribution of the biota sub-compartments to total contamination can be listed in the following order: mycobiota > moss cover > tree floor > grasses and shrubs. An analysis of the radionuclide fluxes between the main compartments (soil and biota) shows that in automorphic landscapes the annual return of ^{137}Cs to the soil is still 2-5 times higher than root uptake. This means that even 8-10 years after the fallout, the self-decontamination process still dominates over root uptake. The opposite situation takes place in the forests of accumulative landscapes, where the annual return of ^{137}Cs to the soil is roughly the same as its accumulation via the root pathway.

Generally, ten years after the fallout the radionuclide inventory in the tree phytomass is concentrated in different tissues depending on the type of landscape. In the automorphic (dry) landscapes, the contamination is determined by external tissue

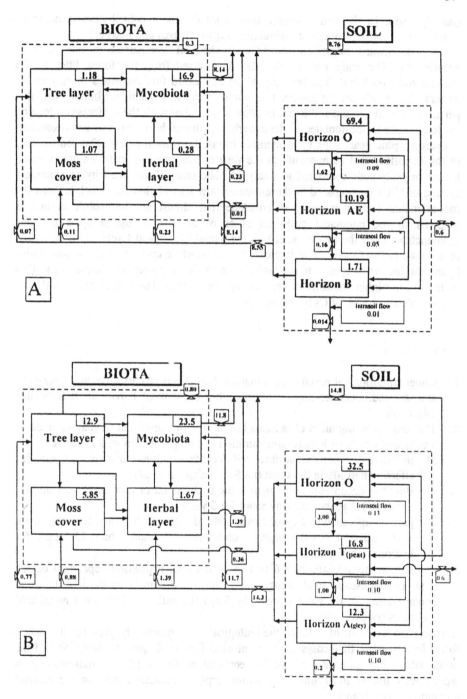

Fig.11. Biogeochemical cycle of ^{137}Cs in forest ecosystem (the intensity of the annual fluxes is expressed as % of total deposition). A - eluvial landscape; B - accumulative landscape.

(mainly bark). In the hydromorphic landscapes the main contribution is made by wood, internal (alive) bark, and assimilating organs (leaves and needles).

Downward migration of ^{137}Cs in the soil compartments has its own peculiarities. The main radionuclide flux is directed from the forest litter to the mineral soil (1.6-3.4%). The intensity of the downward flux decreases sharply in the mineral layers: almost insignificant amounts of the radionuclides (hundredths of a percent) leave the 50-cm soil layer. In other words, almost all the radiocesium leaving the forest litter is accumulated in the underlying mineral horizons. The only exception is hydromorphic peaty soils with intensive intrasoil downward flow. The contribution of the so-called "fast component" to the vertical movement of radiocesium is much higher in peaty soils compared to soddy-podzolic soils. If the intensity of biological cycling of ^{137}Cs is compared to the extremely low flux of this radionuclide from the root-abundant soil layer, we may conclude that almost all the mobile cesium is involved in the biological cycle. In other words, the biological cycle prevents radionuclides from the involvement into the Grand Geological Cycle (from terrestrial ecosystems to the Ocean). The intensity of radionuclide cycling in the accumulative hydromorphic landscapes is much higher than in the automorphic alluvial ones. This is due to the higher intensity of the root uptake in these landscapes along with low irreversible absorption of Cs by the peat layer.

4. Conclusions

1. Under conditions of aerial contamination, forests ecosystems serve as a long-term sink for radionuclides and an effective biogeochemical barrier to radionuclide migration.
2. The long-term migration of radionuclides in the tree level can be characterized by 3 principal stages with increasing duration and different dominant processes:
    a) intensive mechanical surface self-decontamination: duration 1-3 months, effective half life in the stand (half-stay) about 2 weeks;
    b) slow self-decontamination against the background of increasing root uptake: duration 1-3 years, half-stay about 1-1.5 years;
    c) gradual approach to a quasi steady state: duration 3-10 years, half stay about 10-15 years for automorphic conditions and more than 15 years for hydromorphic conditions.
3. Forest biota accumulate from 6 to 43% of the total deposition depending on the type of landscape and ecosystem. The maximum contribution to total contamination of the biota is made by fungi (including underground mycelium) and moss cover.

Intensive involvement of ^{137}Cs to the biological cycle practically prevents its mobile forms from leaving the ecosystem with intrasoil flow to deeper soil horizons and the water table. The intensity of the biogeochemical fluxes of this radionuclide is significantly higher in accumulative hydromorphic landscapes than in the alluvial automorphic landscapes.

5. References

1. Shcheglov, A.I. (1997) Biogeochemical migration of technogenic radionuclides in the forest ecosystems of central regions of East-European plain, *Doctoral Dissertation*, Moscow (available by request).
2. Fedorov, S.F. (1977) *Studies of the water balance in the forest zone in the European territory of USSR*, Nauka, Leningrad.
3. Tikhomirov, F.A. (1993) Radionuclide distribution and migration in EURT forests, *Ecological After-Effect of Radioactive Contamination in South Ural*, Moscow, Nauka, 21-39.
4. Alexakhin, R.M. and Naryshkin, M.A. (1977), *Radionuclide Migration in Forest Biogeocenoses*, Moscow, Nauka.
5. Shcheglov, A.I., Tikhomirov, F.A., Tsvetnova, O.B., et al. Geochemical migration of radionuclides in forest ecosystems within radioactive contamination zone of Chernobyl Nuclear Power Plant (1991), *Soviet Soil Science*, vol. 23, no. 2, 66-75.
6. Shcheglov, A.I., Tsvetnova, O.B., and Tikhomirov, F.A. (1992), Migration of long-lived radioisotopes of Chernobyl fallout in forests soils in the European Part of CIS, *Vestnik MGU*, 17, no. 2, 27-35
7. Kliashtorin, A.L., Tikhomirov, F.A., and Shcheglov, A.I. (1994) Vertical radionuclide transfer by infiltration water in forest soils in the 30-km Chernobyl accident zone, *The Science of the Total Environment*, 157, 285-288.

USE OF STABLE ELEMENTS FOR PREDICTING RADIONUCLIDE TRANSPORT

S. YOSHIDA and Y. MURAMATSU
Environmental and Toxicological Sciences Research Group, National Institute of Radiological Sciences, 4-9-1 Anagawa, Inage-ku, Chiba-shi, 263-8555, JAPAN

1. Introduction

In Japan, about 67% of land area is covered by forests (see Figure 1) and 50 nuclear reactors produce about 30% of the total electricity. The reactors in other East Asian countries are increasing in number. Therefore, forest radioecology becomes more and more important in Japan and other Asian countries. The first radioecological study in Japanese forests was reported in 1960's by Yamagata et al. [1] for ^{137}Cs and ^{90}Sr derived from atmospheric nuclear weapons testing. Distribution and transfer of these two radionuclides in a pine forest were studied and an accumulation in the upper 5 cm layer of soil was observed. After this study, however, Japanese and Asian forests have scarcely been studied in regard to the radionuclides. This is because of the insignificant direct contribution of the forest products to the principal food chain of humans compared with the products from the agricultural systems.

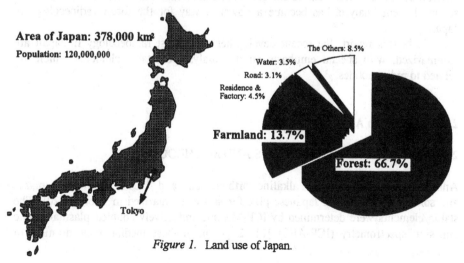

Figure 1. Land use of Japan.

41

I. Linkov and W.R. Schell (eds.), Contaminated Forests, 41–49.
© *1999 Kluwer Academic Publishers. Printed in the Netherlands.*

Many studies on the forest ecosystems were carried out mainly in the European forests after the Chernobyl accident [2-6]. The studies increased in number also in Japan, although the deposition from the accident was low. High concentrations of radiocesium in mushrooms were observed [7-11]. The levels of ^{137}Cs in 124 Japanese mushroom species collected between 1989 and 1991 varied widely, ranging from <3 to 16300 Bq/kg-dry (median: 53 Bq/kg-dry) [9]. The ^{137}Cs in Japanese mushrooms originates mainly from the fallout of nuclear weapons testing.

The experimental measurements in forest ecosystems have provided information for the development of models which explain radiocesium transport through several compartments in a forest ecosystem. Some different types of models were developed as summarized by Schell et al. [12, 13]. However, the fate of long-lived radionuclides in the forest ecosystems is still difficult to predict because of the changing of vertical profiles and availability with time of the radionuclides in soil. Determination of major and trace elements in plants and mushrooms as well as associated soils is one of the most powerful ways to understand the long term behavior of radionuclides and toxic elements in the forest ecosystems. Chemical behavior of radiocesium is expected to be similar to that of stable Cs and the other alkali elements. Myttenaere et al. [14] summarized the relationship between radiocesium and K in forests, and suggested the possible use of K behavior for the prediction of radiocesium behavior.

There are many studies on the behavior of major nutrient elements such as K, Mg and Ca in forests [15, 16] because these elements are directly related to the forest growth. Effect of acid deposition on the behavior of these elements was also studied in the last two decades [17-19]. However, the data for trace alkali and alkaline earth elements such as Rb, Cs, Sr and Ba in forest ecosystems are limited because of the lack of simple analytical methods. Recent analytical techniques, such as neutron activation analysis (NAA) and inductively coupled plasma-mass spectrometry (ICP-MS), can provide the data for stable elements which are related to radionuclides in environmental samples. Due to the low contamination level of Japanese forests, the stable element analysis has become a powerful way for the forest radioecology in Japan.

In this report, the recent developments of forest radioecology in Japan are summarized, with special emphasis on the analyses of stable elements which are related to radionuclides.

2. Concentration And Distribution

2.1. CONCENTRATIONS IN PLANTS AND MUSHROOMS

Analytical results of alkali and alkaline earth elements and ^{137}Cs in mushrooms, plants and soils collected from a Japanese pine forest are summarized in Table 1 [20]. The stable elements were determined by ICP-MS and inductively coupled plasma-atomic emission spectrometry (ICP-AES) [21, 22]. The highest median concentration in mushrooms was found for K followed by Mg, Na, Ca, Rb, Ba, Sr and Cs. For plants, the highest median value was found for Ca followed by K, Mg, Na, Sr, Ba, Rb and Cs. In comparison with the elemental composition of plants, the mushroom composition could be characterized by the high ^{137}Cs, Cs and Rb concentrations and low Ca and Sr concentrations. The concentrations of ^{137}Cs, Cs and Rb for mushrooms were found to be one order of magnitude higher than those for plants growing in the same forest. Accumulations of Cs and Rb in mushrooms were also observed from cultivation experiments in flasks using radiotracers [23, 24]. These results are consistent with the high concentrations of radiocesium in mushrooms reported in many countries [7-10, 25, 26].

The low concentrations for Ca and Sr observed in mushrooms were consistent with the low concentrations of ^{90}Sr in mushrooms after the Chernobyl accident reported by Mascanzoni [27]. Lower accumulation of Sr in mushrooms than those in plants were also observed by radiotracer experiments in the laboratory [23].

TABLE 1. Mean and median cencentration of alkali and alkaline earth elements and ^{137}Cs in mushrooms, plants and soils collected from a pine forest in Tokai-mura, Japan from 1989 to 1991 [20]

	"Cs*	Na	K	Rb	Cs	Mg	Ca	Sr	Ba	Al
	(Bq kg', dry wt)	(mg kg', dry wt)								
Mushrooms (n=29)										
mean	321.0	957	28700	89.80	2.140	1090	535	5.24	8.71	976
median	135.0	1000	27200	87.60	1.010	1050	389	2.88	5.12	556
min	5.4	75	12700	29.50	0.044	683	63	0.50	0.39	102
max	3110.0	1970	51700	214.00	20.000	1580	1990	17.30	28.10	3190
Plants (n=8)										
mean	4.5**	1440	11300	12.10	0.052	2630	20500	74.10	10.40	259
median	3.8**	802	8860	9.35	0.043	1800	16300	70.80	11.10	254
min	<2.0	93	3960	3.04	0.017	1100	2570	13.70	2.73	62
max	8.0	6040	24500	31.90	0.112	5660	58300	151.00	15.30	506
Soils (0-5cm)										
mean	34.7	15900	20100	53.90	1.280	6290	11900	112.0	310.0	48400

*The decay correction was made as to October 1990.
**In the calculation of the mean and median values of ^{137}Cs, values below the detection limits were regarded as measured values.

2.2. RADIAL DISTRIBUTION IN TREE

Behavior of ^{137}Cs and ^{90}Sr as well as related alkali and alkaline earth elements on the radial direction of the annual ring of Japanese ceder (*Cryptomeria Japonica* D. Don) has been investigated by several different research groups [28-30]. The distribution of ^{90}Sr fallout was directly correlated with tree ring age suggests that ^{90}Sr had given a rather direct effect and showed no significant translocation from sapwood to heartwood, whereas ^{137}Cs tends to move from one annual ring to another irrespective of the input from fallout. The similar results were also observed in some other species such as red spruce (*Picea rubens*) and eastern hemlock (*Tsuga canadensis*) collected in USA [31]. The distribution profile of ^{137}Cs is similar to that of stable Cs and of related alkali elements [28].

2.3. RELATIONSHIP BETWEEN ^{137}Cs AND STABLE Cs

Correlations between ^{137}Cs and stable Cs for mushrooms collected from a pine forest in Tokai-mura, Japan [20] and 2 forests in Finland are shown in Figure 2. The data for mushrooms collected from a pine forest in Rokkasho-mura, Japan reported by Tsukada et al. [32] are also plotted in the figure. A good correlation between ^{137}Cs and stable Cs was observed for each site, although several different species of mushrooms are included. This finding suggests that ^{137}Cs is taken up by mushrooms together with stable Cs. The ^{137}Cs/Cs ratios were almost constant for samples collected in the same site. The highest ratio was obtained at Kullaa (^{137}Cs/Cs = 5000 ± 1670 Bq mg^{-1}), followed by Kirkkonummi (733 ± 240 Bq mg^{-1}), Rokkasho-mura (430 ± 110 Bq mg^{-1}) and Tokai-mura (134 ± 36 Bq mg^{-1}). The difference of the ratios between the sites might be attributable to the difference of deposition rate of ^{137}Cs and the difference of the forest type (e.g. geology and soil).

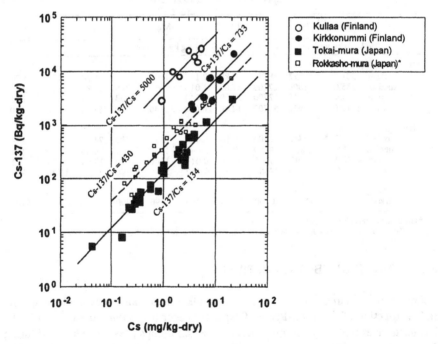

Figure 2. Relationship between stable Cs and ^{137}Cs in mushrooms collected from four different forests in Finland and Japan. The decay correction for^{137}Cs was made as to October 1990.
*Data from Tsukada et al. (1998)

In the pine forest in Tokai-mura, the ^{137}Cs/Cs ratios for 2 plant samples, in which ^{137}Cs was detected, were almost the same as those for mushrooms. The similar ratio, 166 Bq mg[-1], was observed also in the litter layer of the forest floor. These findings suggest that ^{137}Cs, mainly deposited to the forest ecosystem in the 1960s, is cycling within the biological samples in the forest with the constant ^{137}Cs/Cs ratio.

The ^{137}Cs/Cs ratio might be a useful criterion for judging the equilibrium of deposited ^{137}Cs to stable Cs in a forest ecosystem. The ^{137}Cs/Cs ratio might also be useful for the prediction of ^{137}Cs contamination by using stable Cs concentration, although further studies on the factors controlling the ratio are required.

2.4. RELATIONSHIP AMONG ALKALI AND ALKALINE EARTH ELEMENTS

Correlations among alkali elements in plants and mushrooms collected in a Japanese pine forest are shown in Figure 3 [20]. In plant samples, good correlations were observed among K, Rb and Cs. Correlations between these elements and Mg in plant samples were also good. Correlation between Cs and K has also been observed by other authors [14, 33]. These findings suggest that K could be used as an indicator of the behavior of radiocesium in soil-plant systems in forests. Correlations between Ca and Sr in plants were also good ($r = 0.99$), suggesting the possibility of using the behavior of Ca to predict the behavior of [90]Sr. No correlations were observed between Na and the other elements.

Cesium was not correlated with K in mushrooms in contrast with plant samples. The K concentration seemed to be controlled within a narrow range regardless of the mushroom species, while the concentrations of Cs in mushrooms varied very widely. These findings suggested that the mechanism of Cs uptake was completely different from that of K. Rubidium showed intermediate behavior between K and Cs. There was a correlation between Rb and Cs ($r = 0.82$).

3. Transfer Factors

In order to compare the concentrations in mushrooms or plants with those in soil, the tentative transfer factor defined as "median concentration in mushrooms or plants on a dry weight basis" divided by "concentration in the surface 0 - 5 cm soil on a dry weight basis" was calculated for each element and ^{137}Cs as shown in Table 2 [20, 25]. The transfer factors of higher than 1 were observed for ^{137}Cs, K and Rb in mushrooms and for Ca in plants. The transfer factors between 0.1 and 1 were observed for Cs and Mg in mushrooms and ^{137}Cs, K, Rb, Mg and Sr in plants. The transfer factors for the others were lower than 0.1. The transfer factors of ^{137}Cs were higher than those of stable Cs for both plants and mushrooms. The surface 0 - 5 cm soil is a mixture of organic materials and minerals (sand). Stable Cs is originally contained in the mineral components and this stable Cs is difficult for plants and mushrooms to take up.

Tsukada et al. [32], who investigated ^{137}Cs and stable Cs in 21 different species of

46

fungi, report on transfer factors for ^{137}Cs due to weapons fallout being on the average a factor of 2.4 larger compared to the corresponding transfer factors for stable Cs. The transfer factor was estimated by using the concentration of Cs in the whole layer rich in organic material such as litter and humus. The higher transfer factor for fallout ^{137}Cs compared to stable Cs observed in Tsukada et al. might be attributable to the differences in physico-chemical forms of the nuclides, e.g. existence of the strong bound fraction of stable Cs with clay minerals, as suggested by these authors.

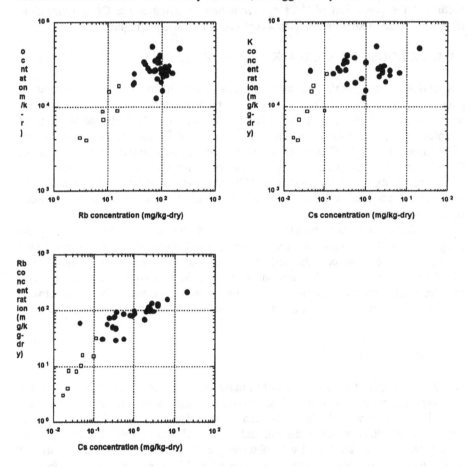

Figure 3. Relationship among K, Rb and Cs in mushrooms ( ) and plants ( ) collected from a pine forest in Tokai-mura, Japan [20].

TABLE 2. Tentative transfer factors of alkali and alkaline earth elements and ^{137}Cs for mushrooms and plants [20]

Sample	[137]Cs	Na	K	Rb	Cs	Mg	Ca	Sr	Ba	Al
Mushrooms	3.9	0.06	1.4	1.6	0.80	0.2	0.03	0.03	0.02	0.010
Plants	0.2	0.05	0.4	0.2	0.03	0.3	1.40	0.60	0.04	0.005

TABLE 3. Transfer factors for mushrooms and berry plant for stable [133]Cs and Chernobyl [134]Cs, both calculated with respect to the layers from which radiocesium is taken up (at Hochstadt, Germany) [34]

Species	Horizon	Cs-134*	Cs-133
Lepista nebularis	L&Of	0.5 ± 0.10	1.0 ± 0.4
Xerocomus badius	L&Of&Oh	21.8 ± 2.30	13.1 ± 2.8
Hydnum repandum	Oh	43.6 ± 7.40	24.2 ± 1.9
Russula cyanoxantha	Oh&Ah	Oh: 2.9 ± 1.90 Ah: 15.3 ± 10.2	Oh: 6.2 ± 3.8 Ah: 10.5 ± 6.5
Vaccinium myrtillus (berry leaves)	L&Of	2.5 ± 0.50	3.9 ± 1.5

*The decay correction for Cs-134 was made as to May 1986.

In contrast, the strong bonding of stable Cs in the organic layers might be negligible, if the mineral content is low and, on favorable conditions, the physico-chemical properties of stable elements and radionuclides are expected to be similar. Four different mushroom species and one berry plant collected from a forest in Hochstadt, Germany, were analyzed for [134]Cs and stable Cs [34]. The stable Cs was analyzed in National Institute of Radiological Sciences, Japan, by ICP-MS. This site has been under investigation since 1987, and the soil layers from which certain species of mushrooms take up radiocesium were estimated by using the ^{137}Cs/[134]Cs ratios [35]. Transfer factors for [134]Cs and stable Cs were calculated on the basis of the soil layers, from which the corresponding species take up radioactive [134]Cs (Table 3). Most species take up [134]Cs from organic layers (L, Of and/or Oh). The resulting transfer factors for stable Cs were close to the corresponding transfer factors for [134]Cs, indicating that bio-availabilities of Chernobyl [134]Cs and stable Cs are similar. With this result, it is possible to predict the future contamination of radiocesium in understory vegetation of forest ecosystems; changes will occur due to physical decay of radiocesium, and due to migration of radiocesium to or from the rooting zone of the corresponding species. There will be no future change (aging effect) of the bio-availability of radiocesium in the organic layers for the investigated species and site.

Analyses for stable elements have provided much information on the behavior of elements which are related to radionuclides in forest ecosystems. Such studies could be useful in predicting the migration of long-lived radionuclides in contaminated forests.

4. References

1. Yamagata, N., Matsuda, S. and Chiba, M. (1969) Radioecology of cesium-137 and Strontium-90 in a forest, *J. Rad. Res.* 10, 107-112.
2. Heinrich, G., Muller, H. J., Oswald, K. and Gries, A. (1989) Natural and artificial radionuclides in selected Styrian soils and plants before and after the reactor accident in Chernobyl, *Biochem. Physiol. Pflanz.* 185, 55-67.
3. Schimmack, W., Förster, H., Bunzl, K. and Kreutzer, K. (1993) Deposition of radiocesium to the soil by stemflow, throughfall and leaf-fall from beech trees, *Radiat. Environ. Biophys.* 32, 137-150.
4. Wirth , E., Hiersche, L., Kammerer, L., Krajewska, G., Krestel, R., Mahler, S. and Römmelt, R. (1994) Transfer equations for cesium-137 for coniferous forest understorey plant species, *Sci. Total Environ.* 157, 163-170.
5. Thiry, Y. and Myttenaere, C. (1993) Behaviour of radiocaesium in forest multilayered soils, *J. Environ. Radioactivity* 18, 247-257.
6. Bunzl, K., Kracke, W. and Schimmack, W. (1995) Migration of fallout $^{239+240}$Pu, ^{241}Am and ^{137}Cs in the various horizons of a forest soil under pine, *J. Environ. Radioactivity* 28, 17-34.
7. Muramatsu, Y., Yoshida, S. and Sumiya M. (1991) Concentrations of radiocesium and potassium in basidiomycetes collected in Japan, *Sci. Total Environ.* 105, 29-39.
8. Yoshida, S. and Muramatsu, Y. (1994) Accumulation of radiocesium in basidiomycetes collected from Japanese forests, *Sci. Total Environ.* 157, 197-205.
9. Yoshida, S. and Muramatsu, Y. (1994) Concentration of radiocesium and potassium in Japanese mushrooms, *Environ. Sci.* 7, 63-70.
10. Yoshida, S., Muramatsu, Y. and Ogawa, M. (1994) Radiocesium concentrations in mushrooms collected in Japan, *J. Environ. Radioactivity* 22, 141-54.
11. Sugiyama, H., Shibata, H., Isomura, K. and Iwashima, K. (1994) Concentration of radiocesium in mushrooms and substrates in the sub-alpine forest of Mt. Fuji Japan, *J. Food Hyg. Soc. Japan* 35, 13-22.
12. Schell, W.R., Berg, M.T., Myttenaere, C. and Massey, C.D. (1994) A review of the deposition and uptake of stable and radioactive elements in forests and other natural ecosystems for use in predictive modeling, *Sci. Total. Environ.* 157, 153-161.
13. Schell, W.R., Linkov, I. Myttenaere, C. and Morel, B. (1996) A dynamic model for evaluating radionuclide distribution in forests from nuclear accidents, *Health Phys.* 70, 318-335.
14. Myttenaere, C., Schell, W.R., Thiry, Y., Sombre, L., Ronneau, C. and van der Stegen de Schrieck, J. (1993) Modelling of Cs-137 cycling in forests: recent developments and research needed, *Sci. Total Environ.* 136, 77-91.
15. Eaton, J.S., Likens, G.E. and Bormann, F.H. (1973) Throughfall and stemflow chemistry in a northern hardwood forest, *J. Ecol.* 61, 495-508.
16. Likens, G.E., Bormann, F.H., Pierce, R.S., Eaton, J.S. and Johnson, N.M. (1977) Biogeochemistry of a forested ecosystem. Springer-Verlag, New York.
17. Mayer, R. and Ulrich, B. (1977) Acidity of precipitation as influenced by the filtering of atmospheric sulphur and nitrogen compounds - its role in the element balance and effect on soil, *Water, Air and Soil Pollut.* 7, 409-416.
18. Yoshida, S. and Ichikuni, M. (1989) Role of forest canopies in the collection and neutralization of airborne acid substances, *Sci. Total Environ.* 84, 35-43.
19. Liechty, H.O., Mroz, G.D. and Reed, D.D. (1993) Cation and anion fluxes in northern hardwood throughfall along an acidic deposition gradient, *Can. J. For. Res.* 23, 457-467.
20. Yoshida, S. and Muramatsu, Y. (1998) Concentrations of alkali and alkaline earth elements in mushrooms and plants collected in a Japanese pine forest, and their relationship with ^{137}Cs, *J. Environ. Radioactivity* 41, 183-205.
21. Yoshida, S., Muramatsu, Y., Tagami, K. and Uchida, S. (1996) Determination of major and trace elements in Japanese rock reference samples by ICP-MS, *Intern. J. Environ. Anal. Chem.* 63, 195-206.
22. Yoshida, S. and Muramatsu, Y. (1997) Determination of major and trace elements in mushroom, plant and

soil samples collected from Japanese forests, *Intern. J. Environ. Anal. Chem.* 67, 49-58.

23. Ban-nai, T., Yoshida, S. and Muramatsu, Y. (1994) Cultivation experiments on uptake of radionuclides by mushrooms, *Radioisotopes* 43, 77-82 (in Japanese).

24. Ban-nai, T., Muramatsu, Y., Yoshida, S., Uchida, S., Shibata, S., Ambe, S., Ambe, F. and Suzuki, A. (1997) Multitracer studies on the accumulation of radionuclides in mushrooms, *J. Radiation Res.* 38, 213-218..

25. Haselwandter, K., Berreck, M. and Brunner, P. (1988) Fungi as bioindicators of radiocesium contamination: pre- and post- Chernobyl activities, *Trans. Br. Mycol. Soc.* 90, 171-174.

26. Baldini, E., Bettoli, M. G. and Tubertini, O. (1989) Further investigation on the Chernobyl pollution in forest biogeocenoses, *Radiochim. Acta* 46, 143-144.

27. Mascanzoni, D. (1990) Uptake of ^{90}Sr and ^{137}Cs by mushrooms following the Chernobyl accident. In Proc. CEC Workshop, Transfer of Radionuclides in Natural and Semi-Natural Environments, eds. G. Desmet, P. Nassimbeni & M. Belli. Elsevier Applied Science, London, pp. 459-467.

28. Katayama, Y, Okada, N., Ishimaru, Y., Nobuchi, T. and Aoki, A. (1986) Behavior of radioactive nuclides on the radial direction of the annual ring of Sugi, *Radioisotopes* 35, 636-638 (in Japanese).

29. Kohno, M., Koizumi, Y., Okumura, K. and Mito, I. (1988) Distribution of environmental cesium-137 in tree rings, *J. Environ. Radioactivity* 8, 15-19.

30. Chigira, M., Saito, Y. and Kimura, K. (1988) Distribution of ^{90}Sr and ^{137}Cs in annual tree rings of Japanese cedar, Cryptomeria Japonica D. Don., *J. Radiat. Res.* 29, 152-160.

31. Momoshima, N. and Bondietti, E. A. (1994) The radial distribution of ^{90}Sr and ^{137}Cs in trees, *J. Environ. Radioactivity* 22, 93-109.

32. Tsukada, H., Shibata, H. and Sugiyama, H. (1998) Transfer of radiocaesium and stable caesium from substrata to mushrooms in a pine forest in Rokkasho-mura, Aomori, Japan, *J. Environ. Radioactivity* 39, 149-160.

33. Ronneau, C., Sombre, L., Myttenaere, C., Andre, P., Vanhouche, M. and Cara, J. (1991) Radiocaesium and potassium behaviour in forest trees, *J. Environ. Radioactivity* 14, 259-268.

34. Rühm, W., Yoshida, S., Muramatsu, Y., Steiner, M. and Wirth, E. (1998) Distribution patterns for stable ^{133}Cs and their implications with respect to the long-term fate of radioactive ^{134}Cs and ^{137}Cs in a natural ecosystem, *J. Environ. Radioactivity* (submitted).

35. Rühm, W., Kammerer, L, Hiersche, L. and Wirth, E. (1997) The ^{137}Cs/^{134}Cs ratio in fungi as an indicator of the major mycelium location in forest soil, *J. Environ. Radioactivity* 35, 129-148.

RADIOCAPACITY OF FOREST ECOSYSTEMS

Yu. KUTLAKHMEDOV
Institute of Radioecology, UAAS, 14 Tolstoy str., Kiev, 252033, UKRAINE

V. DAVYDCHUK
Institute of Geography, NASU, 44 Volodimirskaya,, Kiev, 252034, UKRAINE

G. ARAPIS
Agricultural University of Athens, 75 Iera Odos Athens, 11855, GREECE

V. KUTLAKHMEDOVA-VYSHNYAKOVA
Institute of Radioecology, UAAS, 14 Tolstoy str., Kiev, 252033, UKRAINE

1. Introduction

Vast territories have been exposed to radioactive fallout from to the Chernobyl accident and require integrated assessment of the ability of ecosystems to retain and accumulate radionuclides. The fate of radionuclides in an ecosystem is determined by its capacity to retain radionuclides, defined as radiocapacity. Radiocapacity provides the absolute quantity of radionuclides that can be retained in an ecosystem without damaging its living ability. The estimation of this parameter for different types of ecosystems, allows an assessment to be made of many properties of the ecosystem, a prediction of the fate of the radionuclides and their public health impact.

This paper presents our approach to evaluation of the radiocapacity of forest ecosystems. Coniferous and mixed forests occupy about 40 % of the Chernobyl Nuclear Power Plant (ChNPP) Exclusion zone and contain more than 50 % of the total radionuclide inventory. In comparison with data obtained after the Kyshtym accident suggests that the radionuclide deposition in forest ecosystems is 3-7 times higher than that in open landscapes.

The main approach of our research is to estimate the contribution of a radionuclide source (forest ecosystems) to the amount and rate of radionuclide released to the Dnieper-river and into the Black Sea. It is important to investigate and to estimate the surface radionuclide transport on river slopes during normal conditions as well as in the case of forest fires. Our data have shown that evaluation and prediction of the radiocapacity of large ecosystems is a very useful way of studying ecosystem radioecology. The radiocapacity can be used to quantify radionuclide accumulation in the ecosystem without damaging its basic characteristic. The factor of radiocapacity determines what proportion of which radionuclides can be retained in the ecosystem for a long time, and consequently, what proportion of these can easily migrate to other environments (for example to the chain of the Dnieper reservoirs and into the Black Sea).

51

I. Linkov and W.R. Schell (eds.), Contaminated Forests, 51–61.
© 1999 *Kluwer Academic Publishers. Printed in the Netherlands.*

2. Peculiarities of radionuclide behavior in forest ecosystems

Forests can accumulate large amounts of radionuclides and can prevent radionuclide transfer beyond the boundaries of the polluted territory. Therefore, the forests influence the migration of radionuclides on a global scale. Research of radionuclide behavior in a forest ecosystem is an important problem of radioecology, especially in connection with large-scale radionuclide contamination after atmospheric releases as a result of large nuclear accidents, such as Chernobyl or Kyshtym.

The comparative level of radionuclide contamination of different landscapes can be illustrated by the data obtained after the Kyshtym accident (Table 1) [1]. Contamination of arable lands is assumed to be 1. It is evident that the contamination of the forest ecosystems is 3-7 times higher than in arable types of landscapes.

TABLE 1. Relation of contamination of various elements in a landscape
in comparison with arable lands (is accepted for 1)

Type of a landscape	Contamination (relative units)
Clearing	1.8
Meadow	7.3
Forest:	
Deciduous	3.2
Pine	4 - 6
Birch	6
Mixed	7

The following questions are high priority for understanding contaminated forest ecosystems:
1. How does radioactive contamination effect the survival and function of forest ecosystems (especially radiosensitive pine forests)?
2. How important are forest ecosystems in retaining of radionuclides and in protecting neighboring territories from possible secondary contamination?
3. What is the acceptable radionuclide deposition for possible utilization of contaminated forests?

Let us consider radionuclide interception by forests from atmospheric transport. The wind flow over arable land splits at the forest border: a part of it goes over the canopy and a part penetrates into the forest where it is filtered by the canopy. The large surface area of the tree canopy allows an effectively sorption and retention of radionuclides. A significant portion of radionuclides deposit on the border of forest at the first 500 m. This is the so called the border effect. Radionuclides are especially well retained by single standing trees (up to 30 times higher retention) than by a larger forested area [2].

A retention coefficient is recommended for an evaluation of the ability of the forest to retain radionuclides. The retention coefficient is determined by the ratio of the quantity of radionuclides retained by forest to the total radionuclide deposition. This retention coefficient varies within a broad range depending on the type and age of the plot, seasonal and meteorological conditions and the chemical form of the radionuclide.

The forest canopy is particularly effective at retaining fallout radionuclides in the form of particles and gas (especially I^{131}). Deciduous trees are capable of retaining 10-20% of the annual quantity of radionuclides in atmospheric precipitation. The canopy of a coniferous trees is able to retain more (20-30 %). The value of the retention coefficient depends on both duration and intensity of rain and snow. This factor for radionuclides in forests varies from 0.2 to 1.0.

Part of the radionuclides deposited on the forest canopy can penetrate into internal tissues of the plant. Surface contamination and absorption of radionuclides (such as P^{32}, K^{40}, Co^{60}, Sr^{90}, Ru^{106}, Rh^{106}, Cs^{137}, Ce^{144}) is an important pathway for radionuclide retention. Some radionuclides (P^{32} and K^{40}) are capable of moving freely inside plants and to accumulate in different organs. The mobility of other radionuclides is relatively insignificant.

A large portion of the radionuclides become concentrated in the forest litter shortly after deposition because of weathering from trees. The time required to remove about 95 % of radionuclides is estimated to be one year for deciduous, and three years for coniferous forests.

Both equilibrium and dynamic models are widely used for the description of radionuclide migration in forests. Several compartmental models for forest ecosystems have been developed. The FORESTPATH model developed by Schell and Linkov [2-3] will be used in our work.

Mushrooms and berries are very important components of the forest ecosystems. Concentration of radionuclides in mushrooms is small shortly after the release. After a few years, radionuclide contamination of mushrooms becomes significant. Radionuclide concentration in mushrooms currently reaches 370 kBq/kg^{-1} in the ChNPP 30 km zone. Forest mushrooms and berries contribute significantly to the individual and collective doses for the population living in the contaminated areas of Ukraine and Belorus (data for the Rovno region of Ukraine are presented in [5]) Therefore special monitoring of radionuclide concentrations in mushrooms is required to limit the radiation dose to the population.

3. Problem of a radiocapacity of terrestrial ecosystems

3.1. EVALUATION OF RADIOSENSIBILITY OF FOREST ECOSYSTEMS

The radiocapacity of a terrestrial ecosystem can be defined as its ability to retain and accumulate radionuclides without serious damage to the ecosystems function and well-being. It is obvious that the radiocapacity of a terrestrial ecosystem decreases in the case of its destruction or illness [6]. The question is: what level of radionuclide deposition and irradiation dose is necessary to cause such damage?

It is known that conifers are the most radiosensitive forest species. Loss and depression of young tree growth begins at the dose of 10 Gy; depression of adult trees begins at doses 50 Gy (Table 2) [4]. It is easy to estimate that the deposition on the pine ecosystem of the Red Forest in the Chernobyl Exclusion Zone was about 2000 Ci/km^{-2} (or 370 kBq/kg^{-1}). The decline and/or destruction of the ecosystem is very probable in this cause for the Red Forest.

TABLE 2. Radiosensitivity of the main types of ecosystems to ionizing radiation (dose in rad)

Ecosystem Type	Chromosome Number	Chromosome Volume (mk³)	Dose depressing growth	Lethal Dose
Coniferous forests	24	9 – 48	200-1000	500- 2700
Birch-maple forests	24- 48	2-21	400-3800	110-10000
Oak-chestnut	24	6-48	200-2500	500-6500
Leaf-fall forests	24-48	2-7	1300-3800	3600-1000
Grassy	40	6,4	2300	9200
Agrocoenosis (field culture)	20-42	14	1000-5800	4-10000

TABLE 3. Pine growth reduction resulting from application of 1 Ci/km² (200 Bq/kg¹) to soil

Response	Deposition Density Ci/km²	Radioactivity uCi/kg
Complete growth depression	3500-4000	17-20
Dying of branches, possible recovery	1500-2000	8-10
Increased growth of branches	500-700	2.5-3.5
Formation haggen broom	300-500	1.5-2.5
Violation of monopodial branching at trees	200-500	1 - 2.5
Tolerated doses	100	0.5

Response of a pine forest to different radiation depositions is shown in a Table 3 [4]. Significant damage to a forest ecosystem is expected at the soil deposition of 500-1000 Ci/km² or 10^{-5} Ci/kg¹. This value can be used to characterize the radiocapacity of the terrestrial ecosystems. The total radiocapacity of the ecosystem is then the product of surface deposition density (500 Ci/km²) and the total area of the ecosystem [7].

Contaminated ecosystems can retain and accumulate radionuclides. It is known that forests can hold up to 95-97 % of the initially deposed radionuclides for a long time. As low as 3-5 % of radionuclides leave the forest ecosystem annually (primarily by the surface runoff). Therefore the radiocapacity factor, i.e., the proportion of radionuclides retained in the forest ecosystems to the total deposited is estimated to 0.95-0.97.

At the same time, forests are potential sources of radionuclides for neighboring ecosystems. The following radionuclide pathway is common for the territory of Ukraine: forest - meadow - floodplain - river. Our field research in the floodplain of the Uzh-river (Novoselky testing area) revealed that the forest ecosystems serve as a source of radionuclides for the aquatic ecosystem. The major transport process is by surface runoff. This process results in a significant increase in radionuclide deposition in the marginal terrace of the river (Fig 1.).

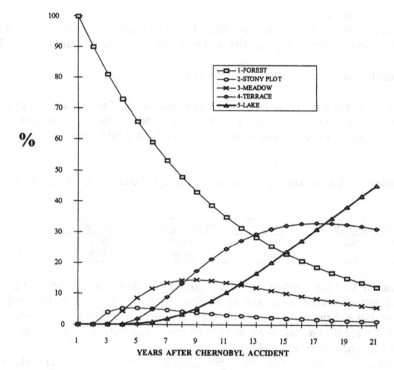

Fig. 1. Radiocapacity of Slope Ecosystem

The accumulation ability of forest ecosystems can be easily changed by forest fires. The forest fires occurring in the Chernobyl Exclusion Zone in 1992 affected areas with high radionuclide deposition. Sharp decrease in the radiocapacity and the corresponding increase in radionuclide migration are expected due to these forest fires. Our monitoring of the slope system of "Burned forest" revealed redistribution of radionuclides as fast as two years after the fire. The processes of radionuclides in surface runoff are well observable in the lower part of the slope. These observation suggests that the radiocapacity of the forest ecosystem can be decreased by extensive forest fires because of the surface runoff and secondary redistribution. The Exclusion Zone can become a dangerous source of secondary radioactive contamination of the air and water environment in the Ukraine.

3.2. RADIOCAPACITY OF FOREST ECOSUSTEMS IN THE CHERNOBYL EXCLUSION ZONE

The following parameters influence the radiocapacity of forests:
- Type forest ecosystem (coniferous, mixed, leafed)
- Biomass of the forest ecosystem compartments

- Landscape characteristics
- Chemical form of the deposited radionuclides. (In the Chernobyl Exclusion Zone, the fuel particles have dissolution properties different from regions far from the NPP accident.)
- Total radionuclide inventory and the area of the ecosystem.

Knowledge of these parameters should be sufficient to calculate the amount of a radionuclide leaving the forest ecosystem and accumulated in different ecosystem compartments. The biomasses of common ecosystems of the Chernobyl Exclusion Zone are given in Table 4.

TABLE 4. Biomass of forest ecosystems in the Chernobyl Exclusion Zone (t/km^2 dry weight).

Type of a forest	Young	Middle-aged	Matured
Coniferous	15000	23000	25000
Mixed	22000	32000	38000

Immediately after the accident, the largest portion of radionuclides (82-92 %) are concentrated in forest litter while the remaining fraction is in the upper soil layer (0-5 cm). According to our data and the data collected by the Ukrainian Forest Service the average accumulation of ^{137}Cs in the biomass is 2.6 nCi/kg^1 for coniferous and 7 nCi/kg^1 for mixed forest where the initial deposition is 1 Ci/km^2.

Our data and information from other publications show that the annual surface runoff from a forest located on a slope of 1-2 degrees does not exceed 0.2% of the total deposition. For a forest on a slope of 3-4 degrees, this parameter is estimated to be 0.5%. For a steeper slope (6-8 degrees) the runoff is about 1% (Table 5). This evaluation can be an initial basis for the conservative estimations that should be refined by the field measurements.

Table 5 shows that 8 years after the Chernobyl accident the general runoff from the contaminated (10 Ci/km^2) territory varies form 1.5 to 7.4% of the total deposition. This means that no significant loss of radionuclides from the forest ecosystem is expected in the near future. For the forest located in flat areas, this is determined by high radiocapacity of forest ecosystems. At higher inclination of the forest area, surface runoff of radionuclide is controlled by vertical migration within the ecosystems and accumulates in areas with lower elevation. Therefore radionuclides do not necessarily migrate towards the Kiev reservoir even from forests located in the hills since they can be captured within the ecosystem.

Landscape maps of the Chernobyl exclusion Zone, overlayed with the radionuclide deposition maps allows estimates to be made of radiocapacity of the forest in the Chernobyl Exclusion Zone (Table 6). A high radiocapacity is characteristic of the Chernobyl forests.

SLOPE SYSTEM "NOVOSELKY"

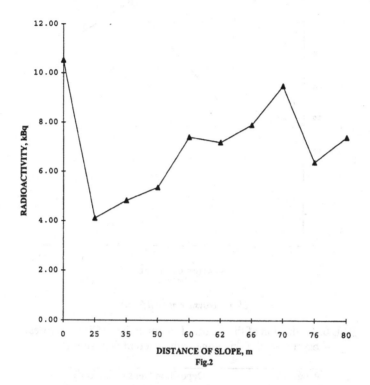

Fig. 2. Slope system "Novoselky"

TABLE 5. Radionuclide removed by the surface runoff from an ecosystem (area 1 km^2, and deposition 10 Ci km^{-2}) (% of total inventory) and radiocapacity (%) for different slope steepness.

	Coniferous forest			Mixed forest		
Slope, degrees	1-2	3-4	6-8	1-2	3-4	6-8
Year						
1986	0.05	0.2	0.3	0.12	0.3	0.6
1987	0.12	0.3	0.6	0.16	0.4	0.8
1988	0.2	0.5	1	0.2	0.5	1
1989	0.2	0.5	1	0.2	0.5	1
1990	0.2	0.5	1	0.2	0.5	1
1991	0.2	0.5	1	0.2	0.5	1
1992	-	-	-	-	-	-
1993	-	-	-	-	-	-
Total loss (8 years)	1.5	3.5	7	1.6	3.7	7.4
Radiocapacity	98.5	96.5	93.0	98.4	96.3	92.6

58

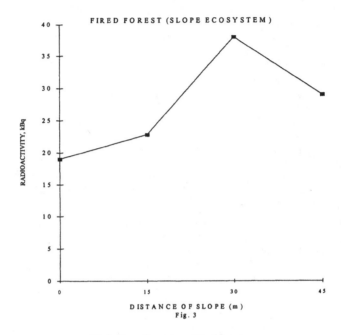

FIRED FOREST (SLOPE ECOSYSTEM)

Fig.3. Slope Ecosystem at fired forest

TABLE 6. Evaluation of the runoff of radionuclides (Ci) from forest ecosystems of 30-km zone with different deposition

Parameter	Deposition Density (Ci/km^2)		
	5-15	15-40	>40
General stock, Ci	5630	8700	34600
Runoff, Ci	104,3	108,2	526.8
Radiocapacity	98.2	98.8	98.5
Average radiocapacity		98.5	

3.3. RADIOCAPACITY OF MOUNTAIN LANDSCAPES

The radiocapacity for an experimental plot on a plain (with a small slope) is high for forests in the Chernobyl Exclusion Zone. Some 98-99% of the total radionuclides initially deposited are strongly retained in the ecosystem. For landscapes with 4-8 degrees incline the total loss is about 3-5 % over the first eight years following the accident.

The radiocapacity is clearly different for mountain landscapes with inclines of 20-30 degrees. Table 7 presents estimated radiocapacities for the coniferous forest ecosystem on a uniform mountain slope (20-30 degrees). The total area of the forest is 1 km^2, radionuclide contamination is assumed to be 10 Ci/km^2. Radionuclides migrate from the forest to a rocky area of 1 km^2. There is then an undisturbed meadow (1km^2) covered by lucerne, corn, etc. A small lake (1 km^2 area and 10 m deep) is the radionuclide sink.

TABLE 7. Basic parameters of the slope ecosystems (^{137}Cs)

Ecosystems	Slope (degrees)	Biomass (ton/km^2)	Biomass Transport (%)
Forest	10	15000	1
Rocky area	30	-	-
Meadow	15	50	5
Terrace	5	500	20

If forest ecosystems are the most contaminated, a general strategy of countermeasures can be developed for different portions of the slope ecosystems. During the first 5 years after contamination, the most effective countermeasure is to collect radionuclides at the forest boundary. For the subsequent 10 years, remedial policies should be applied to the forest as well as to the meadow ecosystems. In the long run, the entire slope ecosystem must be cleaned.

One of the possible effective countermeasures is to establish a series of special protective terraces at the forest boundary. Radionuclides removed from the forest will then be collected at the terrace. Construction of such protective terraces can be effective countermeasures even 12 years after the Chernobyl Accident. Our estimations show that such protective terraces can intercept up to 90 % all radionuclides leaving the forest ecosystem.

We have studied a slope ecosystem with a current cesium contamination of 3-4 Ci/km^{-2}. Both the initially contaminated upper polluted part of the slope (2-3 km^2) and downhill area (about 5 km^2) are arable land with quite homogenous contamination. If a protective terrace were created during the first years after the accident, the lower arable land would be protected from some 3Ci of Cs.

3.4. RADIOCAPACITY OF BURNED FOREST ECOSYSTEMS IN THE CHERNOBYL EXCLUSION ZONE

Forest fires are frequent in the Chernobyl Exclusion Zone. A major fire occurs once every 3-4 years. Experimental data for a burned forest in the Chernobyl Exclusion Zone shows that the radionuclide runoff from the slopes increases by more than one order of magnitude after forest fires. Using the probability of forest fires, the total forest and radionuclide inventory in the Chernobyl Exclusion Zone, the radionuclide runoff from forest ecosystems on different slopes can be estimated. The results for three levels of Cs^{137} deposition are shown in Table 8. Over the next 20-30 years, forest fires will increase the radionuclide runoff from forests by 136 Ci.

Table 9 summarizes data on the influence of forest fires on radionuclide accumulation in Chernobyl forests. The radiocapacity of burned forests decreases by up 0.9%. These model estimations are supported by our 1994 studies of a forest in the Kupovato district of the Chernobyl Exclusion Zone burned in 1992.

TABLE 8. Radionuclide runoff from the Chernobyl forests after forest fires.

Slope (degrees) Contamination (Ci/km²)	Radionuclide Runoff (Ci)				Total Runoff (Ci)
	0-1	1-2	2-4	4-8	
5-15	0	5.0	43.4	-	48.4
15-40	0	5.8	9.7	-	15.5
> 40	0	50.4	4.8	16.6	71.8

TABLE 9. Annual radionuclide runoff from Chernobyl normal and burned forest ecosystems

Contamination		Radionuclide Runoff (Ci)			Total Runoff (Ci)
		5-15	15-40	>40	
Normal Forest	Total Inventory, Ci	5630	8700	34600	
	Annual Runoff, Ci	104.3	108.2	526.8	739.3
	Radiocapacity	98.2	98.8	98.5	
Burned Forest	Annual Runoff, Ci	152.7	123.7	598.6	875
	Radiocapacity	97.3	98.6	98.3	

4. Conclusions

Coniferous ecosystems can sustain their development if radionuclide deposition density is less than 500 Ci/km². This value can be accepted as the lower limit for the radiocapacity of terrestrial ecosystems. Forest ecosystems can efficiently retain radionuclides; the factor of radiocapacity (fraction of retained radionuclides) is about 0.95-0.97 for forests. This high retention capacity makes contaminated forests a long term sink for radionuclides and, at the same time, a source for neighboring meadow and aquatic ecosystems.

In the present paper we calculate the loss of radionuclides from forest ecosystems in the Chernobyl Exclusion Zone and estimate radiocapacity of these ecosystems. Over the last 12 years, less than 2 % of the initial deposition have been lost from the forest ecosystems. Therefore, the factor of radiocapacity for the Chernobyl forests exceeds 98 %. In extreme environmental conditions (forest fires, unusually high snow melting, high spring flood, extreme soil freezing) the annual surface runoff can increase to 3-5 % or even to 15-20 % of the total deposition. These conditions are very dangerous since they lead to secondary radionuclide resuspension and significant contamination of neighboring areas. These processes are especially dangerous in floodplain and meadow ecosystems.

5. References

1. F. Tikhomirov. Ionizing radiation's influence on ecological systems. M., Atomizdat, 1972, 248 pp. (in Russian).

2. W.Schell, I.Linkov and others. A forest and natural ecosystem model for short and long term distribution of radionuclide from nuclear accidents. in: Proceedings of an international symposium IAEA, 830-836 p.
3. W. Schell, I. Linkov. Human radiation dose resulting from forests contaminated by radionuclides: Generic model and applications to the Chernobyl ecosystems. in: Science and Total Environment (in press).
4. Kyshtym accident by the large plan. Priroda V5, 1990. pp 46-76.
5. Yu. Kutlakhmedov, A. Jouve., V. Davydchuk and others. Strategy of decontamination. ECP-4. Final Report. EUR 16530 en.
6. Antropogenic radionuclide anomaly in plants. Edited by D. Grodzinsky. Kiev. Lybid, 1991, p. 264.
7. Yu. Kutlakhmedov and V. Korogodin. Fundamentals of Radioecology. Kiev, High School, 1997 (in press).

WOODY BIOMASS PRODUCTION AND RADIOCAESIUM ACCUMULATION RATE IN PINE (*Pinus sylvestris* L.) FROM A CONTAMINATED FOREST IN THE VETKA AREA

Y. THIRY
Radiation Protection Research Unit, SCK-CEN, Boeretang, 200, 2400 Mol, BELGIUM

T. RIESEN
Division for Radiation Protection and Waste Management, PSI, 5232 Villigen PSI, SWITZERLAND

N. LEWYCKYJ
Radiation Protection Research Unit, SCK-CEN, Boeretang, 200, 2400 Mol, BELGIUM

1. Introduction

After the Chernobyl accident, radioactive deposits affected wide forest and agricultural areas in Europe. In Belarus, 23% of the national forest cover are contaminated to radiocaesium levels >37 kBq.m^{-2} (1 Ci.km^{-2}). The forest ecosystem as other perennial culture with a high biomass density helps in the long-term stabilisation of the contamination by acting as a reservoir of radionuclides. Nevertheless, the radiocaesium tends to accumulate in the vegetation woody biomass leading to increasing levels of radiopollutants in woody products with time.

In large areas including slightly contaminated forest, nominally those with from 1 to 15 Ci.km^{-2} of ^{137}Cs, wood may be too contaminated for traditional harvesting and processing in a few years. The evolution of this problem which affect more than $1.2 \ 10^6$ ha of contaminated forest in Belarus is considered a priority. In this regard, the prognosis on radiocaesium distribution and accumulation in wood are needed to account for the radioecological and economical limitations associated with the future management and economic value of contaminated forest areas.

In this study, we considered a 17 year old scots pine stand located near Vetka (Belarus) and affected by radiocaesium and radiostrontium deposition levels of about 30 and 2 Ci.km^{-2}, respectively, following the Chernobyl accident. Our purpose was to investigate the time-trends in cation and radiocaesium distribution in annual rings in order to characterize and quantify the rate of radiocaesium accumulation in trunk biomass as a result of growth.

I. Linkov and W.R. Schell (eds.), Contaminated Forests, 63–70.

2. Materials and method

In December 1997, a sampling trip was made in a pine (*Pinus sylvestris* L.) forest situated near Vetka (Belarus) and contaminated by Chernobyl deposition. We investigated a 17 years old plantation established on a soddy-podzolic sandy soil. The mean surface soil contamination by ^{137}Cs was 1.46 10^6 Bq.m^{-2}. The CBH (circumference at breast height) of all trees were measured in three circular plots with a radius of 6 m. Breast height stem sections of scots pine were taken from three freshly cut trees of average circumference category in each plot. To avoid physical or chemical alteration of the fresh wood, each bole section was wrapped in a polyethhylene bag and stored at 4°C. Before the wood samples were extracted for chemical analysis, a 2-cm disc was cut from each bole section with a band saw. The surface of each disk was polished and scanned to reveal annual growth rings. The cross dating was made *de visu*. The mean annual ring width data was obtained from measurement within 0.25 mm along four radii and used to reconstruct the individual dendrochronological curves. The last ring (1997) and outer bark were separated and prepared separately for analysis. Samples of annual tree-rings formed before 1997 were cut off from the half of each disc with a chisel. All fresh materials sampled from wood and bark were weighted and then dried at 105°C for 24h for a dry weight. After burning at 550°C, the ash samples were dissolved in 10 ml of 2N HNO$_3$. The acid solutions were measured for their ^{137}Cs concentrations using a Ge(Li) gamma-ray spectrometer and the radioactivity was decay-corrected to 26 April 1986. Potassium and calcium content in wood samples were measured from aliquots of acid solutions using of atomic absorption spectrometry.

3. Results and discussion

3.1 STAND CHARACTERISTICS

Figure 1 describes the mean frequency distribution of the trees in the stand as a function of CBH categories. Table 1 shows the mean stand characteristics according to the stand density measured in each plot and to the dendrochronology parameters observed for one tree sampled in the mean CBH category of each plot.

With reference to the English production forecast table for scots pine, the stand belongs to the class 2 (of 6) of fertility with a potential mean annual increment of 12 m^3.ha^{-1}. The mean density and height parameters are also in good agreement with those of the mean stand described in the yield table for the same age class.

TABLE 1. Mean stand characteristics

Dendrochronology parameter	Mean value (standard deviation)
Height (m)	7.78 (0.55)
CBH under bark (cm)	24.9 (1.7)
Stand density (number trees ha^{-1})	3892 (386)

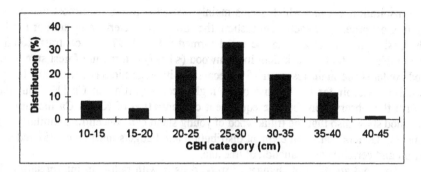

Fig. 1. Mean tree frequency distribution as a function of CBH categories

3.2 QUALITATIVE ASPECT OF THE RADIOCAESIUM AND CATION DISTRIBUTION IN STEM WOOD

The dynamics of radiocaesium and nutrient accumulation in the trunk wood has been studied from extraction and chemical analysis of growth ring samples taken on a stem section sampled at breast height of three representative trees. Figure 2 shows the radial distribution of radiocaesium, potassium and calcium concentrations compared to the relative water content for each tree.

Between each tree, the discrepancy in element content distribution is very low. The concentrations of ^{137}Cs ranged from 2.5 to 7.5 Bq g^{-1} in the wood formed till 1996, increase drastically in the ring formed in 1997 and decrease in the bark (indicated as 98) (Fig.2). The same general pattern is observed for K and Ca which show comparable mean concentration ranged from 300 to 800 µg.g^{-1} in the wood formed till 96. Compared to ^{137}Cs and K, the Ca concentration in bark (4000 µg.g^{-1}) is however higher than in stemwood (<1997).

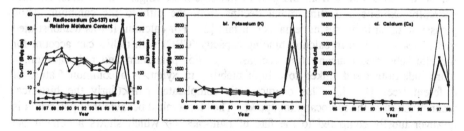

Fig. 2. Radial distribution of (a) radiocaesium, (b) potassium and (c) calcium concentrations in comparison to the relative water content measured in the wood section for each tree

The last ring formed in 1997 shows similarly higher content in all elements than those measured in the previous rings. The similar broad peak for ^{137}Cs, K and Ca concentrations observed in the newly formed ring is also associated with the highest water content. These observations suggest that the cations upward movement follows the transpiration streams which seems mainly restricted to the outermost one annual ring in scots pine. For each stem section, the bark is characterized by similar level of radiocaesium and potassium to the wood formed before 1997. In contrast, calcium shows higher content in bark than in stemwood (<1997). In mature forest stands, the wood contains the main reservoir of K because of its huge biomass, while the bark is the main reservoir for Ca because of its higher concentration in Ca [1]. Our data confirm this observation but the equivalent concentration of K and Ca measured in stemwood show also that the trunk wood of young stand can concentrate similarly both elements. Generalization of element distribution and budgets must be considered with caution and certainly as a function of tree age.

To avoid an effect of changes in wood density with radius in interpretating the cation patterns, the cations content were expressed on a volumic basis. According to the former data expression, figure 3 focus on the mean trends in element concentration for the wood formed between 1986 and 1996.

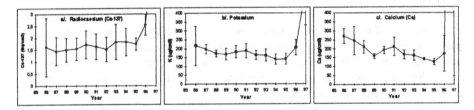

Fig. 3. Mean radial distribution of (a) radiocaesium, (b) potassium and (c) calcium concentrations measured in the wood formed till 1996 and expressed on a volumic basis (Bq cm⁻³ or μg cm⁻³)

In the wood xylem previously formed (1986-1996), the distribution of ^{137}Cs concentration expressed on a volumic basis is uniform with a slight increasing trend in function of radius (Fig.3). By contrast, concentrations of Ca and of K to a lesser extent, decrease slowly from the pith to the outermost rings. In conifers, calcium generally exhibits a decline radial decline from the pith to the cambium while potassium often increases or remains constant [2]. For calcium, it was demonstrated that a steady decline in calcium binding capacity (CBC) with radius can account for the often-observed concentration decrease [3]. For potassium and radiocaesium, a more indefinite trend linked to a high mobility in xylem is a common feature of different trees [4, 5]. This characteristic complicates particularly the historical reconstruction of the radiocaesium uptake. The biochemical behaviour of potassium is however mostly compared to the fate of radiocaesium which shows a comparable mobilization in forest trees [6]. In the wood section studied, the radial pattern observed for ^{137}Cs in the wood formed till 96 is different compared to the monotonous decrease of wood K concentrations. This differential radial pattern shown by ^{137}Cs might thus tentatively be ascribed to an increase of soil radiocaesium availability. An alternative hypothesis is that an increase in ^{137}Cs immobilization in wood could also be the result of a prevailing transient retranslocation phenomenon from the crown to the stem. However, the high ^{137}Cs content in connection with a high water content in the last ring formed supports the assumption of an enhanced uptake from the soil.

3.3 QUANTITATIVE ASPECT OF THE RADIOCAESIUM AND CATIONS DISTRIBUTION IN STEM WOOD

For each wood section, ring width measurements allowed the reconstruction of individual dendrochronological curve. Height and diameter growth data were coupled to calculate the annual volume increment at the scale of the entire bole. According to the assumption of a steady radial distribution of each cation concentration with height, the same calculations were made for ^{137}Cs, K and Ca cations respectively. The comparison between the mean volume growth and the mean cumulative occurrence of ^{137}Cs and of K and Ca in the trunk wood formed before 1996 is shown in figure 4

The ^{137}Cs, K and Ca mean distribution follows the mean volume growth with a comparable exponential function indicating that radiocaesium is also mainly absorbed by roots before to be redistributed and included as the other major nutrients inside the growing trunk biomass. The wood formed till 96 accumulates a total of 43800 Bq of ^{137}Cs but the last formed ring (97) contains more than 40% (i.e. 29700 Bq) of the total activity of the stemwood. In the pine stand studied, the average global ^{137}Cs activity accumulated in the stemwood of the mean tree (86-97) is 73500 Bq. According to the mean activity measured in wood (6540 Bq.kg^{-1}) and in soil (1.46 10^6 Bq.m^{-2}), the total accumulation of radiocaesium in the trunk is associated with a current mean transfer factor of 4.54 10^{-3} m^2.kg^{-1} which is close to other values estimated for scots pine in similar contaminated forest area in south-east Belarus [7] [8].

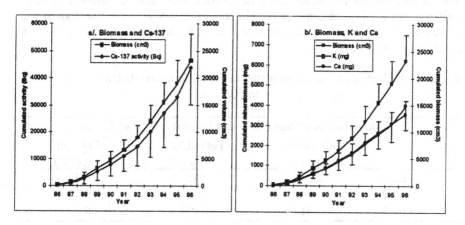

Fig 4.. Comparison between the mean cumulative volume growth (cm^3) and the mean cumulative occurrence of (a) ^{137}Cs and (b), K and Ca measured in the trunk wood formed till 1996

As showed previously, the mobility of cations in xylem can be high and does not allow direct comparisons of the cation concentrations in individual tree rings with historical changes in soil availability, especially for potassium and radiocaesium. However, it was possible to correlate the cumulated amounts of each cation with the cumulated biomass increments that followed an almost simple linear relationship for ^{137}Cs and K as shown in Figure 5.

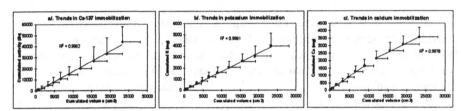

Fig. 5. Relationship between the cumulated annual amounts of (a) ^{137}Cs,(b) K and (c) Ca and the cumulated biomass annual increments in the trunk wood formed till 1996

For Ca, the appearance of a second slope is probably associated with a sharp decrease of the Ca content in the last 5 to 6 rings which represent an important contribution to the volume growth. The slope of the linear regression was estimated and used to deduce a minimal rate of cation accumulation in stemwood as a function of volume growth. For Ca, we considered only the second slope because cumulative calcium fate in the last year of wood formation is undoubtedly more representative for the calcium accumulation in the immediate few years. Table 2 indicates the mean accumulation rate element estimated for individual trees.

In these young trees, the calculated accumulation rate is higher for potassium than for calcium but this trend would be reversed with time as observed in other conifer stands [7] [9]. Because of the biochemical analogy between potassium and radiocaesium, the inevitable accumulation of radiocaesium could be significant especially in young stand.

TABLE 2. Individual and mean calculated values of the accumulation rate of ^{137}Cs, K and Ca

	Annual accumulation rate values in the newly formed biomass		
	Radiocaesium (Bq cm^{-3})	Potassium (mg cm^{-3})	Calcium (mg cm^{-3})
Tree 1	2.27	0.15	0.12
Tree 2	1.64	0.18	0.13
Tree 3	1.51	0.15	0.13
Mean	1.81	0.16	0.13
(St.dev.)	(0.41)	(0.02)	(0.008)

The mean current accumulation rate for ^{137}Cs is estimated at 1.81 Bq.cm^{-3} of new biomass and per year. With reference to the English production forecast table, a similar young pine stands would produce annually 9.9 m^3.ha^{-1} of new biomass. According to the volume growth observed for the last 5 years of the mean trees and according to the mean tree density observed, the current annual volume increment calculated for the stand was 14.8 m^3.ha^{-1}. Based on these two different growth estimations, the contamination of the woody compartment of this stand could increase in the next few years from 17.9 to 26.7 10^6 Bq.ha^{-1}.year^{-1}, respectively. Till now, the trunk wood compartment of the pine stand had accumulated an average amount of radiocaesium equivalent to 2% of the soil activity. The minimal annual increase in the stemwood contamination should correspond to a soil-to-wood flux ranged between 0.12 and 0.18 %.ha^{-1}.year^{-1} of the existent soil radiocaesium pool.

4. Conclusions

A dendrochemical approach was performed to reveal the tendency in radiocaesium distribution and accumulation in trunk wood of a young pine stand in connection with the biomass growth yields. It was shown that, as for potassium, the radiocaesium pattern in wood is consistent with a high mobility in pine tree and with a significant accumulation resulting from a prevailing root uptake process. According to a decreasing potassium accumulation in wood of older stands and to a similar feature for radiocaesium, the radiocaesium accumulation process could thus be reduced with time in a growing young stand or in an existing older stand. Until now, only few reference values on radionuclides accumulation in the wood compartment of contaminated forest ecosystems are available in the litterature; these values are mostly used to elaborate generic models in spite of the fact that these data are site-specific.

Nutrient or radionuclide accumulation can not be dissociated from stem biomass which is a cumulative component and is largely age, species and soil dependent. It is thus primordial to multiply such an approach in different edaphic forest types and for different age of tree to provide relevant radiocaesium flux data and to increase our ability to make prognosis using convenient modelling tools.

5. References

1. Duvigneaud, P. and Denayer-De Smet, S. (1973) In: Analysis of temperate forest ecosystems, Reichle, D.E. (Eds.), Springer-Verlag, Berlin, pp 199-224.
2. Arp, P.A. and Manasc, J. (1988) Red spruce stands downwind from a coal-burning power generator: tree ring analysis. Can. J. For. Res. 18:251-264.
3. Momoshima, N. and Bondietti, E.A. (1990) Cation binding in wood: applications to understanding historical changes in divalent cation availability to red spruce. Can. J. For. Res. 20:1840-1849.
4. Kohno, M., Koizumi, Y., Okumura K. and Mito, I. (1988) Distribution of environmental Cesium-137 in tree rings. J. Environ. Radioactivity, 8:15-19.
5. Kudo, A., Suzuki, T., Santry, D.C., Mahara, Y., Miyahara, S. and Garrec, J.P. (1993) Effectiveness of tree rings for recording Pu history at Nagasaki, Japan. J. Environ. Radioactivity, 21:55-63.

6. Ronneau, C., Sombré, L., Myttenaere, C., André, P., Vanhouche, M. and Cara, J. (1991) Radiocésium and potassium behaviour in forest trees. J. Environ. Radioactivity, 14(3):259-268.
7. Dvornik, A.M. and Zhuchenko, T.A. (1994) Predictive model of pine wood contamination (Forestlive). Paper presented at 6th VAMP meeting, Terrestrial Working Group, IAEA, Vienna, November 1994.
8. Thiry, Y., Ronneau, C. and Myttenaere, C. (1992) Comportement du radiocésium en écosystème forestier contaminé lors d'un accident nucléaire. In Proc. IRPA8, Réalisations Mondiales au niveau de la Protection de la Santé du Public et des Travailleurs contre les Rayonnements, Montréal, Canada, May 17-22, Vol II, 1689-1692.
9. Ranger, J., Marques, R., Colin-Belgrand M., Flammang, N. and Gelhaye, D. (1995) The dynamics of biomass and nutrient accumulation in a Douglas-fir (*Pseudotsuga menziesii* Franco) stand studied using a chronosequence approach. Forest Ecology and Management, 72:167-183.

THE DIRECTION AND INTENSITY OF ^{137}CS FLUXES IN FOREST ECOSYSTEMS

V.P. KRASNOV

Laboratory of Forest Radioecology of Polesskaya Forest Scientific Research Station of UkrSRIFA, 262004, Zhitomir, pr. Mira, 38, Polesskaya FSRS. UKRAINE

Abstract

Immediately after the accident at the Chernobyl nuclear power plant, 60–95% of radioactivity in the forest biogeocenosis was retained by the tree canopy. In the following years, radionuclides migrated at different rates to the lower layers of the phytocenosis and to the soil where they have been redistributed in the various soil horizons. Contamination of the biota now occurs primarily through root uptake and resuspension of contaminated particles. In this paper we present our data on radionuclide fluxes in Chernobyl forests

1. Radionuclide migration in soil

As a result of the Chernobyl accident, 60–95% of the total radioactivity in the forest biogeocenosis was initially concentrated in the tree canopy. The migration of radionuclides became apparent in the first few weeks and by 1987, more than 90% of the radionuclides were in the soil surface layers. The migration process began during the vegetative season of 1986 with the removal of leaves, needles, parts of bark and branches to the soil surface and the weather conditions (rain, wind and at a later stage, snow-thawing) enhanced this effect.

In the following years, the rate of ^{137}Cs migration in the forest biogeocenosis was determined by the speed of its vertical redistribution in soil horizons. The species composition of the growing plantations had a great influence on the rate at which these processes occurred. In broadleaf plantations, where leaf-fall occurs annually and decomposition takes place in a relatively short time, most of the ^{137}Cs activity became concentrated in the mineral soil within a few years. The redistribution of ^{137}Cs in coniferous plantations proceeds much more slowly. Coniferous needles contaminated by direct deposition were only completely shed in 1988. As a result, the bulk of ^{137}Cs was concentrated in the upper layers of forest litter from 1988 to 1990 (Fig. 1). The

71

I. Linkov and W.R. Schell (eds.), Contaminated Forests, 71–76.
© *1999 Kluwer Academic Publishers. Printed in the Netherlands.*

stand under study is a 50-year old pine plantation mixed with birch. The density of the ^{137}Cs ground deposition is 320 kBq m^{-2} (8.6 Ci km^{-2}).

In 1990, the magnitude of ^{137}Cs specific activity in undecomposed litter was 2.0 times greater than in the semi-decomposed layer and 2.8 times greater than in the wholly decomposed one. In 1994 the highest concentration of ^{137}Cs was found in the semi-decomposed Of layer of the forest litter. By 1997, the wholly decomposed Oh layer contained the maximum concentration of ^{137}Cs. Radionuclide concentration in mineral soil is gradually increasing during these years.

2. Radionuclide Fluxes in Vegetation

Specific activities of ^{137}Cs in some components of the forest biogeocenosis in various years are given in TABLE 1

TABLE 1. Radiocontamination of some species of vascular plants and mushrooms in various years (density of ground deposition ^{137}Cs about 320 kBq m^{-2})

Species	Parts and organs of plants and mushrooms	Specific activity (Bq/kg) ^{137}Cs in years		
		1991	1995	1997
Pinus sylvestris L.	Conifer	6043	13103	15068
	Bark, external part	4810	4360	4280
	Bark, internal part	1764	3830	3970
	Wood	408	1128	1240
Vaccinium myrtillus L.	Berries (f.w.)	3820	2090	2020
	shoots	19928	11770	10900
Convallaria majalis L.	Flowers	2328	2230	2150
	Leaves	1791	1652	1590
Suillus luteus	Fruit bodies	8965	6307	4164
Leccinum scabrum	Fruit bodies	4007	3839	3001

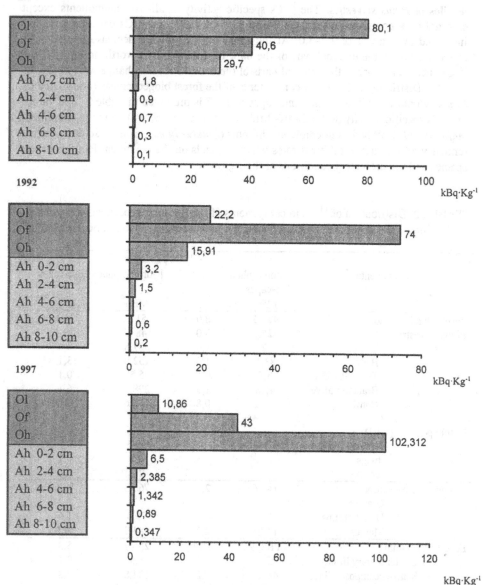

Fig. 1. *Dynamics of ^{137}Cs specific activity in soil horizons of soddy-podzol sandy-loam soil in 1990-1997*

From the Table 1 it can be seen that the highest levels of ^{137}Cs are contained in the needles of *Pinus sylvestris*. The ^{137}Cs specific activity in all tree components except external bark continues to increase. From 1991 to 1997, the ^{137}Cs concentration has increased by a factor of 2.2–3.0. At the same time, a gradual decrease of the ^{137}Cs concentration in the external part of the bark is observed. It is worth noting that the ^{137}Cs specific activity in the external parts of the cortex does not change with time

Distribution of ^{137}Cs in components of the forest biogeocenosis (dominated by *Pinus sylvestris* L.) in different landscapes in 1997 is presented in Table 2. The values of ^{137}Cs specific activity of the dwarf-shrubs (*Vaccinium myrtillus*), herbs (*Convallaria majalis*) and fruit bodies of edible mushrooms (*Boletus luteus, Boletus scabra*) have a tendency to decrease at different rates which depends on the biological features of the species and the ecological conditions of their growth.

TABLE 2. Distribution of ^{137}Cs in components of the forest biogeocenosis (dominated by *Pinus sylvestris* L.) in automorphous and hydromorphous landscapes in 1997

Components	^{137}Cs per 1ha			
	automorphous lanscapes		hydromorphous landscapes	
	MBq	%	MBq	%
Forest stand: total	425,5	8,0	1517,6	31,1
Pinus sylvestris (total)	425,5	8,0	1480,0	30,4
Wood	212,9	4,0	407,8	8,4
Bark	83,1	1,6	152,0	3,1
Twigs dried up	3,5	0,1	5,5	0,1
Branches alive	81,8	1,5	298,1	6,1
Conifer	44,2	0,8	616,6	12,7
Betula pendula (total)			37,6	0,7
Wood + bark			30,6	0,6
twigs			4,8	0,1
Leaves			2,2	0,0
Underwood vegetation	147,6	2,7	760,3	15,6
Grasses				
Dwarf-shrubs			75,4	1,6
Mosses	147,6	2,7	684,9	14,0
Forest litter (total)	1160,9	21,9	256,7	5,3
Contemporary litter	5,6	0,1	5,9	0,1
Semi-decomposed litter	487,4	9,2	134,8	2,8
Decomposed litter	667,9	12,6	116,0	2,4
Mineral part of the soil (0-30 cm)	3579,8	67,4	2340,0	48,0
Total in the biogeocenosis	5313,8	100,0	4874,6	100,0

3. ^{137}Cs Distribution in automorphous and hydromorphous landscapes

In automorphous landscapes where there is a prevalence of pine plantations on turf-soddy-podsol sandy soil, most of ^{137}Cs is concentrated in the mineral layers (67.4%). In spite of having the highest ^{137}Cs specific activity, the forest litter contains only 21.9% of the total forest content because of the low density of the substratum. In general, soil is the major repository for ^{137}Cs accounting for 89.3% of the total content.

Due to the considerable biomass of forest stands in tree plantations, which make up the first canopy of the phytocenosis, the ^{137}Cs contribution to the total content (8.0%) is much greater than that of the lower layers of vegetation (2.7%).

In hydromorphous landscapes 31.1% of the total ^{137}Cs content is found in the above-ground part of the trees and 15.6% is found in the understorey. In the latter case, most of the ^{137}Cs is concentrated in the moss canopy. As expected, soil contains a smaller percentage of the total ^{137}Cs content (53.3%) than that of automorphous landscapes. These data show that both the uptake of ^{137}Cs by plants is enhanced and the fixation of ^{137}Cs in soil is weaker in the more moist ecological conditions.

4. Comparison of ^{137}Cs distribution in two pine plantations grown on soils of different nutrient content but similar moisture contents

Table 3 compares the distribution of ^{137}Cs in two pine plantations growing on different soils with similar moisture content. A higher plant uptake of ^{137}Cs is observed in a 50-year old pine plantation growing on soddy-podzol sandy soil (density of ground deposition is 6.7 Ci km^{-2}) than on a similar plantation growing on podzol-soddy soil. In the first case, 15.9% of the total ^{137}Cs activity is in the forest stands compared to 5.5% in the second case.

In understorey vegetation a similar distribution pattern is observed, accounting for 10% of the total ^{137}Cs activity in forests grown on nutrient-deficient soils compared with 0.6% on the richer podzolic-soddy loam soils. It is worth noting that the bulk of the ^{137}Cs activity in understorey vegetation grown on poorer soils is concentrated in the dense moss canopy.

In richer soils, 76.1% of ^{137}Cs content is found in the mineral layers as compared to 65.3% for the poorer soils. This difference can be explained by the different rates of radionuclide redistribution between forest litter and the mineral soil, the degree of its absorption by organic substances and its fixation by clay minerals. On podzol-soddy, soddy and grey forest soils these processes take place at a faster rate and as a result, have a great influence on the lower rate of ^{137}Cs migration into the phytocenosis components.

5. Conclusion

In a pine plantation, the maximum concentration of ^{137}Cs was found in the surface layers of forest litter during the years 1987–1989; in the period 1990–1992, it was in the partly

TABLE 3. Distribution of ^{137}Cs in the components of the forest biogeocenosis (dominated by Pinus sylvestris L.) growing on two different soils (1997)

Components	Activity ^{137}Cs per 1 ha			
	soddy-podzolic sandy-loam		podzolic-soddy loam	
	MBq	%	MBq	%
Forest stand: total	557,4	15,9	228,7	5,5
Pinus sylvestris (total)	557,4	15,9	194,2	4,7
Wood	221,7	6,3	64,5	1,5
Bark	81,1	2,3	32,6	0,8
Twigs dried up	6,9	0,2	1,3	0
Branches alive	176,7	5,1	18,2	0,5
Conifer	71,0	2,0	77,6	1,9
Betula pendula (total)			34,5	0,8
Wood			16,3	0,4
Bark			11,4	0,3
Twigs dried up			0,4	0
Branches alive			4,7	0,1
Leaves			1,7	0
Underwood			19,5	0,5
Understorey vegetation, total	349,5	10,0	25,1	0,6
Herbaceous plants			1,1	0
Dwarf-shrubs	9,1	0,3		
Mosses	340,4	8,8	24,0	0,6
Forest litter, total	310,4	8,8	723,4	17,3
Contemporary litter	1,8	0	5,1	0,1
Semi-decomposed litter	143,8	4,1	130,9	3,1
Decomposed litter	164,8	4,7	587,4	14,1
Mineral part of the soil (0- 30 cm)	2290,9	65,3	3173,0	76,1
Total:	3508,2	100,0	4169,7	

decomposed Of layer and in the following years, it was in the wholly decomposed Oh layer. Under similar soil conditions, the bulk of ^{137}Cs was found in the upper mineral soil layers in broadleaf forests and in the decomposed part of soil litter in coniferous forests. The distribution of radionuclides in forest biogeocenoses depends on many factors including soil conditions, forest composition and stand age. The dwarf-shrubs, mushrooms, mosses and lichens are contributing to the redistribution of ^{137}Cs.

PECULIARITIES OF ^{137}Cs VERTICAL MIGRATION IN PINE ECOSYSTEM WITH STEM FLOW, THROUGHFALL, LITTERFALL, AND INFILTRATION

A.L. KLYASHTORIN

Radioecology Laboratory, Faculty of Soil Science, Moscow State University, 199899 Moscow, RUSSIA

Abstract

The study is based on long-term monitoring of key plot located in the exclusion zone of Chernobyl NPP (ChNPP). Concentrations and annual fluxes of ^{137}Cs in the stem flow, crown water (throughfall), litterfall, and lysimetrical water are estimated. In spite of high caesium concentration, the contribution of stem flow to the overall migration process is revealed to be much less than that of throughfall because of larger area of the tree crown compared to the trunk. Total contribution of the crown and stem fluxes to the migration of ^{137}Cs is about 0.05% of the total deposition. Annual influx of ^{137}Cs with litterfall to the soil is about 0.1% of its deposition. Maximum concentration of ^{137}Cs in the lysimetrical water takes place in the lowest horizon of forest litter. In the deeper soil horizon this radionuclide is strongly fixed by mineral soil components. Annual output of ^{137}Cs from the forest litter and 0-20 cm layer is about 0.1% and 0.007% of the deposition, respectively. Annual influx of radiocaesium to the soil surface from the tree floor almost completely compensates its losses from the forest litter for vertical infiltration and is much higher than the radionuclide outflow from the 0-20 cm soil layer.

1. Introduction

Vast area of forests in Ukraine, Belarus, and European part of Russia were exposed to the radioactive contamination due to the accident at Chernobyl Nuclear Power Plant (ChNPP) in 1986. In the first years after the accident, researcher's attention was concentrated mainly on the estimation of the deposition over different areas, peculiarities of primary distribution of the radionuclides, development of radioecological monitoring network, and short-term forecasting of the radionuclide behaviour in the contaminated territory. Nowadays, the top priority is given to the studies of biological cycling of long-lived radionuclides (^{137}Cs, ^{90}Sr, etc.) and development of this data base for long-term forecasting of radionuclide behaviour in the environment.

I. Linkov and W.R. Schell (eds.), Contaminated Forests, 77–84.

This kind of information is of particular importance for forest ecosystems (biogeocenoses). The latter are known to possess complex inner arrangement (multi-layer structure, various species composition, variable seasonal dynamics) and occupy more than 30% of the contaminated territory.

In our previous publications it was shown that 6-8 years after the accident the radionuclides in forest ecosystems come to the so-called quasi-steady state in the "soil-plant" system, with prime significance of biological turnover [5]. The contribution of individual processes constituting the biological cycling of the radionuclides varies widely and depends on physico-chemical properties of the initial fallout and type of ecosystem.

Thus, some publications suggest that at least one half of the total downward flux of chemical elements in the forest ecosystems is represented by nutrients leached from the alive plant tissues by stem flow and crown water (throughfall) [2, 4]. Possible contribution of these processes to the long-term secondary contamination of the soil surface also has been estimated [1, 7].

Another important process of radionuclide recycling - annual litterfall - has still not been investigated in detail, despite the fact that radionuclides, released from the decomposing leaves and needles are in contact with the root-abundant zone and thus can be immediately re-accumulated by the trees.

Fresh litter serves also as a feeding substrate for the soil invertebrates and the radionuclides may enter the trophic food chain by this way. In this respect, the availability of the radionuclides contained in the fresh litter is close to that in ionic and colloidal solutions.

The purpose of our work was to estimate basic parameters of ^{137}Cs migration in the litterfall, throughfall, and stem flow in typical pine ecosystem of Ukrainian Poles'e, and collate these parameters with the radionuclide migration by infiltration.

2. Materials and Methods

The investigated key plot was established 6.5 km to the South-West from the accidental unit, on the first terrace of the Pripyat-river, in a pine ecosystem of 50-60 years old. Herbal layer in the ecosystem constituted mainly of green moss. In 1989, the total deposition of gamma-irradiating radionuclides in the plot was of 18000 kBq/km^2, with ^{137}Cs deposition of 2430 kBq/m^2 (Table 1).

TABLE 1. Basic soil properties in the key plot in the South-East Exclusion Zone, Ukrainian Poles'e.

DEPOSITION Bq/M^2		TYPE OF FOREST	SOIL TYPE	SOIL PROPERTIES			
Total	Cs-137			Humus content, %	Exchangeable Ca^{2+} & Mg^{2+}	pH KCl	Clay content, %
18000	2430	Pure pine forest, 50-60 years old	Soddy-podzolic	1.3	14-17% of total exch. cations.	4.2	< 5

Soil cover at the plot is represented by soddy-podzolic sandy soil with low clay (5%) and humus (1.3%) content. The proportion of Ca and Mg cations in the soil absorbing complex is low (14-17%), and total soil acidity is rather high (pH$_{KCl}$ is of 4.2). The ecosystems of this kind are typical for the exclusion zone and Ukrainian Poles'e as a whole [6, 8]. The key plot is a part of a monitoring network that was established in the frame of ECP-5 and other international projects [8, 11].

To collect throughfall, special cone samplers (0.015 m^2 in area) were established under the "average" trees at the distance of one half of the crown radius from the trunk using 5-fold replication. The trunks of the model trees were equipped with special stem flow collectors using 3-fold replication.

Litter-traps (1 m^2 in area, 5-fold replication) were used to collect litterfall. Soil lysimeters (0.12 m^2 in area) filled with neutral drainage were established in different layers of the soil and forest litter to intercept and collect infiltrating water. The lysimetrical method and studies have been described in detail in our previous publications [6].

Samples of throughfall, lysimetrical waters, stem flow, and litterfall were collected monthly (sometimes once for two months). Native or concentrated liquid samples (throughfall, lysimetrical waters, and stem flow) were analyzed using gamma-spectrometer Nokia with Li-Ge detector. The concentration of gamma-irradiating nuclides in the litterfall was measured using the same equipment, after preliminary drying and milling of the samples.

3. Results and Discussion

3.1 ^{137}CS IN THE THROUGHFALL AND STEM FLOW

Table 2 presents the data on ^{137}Cs concentration and output from the tree tissues to the forest litter with throughfall and stem flow. Generally, ^{137}Cs concentration in the throughfall does not exceed 9.42 Bq/l, whereas in the stem flow this value is as high as 244 Bq/l.

TABLE 2. Weighted-average annual concentrations and fluxes of Cs-137 to the soil surface with crown and stem flow

SOURCE	Volume of the water l/m^2	^{137}Cs Bq/l	^{137}Cs Bq/m2	^{137}Cs % of total deposition
Crown water	173. 7	5.3	919.4	0.044
Stem flow	0.65	218.77	142.2	0.007

Relatively high concentrations of ^{137}Cs in the stem flow compared to the throughfall can be determined principally by two processes: (1) its sequential extraction from the external bark and (2) extraction from the residual particles of the initial surface contamination. The first process is caused by relatively slow movement of rain water down the stem and gradual enrichment of the water with ^{137}Cs. The water falling through the pine crown contacts the needles for too short a time to be enriched with radiocaesium to the same extent as the stem flow. The second process may take place simultaneously with the first one, but ^{137}Cs is leached from the fallout particles rather than from the plant tissues. The contribution of this pathway is confirmed by the fact that external layers of pine bark in the exlusion zone are still one of the most contaminated components of the stand [11].

Annual influx of ^{137}Cs to the soil with the throughfall and stem flow is 919 and 142 Bq/m^2, respectively These values were calculated using experimental data on the crown cover (75%) and stand density (8 trees per 100 m^2) in the key plot. Therefore, the contribution of the throughfall to the downward ^{137}Cs flux from the stand is higher than the corresponding contribution from the stem flow, in spite of the high ^{137}Cs concentration in the latter.

At the same time, annual flux of ^{137}Cs with the stem flow directly to the stem-adjacent area is of significance (1850 Bq/tree in average). This increases the heterogeneity in the distribution of mobile and available forms of ^{137}Cs over the territory and enriches the stem-adjacent soil areas with this nuclide. The latter may influence the local root uptake, and in this case the TF calculated based on the average content of ^{137}Cs in the soil can be significantly underestimated (since actual TF from the stem-adjacent area is much higher than the average over the territory).

Relative annual ^{137}Cs influx to the soil with the throughfall and stem flow is now of 0.044% and 0.007% of the radionuclide deposition in the ecosystem (including soil), respectively. It is obvious that this value was significantly higher in the first years after the accident (see below). The dynamics of ^{137}Cs influx to the soil with the throughfall and stem flow depends on seasonal variation of radionuclide content as well as variation in the rainfall rate during the season (Fig. 1).

3.2 [137]CS IN THE ANNUAL LITTERFALL

It is believed that 30-70% of the initial fallout was intercepted by the tree crowns in April-May 1986. However, up to 90% of the deposition was replaced to the forest litter as soon as November 1986 due to the so-called self-decontamination processes (washing-off and leaching by rains, litterfall, etc.) [5]. Although direct data on downward radionuclide migration with water flow and litterfall in this period are absent, a large proportion of radionuclides are likely to be transported by these carriers in the form of radioactive particles.

Further long-term studies revealed that the annual litterfall biomass in the investigated key plot varied from 330 to 530 g/m^2 (dry weight) depending on climatic conditions of the year. [137]Cs content in the litterfall also varied in a wide range (7-50 kBq/kg) depending on the proportion of surface contamination in the above-ground phytomass [8]. Eight years after the accident the radiocaesium content in the litterfall stabilized and at the level of 9.5 kBk/kg, with annual downward flux of about 0.14% of the total ^{137}Cs deposition (Table 3).

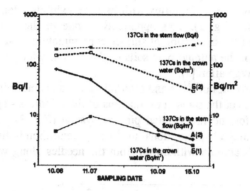

*Fig. 1 Seasonal dynamics of the concentration (left scale)
and monthly influx (right scale) of ^{137}Cs to the soil
surface with the stem flow and crown waters*

TABLE 3. Weighted-average ^{137}Cs concentration and annual fluxes to from the stand soil surface with litterfall

Litterfall biomass (kg/m2 dry weight)	^{137}Cs in the litterfall		
	Bq/kg	Bq/m^2	% of total deposition
0.39	9402	2909	0.139

For the recent 5 years, annual influx of ^{137}Cs to the soil surface with litterfall varied from 0.3 to 0.1% of its deposition in the ecosystem. Taking into account that in the first 2-3 years after the accident this value was significantly higher, we suggest that no more than 2-3% of total ^{137}Cs has been replaced by this pathway from the stand to the soil surface for the period of 1988-1994.

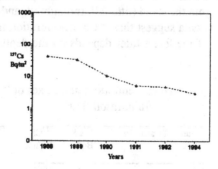

Fig. 2. Long-term dynamics of ^{137}Cs influx to the soil surface with litterfall.

The long-term dynamics of ^{137}Cs influx with litterfall exhibits a tendency to decrease beginning from 1989 (Fig. 2). This apparently is due to the self-decontamination processes, decrease of the proportion of surface contamination, and general processes of caesium "aging" in the automorphic soils of the region, which decreases the ^{137}Cs uptake by the forest vegetation [3].

Seasonal dynamics of ^{137}Cs and litterfall biomass is determined by season and biological peculiarities of the stand. Two maxima of these indices - spring and autumn - are characteristic for the investigated pine ecosystem (Fig. 3). Interestingly, the litterfall biomass is inversely proportional to ^{137}Cs concentration in the litterfall, which is likely due to ^{137}Cs autumn withdrawal from the needles along with nutrients [12, 13].

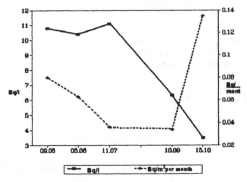

Fig. 3. Seasonal dynamics of ^{137}Cs content and influx to the soils with litterfall

3.3 ^{137}Cs IN THE INFILTRATION FLOW (LYSIMETRICAL WATERS)

The regularities of ^{137}Cs and other radionuclide migration from different soil layers by infiltration have been considered in detail in our previous publications [6, 9]. In this work we would like to estimate the contribution of this pathway relative to caesium influx to the soil surface from the stand (litterfall, throughfall, and stem flow).

The experimental data suggest that ^{137}Cs concentration in the lysimeter waters from different sub-layers of the forest litter depends on the depth of the corresponding layer (Table 4).

Table 4. Weighted-average annual concentrations and fluxes of ^{137}Cs to the soil surface with intrasoil flow

LAYER (cm)	Volume of water (l/m²)	[137]CS Bq/l	[137]CS Bq/m2	[137]CS % of total deposition
Ol+ Of (0-3 cm)	150	12.6	1906	0.091
Ol+ Of + Oh (0-4 cm)	143.5	15.9	2283	0.114
O(total) (0-5 cm)	119.6	8.9	1274	0.061
O + mineral soil (0-20 cm)	70.6	2.18	154	0.007

Thus, the weighted-average radionuclide content in the water from the layers 0-3 cm (Ol+Of), 0-4 cm (Ol+Of+Oh) and 0-5 cm (O + O/A1) is of 12.6, 15.9 and 8.5 Bq/l. This suggests that the water is enriched by ^{137}Cs only when it leaches through the first two layers of the forest litter (0-3 and 0-4 cm). The water from 0-5 cm layer contains almost 2 times less concentration of the radionuclide. Therefore, ^{137}Cs starts to be fixed in the layer 4-5 cm, in the lowest sub-horizon of the forest litter and the topmost layer of the mineral soil profile. This fact suggests that even trace admixture of clay minerals in the organic layer sharply enhances interception and irreversible absorption of ^{137}Cs from the soil solution. In the deeper layers the absorption increases manifold, and only 0.007% of total ^{137}Cs deposition leaves annually from the 0-20 cm layer.

Comparing the data of tables 2, 3, and 4 it is possible to conclude that the investigated pine ecosystem is characterized by extremely low annual loss of ^{137}Cs with intrasoil flow (0.007%). The annual nuclide influx to the soil surface from the stand (about 0.2%) exceeds the above value by more then order of magnitude. In terms of the "soil-vegetation" system such a ratio suggests that reutilization of ^{137}Cs by higher plants impacts significantly on its vertical migration. At the same time, the ^{137}Cs concentration in the lysimetrical waters, litterfall and vegetation have stayed relatively stable for the recent years, which is another evidence for approaching quasi-steady state in the investigated "soil-vegetation" system.

4. Conclusions

Concentration of ^{137}Cs in the stem flow exceeds that in the crown waters by a factor of 40. However, the contribution of this flux to the overall migration process is much less than that of crown waters because of the much wider projection of the tree crown compared to the trunk. Total contribution of the crown and stem fluxes to the migration of ^{137}Cs is about 0.05% of total deposition. Annual influx of ^{137}Cs with litterfall to the soil is about 0.1% of its deposition.

Maximal concentration of ^{137}Cs in the lysimeter water takes place in the lowest horizon of forest litter. In the deeper soil horizon this radionuclide is strongly fixed by mineral soil components. Thus annual output of ^{137}Cs from the forest litter is about 0.1% and from 0-20 cm layer - 0.007% of the deposition.

Annual influx of radiocaesium to the soil surface from the tree floor almost completely compensates for its losses with vertical flow from the forest litter and much higher than the radionuclide outflow from the 0-20 cm soil layer.

5. References

1. Bonnet, P.J.P., Anderson, M.A., 1993 Radiocesium dynamics in a coniferous forest canopy: a mid-Wales case study, *Science of the Total Environment* 136, 259-277
2. Karpachevskii, L.O., 1981, *Forest and Forest Soils*, Moscow, Lesnaya Promyshelnnost'
3. Kulikov, N.V., Molchanova, I.V., and Karavaeva, E.N., 1990, *Radioecology of Soil and Vegetation Covers*, Sverdlovsk.

4. Mina, V.N., 1965, Leaching of some compounds from tree plants by rain waters adn the contribution of this process to the biological cycle, *Pochvovedenie*, no. 6, 7-17.
5. Tikhomirov, F.A., Shcheglov, A.I., Kazakov, S.V., and Klyashtorin, A.L., 1989, Radionuclide distribution in the forest landscapes of Ukrainian Poles'e, *All-Russian Conference "Principles and Methods of Landscape-Geochemical Studies of Radionuclide Migration"*, p. 53.
6. Tikhomirov, F.A., Klyashtorin, A.L., Shcheglov, A.I., 1992, Radionuclides in the vertical intrasoil flow in the forests of the exlusion zone of Chernobyl NPP, *Pochvovedenie*, no. 6, 38-44.
7. Schimmack, W., Forster, H., Bunzl, K., Kreutzer, K. 1993, Deposition of radiocesium to the soil by stemflow, throughfall and leaf-fall from beech trees, *Radiation and Environmental Biophysiscs* 32, 137-150.
8. Shcheglov, A.I., Tsvetnova, O.B., and Tikhomirov, F.A., 1992. Migration of the Chernobyl-born long-lived radionuclides in the forest soils of European part of USSR, *Vestnik MGU*, 17, no. 2, 27-35.
9. Shcheglov, A.I., Tikhomirov, F.A., Tsvetnova, O.B., et al., 1991, Radionuclide distribution and migration in forest ecosystems, Moscow, *Dep. VINITY* 18.04.91, no. 1656-1391.
10. Shcheglov, A.I., Tikhomirov, F.A., Tsvetnova, O.B., Klyashtorin, A.L, Mamikhin, S.V., 1996, Biogeochemistry of the Chernobyl-born radionuclides in the forest ecosystems of European part of CIS, *Radiation Biology and Radioecology*, v. 36, no. 4, 469-477.
11. *Behaviour of radionuclides in natural and semi-natural environments* (editors M.Belli and F. Tikhomirov), 1996, Luxembourg.
12. Witherspoon, J.P., 1962, Cycling of caesium-134 in white oak trees on sites of contrasting soil type and moisture, *Radioecology*, eds. V., Schults and A. Klement, New-York, Reinhold, 127-132.
13. Witherspoon, J.P., Taylor, F.G., 1969, Retention of fallout simulant containing ^{137}Cs by pine and oak trees, *Health Phys.*, 17, 825-839.

PREPARATION OF SOIL THIN SECTIONS FOR CONTAMINANT DISTRIBUTION STUDIES

K.A. HIGLEY and O.G. POVETKO
Department of Nuclear Engineering, Oregon State University, Corvallis OR, 97330 USA

1. Introduction

Dynamic models of the transport and fate of radionuclides in ecosystems include many parameters that can not be easily estimated and often their estimates vary by many orders of magnitude. Estimation of these parameters requires knowledge of such factors as physical and chemical properties of soil, radionuclide depth distribution, concentration, bioavailability, migration mechanisms, root uptake, leaching into the deep soil and other factors. The physical location of radioactive contaminants in soils is an important factor in understanding contaminant migration mechanisms and estimation of models parameters. The common approach in parameters estimation is the assumption that the radionuclides are homogeneously distributed either uniformly or with some exponential depth profile under the ground surface. Soil, however consists of various size clay, sand, rock etc. particles bonded together by organic and inorganic materials [4]. The soil pores provide networks for vertical and horizontal transport of water, gases, dissolved nutrients etc. "A crushed or pulverized soil is related to the soil formed by nature like a pile of debris to a demolished building" [2]. The architecture of building can no more be determined from a pile of rubble than the structure or site-specific composition of a soil from a crushed bulk sample. Examination of soil thin sections having the original soil structure intact has developed over last three decades. The production of thin sections from hard rocks for examination under a microscope is one of techniques used in petrological laboratories. The production of thin soil sections from unconsolidated and/or radioactively contaminated soil is not a widely known or established technique and requires some different materials, safety measures and protocols.

2. Thin section preparation

The thin soil section preparation described in this study followed basic steps described in the literature on soil micromorphology [3 and 7]. The emphasis was made

I. Linkov and W.R. Schell (eds.), Contaminated Forests, 85–93.
© 1999 *Kluwer Academic Publishers. Printed in the Netherlands.*

on the steps that were specific to the analysis of radioactively contaminated soils and to the use of related instruments.

2.1. SAMPLE COLLECTION

Tall plants at the sample collection site need to be cutoff but it's not necessary to trim off grass and small plants. The vertical or horizontal soil profile may be exposed prior to the sample collection. The sample collection Kubiena tin is shown in Fig. 1. The tin of any reasonable dimensions may be used as long as it fits the size of drying and impregnating equipment. One of the lids is removed and the open end of the tin is placed onto the prepared surface in the required orientation and position. A knife is used to carve around the outside of the tin

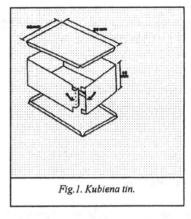

Fig.1. Kubiena tin.

while pressure is applied to the lid. When full, the tin is labeled and placed into a sample box.

2.2. SAMPLE DRYING

Kubiena tins with samples containing non-radioactive soil are placed in the well-ventilated chamber at the 30°C temperature in presence of moisture absorbing silica gel. Tins with radioactively contaminated soil are placed under the well-ventilated hood on the hot plate at 30°C temperature. An additional fan is added to increase ventilation. The loss of moisture is monitored daily by weighing. The samples lose most of their water in 3-5 days.

2.3. RESIN IMPREGNATION

Many resins are available for sample preparation. We use Spurr resin in impregnation. It is prepared and put under 50E3 Pa (0.5 Atm or 15 in Hg) vacuum and 5°C temperature for 30 minutes for release of remaining air. Then, the Kubiena box with the sample is placed in the secondary plastic or foil container and resin is slowly poured between the tin and container walls, so the resin would slowly penetrate the sample from its bottom substituting air in its pores. Afterwards, the container is placed in the vacuum chamber under 85E3 Pa (0.17 Atm or 5 inHg) vacuum for 4-6 hours. Hence, vacuum is released and curing is performed very gradually in five temperature steps - 5°C, 30°C, 40°C, 50°C, 60°C. The samples are left overnight at each step, so the resin gradually increases in viscosity until it hardens. For the samples which will undergo neutron activation analysis (NAA), the Spurr resin slides were used instead of glass slides due to the high induced activity of the glass. For further preparation of resin glass slides the pure resin block is prepared following the above steps for the soil sample.

2.4. SOIL AND RESIN BLOCKS CUTTING

Fig. 2 schematically shows how hardened soil blocks are then cut in slices about 1 cm thick using lapidary saw equipped with a diamond blade. The pure resin block is cut in slices 3 mm thick.

2.5. GRINDING AND POLISHING OF IMPREGNATED SLICES AND MOUNTING ON THE SLIDES

One surface of the prepared slices is ground using Buehler ®PETRO-THIN® thin sectioning system and then, polished manually on lapping plates using loose powders of the following grade sequence: 320, 400, 500 and 600 US grits (from 32 down to 3 μm powder particles sizes, respectively). Then, the slices are mounted either on 5.08 x 4.62 cm (2" x 3") glass or resin slides using epoxy resin and 1.5 kg weight is put carefully on top of the slide. The use of Buehler® mounting wax having a melting point at about 80°C allows later detachment of the thin section from the activated glass slide. It also allows remounting it for further analysis on the glass slide which has not been activated.

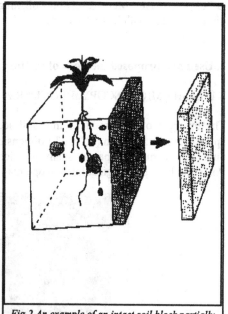

Fig.2 An example of an intact soil block partially impregnated with resin. The shaded area represents the portion of resin-impregnated soil that are trimmed and prepared as a thin section. Note the inclusion of rocks and root segments in the block.

2.6. THIN SOIL SECTIONS GRINDING AND POLISHING

After the resin is completely cured the slides are processed on the surface grinding thin section machine and polished manually on lapping plates using loose SiC powders of the following grade sequence: 320, 400, 500 and 600 US grits. The sections are lapped evenly and can be thinned to 15 μm.

2.7. HEALTH HAZARDS AND SAFETY PRECAUTIONS

In addition to the regular lab safety measures, additional steps are taken to reduce health hazard of the process. The Spurr resin contains one component that is considered carcinogenic. All work with this resin is conducted under the hood. Both machines used in the section preparation, lapidary saw and thin sectioning machine

88

are placed under a hood. In cutting and grinding steps water is used as a coolant, the water splash is confined within an efficient enclosure under the hood. Since the process requires large amount of water the water circulation loop was made in order to recycle the coolant water that may contain radioactively contaminated particles. These particles are filtered out by two filters and filters are periodically disposed as a low-level radioactive waste. Manual section polishing is conducted in the containers which are tightly covered with lids afterwards.

3. Used and proposed methods of the thin section analysis

3.1. LIGHT MICROSCOPY. REGULAR AND POLARIZING MICROSCOPE

The example of the produced thin soil section is shown on Fig.3. Thin sections of approximately 5 x 7 cm (2" x 3") dimensions and 15 - 30 μm thickness are examined under light microscope to identify micromorphological features of the sample. Polarized light petrological microscope may be used.

Fig.3 The photo of the soil thin section in its natural size.

3.2. ELECTRON MICROSCOPY (EM).

For possible analysis using EM, the ultrathin sections 1000 Å thickness sections must be produced by micromilling [1]. Images can be obtained with a scanning electron microscope (SEM). Images can be obtained with a SEM by collecting secondary electrons, backscattered electrons and a combination of these. Table 1 displays visible features of thin sections for various magnification and its potential use in further analysis.

TABLE 1. Relationship of magnification to area scanned on the specimen along with the features visible and potential uses for each magnification (modified from [6-7]).

Magnification	Area scanned	Features visible and use
10x	1 cm^2	Select objects for examination
100x	1 mm^2	Study shape and orientation of sand particles and groups of clay particles
1000x	100 μm^2	Surfaces of sand grains and arrangements of large clay particles visible
10000x	10 μm^2	Individual clay particles visible
100000x	1 μm^2	Small clay particles and iron oxides visible

3.3. NEUTRON ACTIVATION ANALYSIS

Soil sections mounted on the glass or resin slides are activated in the Triga 1 MW nuclear reactor in its thermal column facility. Conditions of irradiation are: thermal neutron flux - 7E10 neutrons cm^{-2}sec^{-1}, irradiation time - 6 hours. Activated samples are analyzed using gamma spectroscopy HPGe detector and by autoradiographic methods. Certain details of interest of the thin sections of 1 mm size are counted by using lead screen with the 1 mm hole in it which is placed against point of interest. Fig. 4 shows a photographic picture of the activated thin soil section, Fig. 5 displays the autoradiographic image of the same section. Figs. 6, 7 show the partial gamma spectrum obtained from the region A of the thin section displayed on Fig. 4 and the partial spectrum obtained from the whole thin section, respectively. Two additional peaks identified as corresponding to the gamma rays emitted by the product of activation ^{137}W (shown in bold face) dominate the partial gamma spectrum of the whole section and can not be seen at all at the spectrum of region A. Spatial element distribution over the thin section may be obtained this way.

3.4. AUTORADIOGRAPHY

The macroautoradiographic methods have been used so far in this study. The Kodak™ XAR X- ray film is placed in contact with the thin soil section. Kodak™ image intensifying screen is placed under the film. This assembly in the light tight container is left in the freezer under -70°C for the time necessary for latent image to be formed. This exposure time is widely vary from minutes to weeks and is determined experimentally. The presence of radioactivity is recognized by an increase in grain density over background levels. Figs.8a,b shows a photographic image and an autoradiography of the thin soil section prepared from the soil contaminated with 1-20 Bq g^{-1} of ^{86}Rb at the moment of exposure.

Measurements of photoemulsion response to date have included visual identification of the soil structure elements that have higher emitting radiation levels and following gamma spectroscopy measurements of these elements. In addition to the

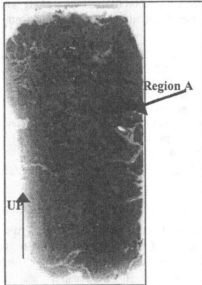

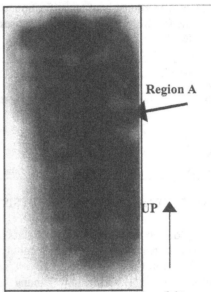

Fig.4. The photo of the soil thin section activated in the experimental 1.1 MW Triga nuclear reactor. Conditions of irradiation: thermal neutron flux - 7E10 n cm^{-2} s^{-1}, irradiation time - 6 hours, time after irradiation - 5 days. Size of the section - approximately 2 x 4 cm.

Fig.5. The autoradiographic of the soil thin section activated in the experimental 1.1 MW Triga nuclear reactor. Conditions of irradiation: thermal neutron flux - 7E10 n cm^{-2} s^{-1}, irradiation time - 6 hours, time after irradiation - 5 days. Size of the section - approximately 2 x 4 cm.

macroscopic autoradiographs, the electron microscopic autoradiographs may be obtained using Ilford™ L4, with a mean crystal diameter of around 1400 A, or Eastman Kodak™ 129-01 permitting resolutions of 500-1000 A [5]. The thin layer of emulsion, in which the beta particles record their passage by the production of individual silver grains may be used to obtain grain density autoradiographs. The stripping film Kodak™ AR-10 is planned to be use. The track emulsion CR-39 may be placed on the thin section surface to obtain track autoradiographs; which the pass through the emulsion layer particle create a track of silver grains.

3.5. ENERGY DISPERSIVE SPECTROMETRY (EDS or EDX)

Energy dispersive spectrometers have a solid-state x-ray detector placed near the specimen to collect x-rays, much like the electron detectors are used to collect backscattered and secondary electrons. Energy dispersive X-ray spectra of soil comprising elements and soil pollutants may be obtained this way. To obtain quantitative measurements, thin sections need to be polished to remove all scratches which could deflect X-rays.

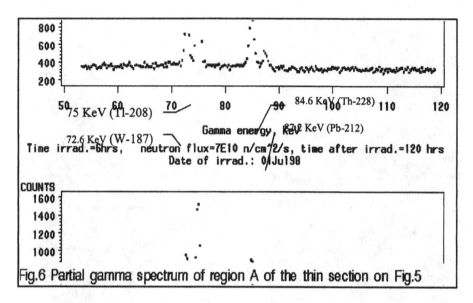

Fig.6 Partial gamma spectrum of region A of the thin section on Fig.5

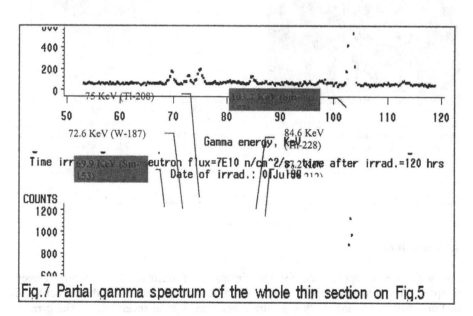

Fig.7 Partial gamma spectrum of the whole thin section on Fig.5

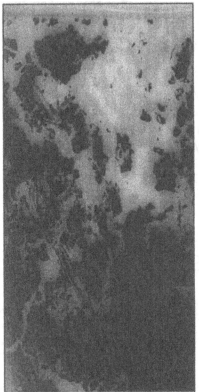

Fig. 8a Photographic image of the thin soil section prepared from the soil contaminated with ⁸⁶Rb at approximate concentration of 1 - 20 Bq g⁻¹ of soil at the moment of exposure. Actual dimensions of the section are approximately 3.5 cm x 2.5 cm.

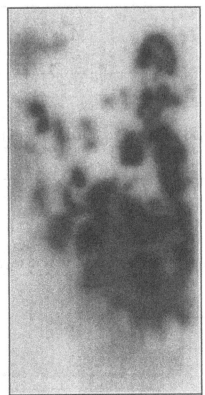

Fig.8b. Autoradiographic image of the thin soil section prepared from the soil contaminated with ⁸⁶Rb at approximate concentration of 1 - 20 Bq g⁻¹ of soil at the moment of exposure. Actual dimensions of the section are approximately 3.5 cm x 2.5 cm.

4. Conclusions

The technique is appropriate for the treatment of unconsolidated, moist, clayey samples. The autoradiographic images of radioactively contaminated and activated sections may be obtained with exposure times from 1 hr to several days. The gamma spectroscopy analysis may be applied to the certain parts of interest of the thin section in order to obtain spatial characteristics of the soil matrix.

5. References

1.   Bresson, L. M. Ion micromilling applied to the ultramicroscopic study of soils. Soil Sci. Soc. Amer. J., 45. 568-573, 1981.
2.   Kubiena, W. L. Micropedology. Collegiate Press, Ames, Iowa, 1938.
3.   Murphy, C. P. Thin Section Preparation of Soils and Sediments. A. B. Academic Publishers, Berkhamsted, UK, 1986.
4.   Miller, R.W.; Donahue, R.L. Soils, An Introduction to Soils and Plant Growth. Englewood Clitffs, NJ Prentice Hall, 1990.
5.   Rogers, A. W. Techniques of Autoradiography. Elsevier/North-Holland Biomedical Press. Amsterdam/New York. 1979.
6.   Smart, P. and N. K. Tovey. Electron Microscopy of Soils and Sediments: Techniques. Clarendon Press, Oxford, 1982.
7.   Vepraskas, M.J. European training course on soil micromorphology. North Carolina State University.

TIME-DEPENDENCY OF THE BIOAVAILABILITY OF RADIOCAESIUM IN LAKES AND FORESTS

E. KLEMT, J. DRISSNER, S. KAMINSKI, R. MILLER, G. ZIBOLD
FH Ravensburg-Weingarten, University of Applied Sciences,
P.O. Box 1261, D-88241 Weingarten, GERMANY

1. Introduction

Lakes and forests are as important to humans as they are providers of drinking water, fish, berries, mushroom, game, and wood. Following the Chernobyl accident, these semi-natural ecosystems have exhibited persistently high contamination levels of radiocaesium, also in the prealpine region of southern Germany. For the past 12 years, the time-dependency of the bioavailability of radiocaesium has been studied for different lakes and forests in order to get an understanding of the persistence and the migration of radiocaesium in semi-natural environments and for radiation protection purposes. As a measure of the time-dependency, the effective half-time and the ecological half-time will be discussed for Cs-137 concentrations and for Cs-137 aggregated transfer factors, respectively.

2. Bioavailability of Radiocaesium in Lakes

The contamination of water and fish of two lakes, Vorsee and Lake Constance, has been studied. They have different limnologic characters and show great differences in the persistence and bioavailability of Cs-137 from the Chernobyl fallout.

Glacially formed, Vorsee is a small (8.98 ha surface area), shallow (2.2 m maximum water depth) eutrophic lake at an altitude of 579 m, which is supplied by a swampy watershed. The watery, several meters thick sediments of this lake consist mainly of organic matter. The initial fallout of Cs-137 onto the lake was about 30 kBq/m^2. In 1994, a Cs-137 inventory in the sediment of about 60 kBq/m^2 with a broad maximum of the activity concentration at a depth of 20 cm to 30 cm was measured. Persistently high levels of dissolved Cs-137 in lake water and fish were observed and attributed to a continuous input of radiocaesium from both sediment and watershed [1].

In the years 1988 to 1998, the Cs-137 activity concentration values have been measured showing a gradual decrease between 300 mBq/l and 60 mBq/l, with maxima in winter and minima in summer (see fig. 1). It was suggested that in summer the

95

I. Linkov and W.R. Schell (eds.), Contaminated Forests, 95–101.

biotic production in the lake is a sink for the Cs-137 dissolved in the lake water whereas all over the year a continuous input of Cs-137 via ion exchange with biogenic ammonium ions takes place [1]. The effective half-time of the Cs-137 activity concentration in the water is determined to be about $T_{1/2}$ = 4.8 years.

In Lake Vorsee, the specific Cs-137 activity of mainly herbivorous fish of the group of small cyprinidae, like roach (Rutilus rutilus), rudd (Scardinius erythrophthalmus), silver bream (Blicca bjoerkna), bleak (Alburnus alburnus), bream (Albramis brama), and tench (Tinca tinca) varies between a maximum value of 1.8 kBq/kg some months after the Chernobyl accident and a present minimum value of 60 Bq/kg. We observe an effective half-time of the Cs-137 activity concentration of these fish of about 3.2 years.

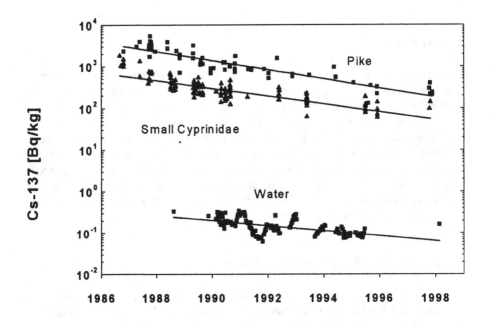

Fig. 1. Contamination of water, pike and small cyprinidae in lake Vorsee

For Pike (Esox lucius), a predator fish, the specific Cs-137 activity is about a factor of 5 higher than that of small cyprinidae. A maximum concentration of more than 4 kBq/kg was measured in late 1987. These findings are due to the enrichment of Cs-137 via the trophic chain. The scatter of the pike activity concentration around the exponential fit-curve is partially due to the positive size effect [2], which means an increase of Cs-137 contamination with fish weight. The effective half-time of the pike Cs-137 activity concentration is determined to be about $T_{1/2}$ = 2.7 years.

It can be assumed that the relatively high effective half-times are due to mainly reversible binding of Cs-137 to the organic matter in sediment and watershed and insufficient irreversible fixation to clay particles.

Lake Constance is a large (572 km² surface area) and deep (254 m maximum water depth) mesotrophic hardwater lake, in which radiocaesium was rapidly removed from the euphotic zone and strongly bound to clay minerals in the sediments [3]. The initial fallout of Cs-137 onto this lake was about 17 kBq/m² [4]. Typical Cs-137 vertical profiles of Lake Constance sediments far away from the tributaries exhibit two well-resolved sharp maxima. An upper one in roughly 2 cm depth, as measured in 1994, can be related to the Chernobyl accident in 1986. A lower one in roughly 8 cm depth can be attributed to the fallout of the atmospheric nuclear weapons testing with maximum fallout in 1963.

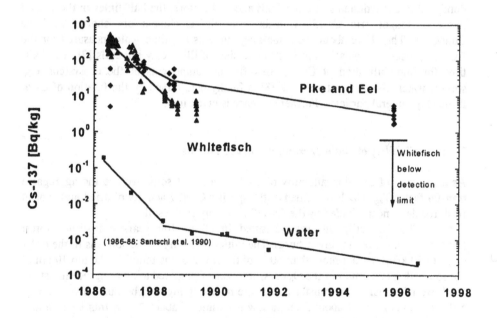

Fig. 2. Contamination of water, pike and eel, and whitefish in Lake Constance

Fig. 2. shows the contamination of Lake Constance water, whitefish (Coregonus wartmanni), and pike (Esox lucius) and eel (Anguilla anguilla). The Cs-137 water concentration, averaged over the vertical distribution, varied between about 200 mBq/l in late 1986 and about 2 mBq/l in 1996 (The data from 1986 to 1988 were estimated after Santschi et al. [5]). As shown by Smith et al. [6], a "double exponential" model for changes in the Cs-137 lake water concentration could be applied to different lakes in Great Britain. The two exponential functions correspond, respectively, to the initial fast flush of activity through the system followed by longer-term transfers from the catchment. This model can also be applied to Lake Constance. Within the first two years after the Chernobyl accident, we observe an effective half-time for the water activity concentration of roughly $T_{1/2} = 0.3$ years. After these two years, the effective half-time of $T_{1/2} = 2.5$ years describes the data reasonably well.

Unfortunately, the specific activity of whitefish was measured mainly during the first phase of the initial fast flush of activity through the system. Here, the effective half-time is roughly $T_{1/2} = 0.4$ years. Afterwards, an upper limit of $T_{1/2} < 2.5$ years can be estimated.

The specific Cs-137 activity in pike and eel increased in the first year after the Chernobyl accident, reached a maximum of more than 100 Bq/kg and then declined again with an initial effective half-time of roughly $T_{1/2} = 0.7$ years. After that, an effective half-time of $T_{1/2} = 2.3$ years is estimated.

Illitic clay mineral particles offering irreversible fixation sites for Cs-ions are dominant in the sediments of Lake Constance. Therefore, the half-times in the second phase of longer-term Cs-137 transfer from the catchment are shorter in Lake Constance. They have about the same magnitude as the time scale measured for the fixation process of potassium at interlattice sites of illitic clay minerals [7]. The half-time for immobilization of Cs-137 in different components of the ecosystem, e.g. surface water, fish, vegetation, milk [8] is finally determined by the fixation of Cs at illitic clay mineral particles. Further evidence is given in 3.

3. Bioavailability of Radiocaesium in Spruce Forests

As a measure of the bioavailability of Cs-137 in forest soils, we use the aggregated transfer factor T_{ag}, which is defined as the specific Cs-137 activity of dried plants or of fresh roe deer meat, divided by the Cs-137 inventory in the soil.

The T_{ag} soil-plant was measured in 80 to 100 years old spruce forest (Altdorfer Wald close to Ravensburg). The initial inventory in the soil was in the order of 12 to 20 kBq/m^2, and more than 80 % of the inventory is found in the top 10 cm of the soil with a maximum in the O_h or A_h horizons. Compared to other spruce forests in south-west Germany, the transfer factors are relatively high [9] because, here, we find a thick humus layer of about 5 cm, a low pH value of about 2.9 in this layer, and a relatively undecomposed raw humus or moder.

In fig. 3. the time-dependency of the T_{ag} is shown for some grazing plants of roe deer: fern (Dryopteris carthusiana), bilberry (Vaccinium myrtillus), clover (Oxalis acetosella), and blackberry (Rubus fruticosus). In brackets the number of sampling sites are noted. Like in other spruce forest sites, T_{ag} is highest for fern. One reason is that the roots of fern are essentially in the horizon where we find the maximum of the Cs-137 activity concentration. In contrast, clover has its roots only in the O_f horizon and therefore, it has access only to a small fraction of the Cs-137 inventory. Consequently, its T_{ag} is one order of magnitude lower. As measured between 1991 and 1997, the ecologic half-times for fern, bilberry, clover, and blackberry are $T_{1/2} = 4.1$ years, 4.0 years, 5.9 years, and 2.3 years, respectively. Like in the case of the lake ecosystems, we observe essentially parallel curves in the time-dependency. The mean ecological half-time, about 4 years, suggests that the time- dependency is not yet dominated by the fixation of Cs-137 in clay minerals but rather by the reversible binding of Cs-137 to organic matter. This assumption is supported by the measured Cs-137 vertical distribution that shows that a large part of Cs-137 is still bound in the humus layer.

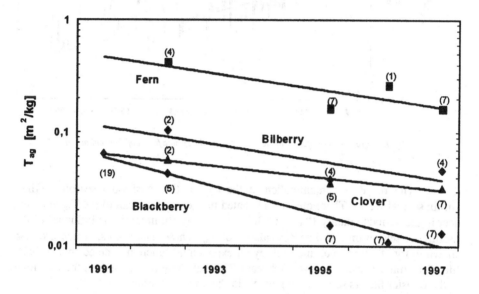

Fig.3. Aggregated transfer factor soil-plant in the spruce forest Altdorfer Wald

Since 1987, we have measured the Cs-137 roe deer meat contamination in the State Forestry Ochsenhausen. The frequency distribution of the activity concentration in roe deer can be described by a log-normal distribution [10]. An average of more than 200 roe deer per year were shot in that forestry, which consists mainly of spruce trees, and which, like the Altdorfer Wald, is a region of relatively high transfer factors [9]. As the places where the roe deer were shot were identified [11] and a map of the Cs-137 inventory of that forestry was established, the aggregated Cs-137 transfer factor soil - roe deer, T_{ag}, could be calculated. Fig. 4. shows the moving 2 weeks geometric mean of the T_{ag}, which is appropriate in the case of a log-normal distribution. The seasonal structure reveals periodic maxima correlated with the mushroom season in fall. As the weather conditions in fall determine the mushroom yield and thus the contamination of the mushroom grazing roe deer, it is not possible to make predictions for future fall periods. Consequently, in highly contaminated forest districts' monitoring of roe deer contamination is still necessary.

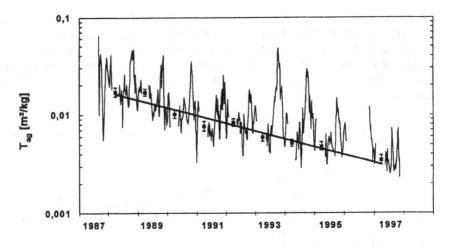

Fig. 4. Aggregated transfer factor soil-roe deer in the spruce forest of Ochsenhausen

The roe deer contamination at the end of the mushroom season declines within several weeks. Therefore, we calculated the geometric mean of the T_{ag} soil - roe deer in the periods January 1st to June 30th, in this way eliminating the influence of the mushroom season on the time dependency of T_{ag}. These data points, which are also shown in fig. 4., are fitted quite well by an exponential function. The ecological half-life of the transfer factor is $T_{1/2} = 3.8$ years. This half-time corresponds to the half-time of the transfer factors soil - grazing plants in the forest Altdorfer Wald.

4. Summary

After the Chernobyl accident, the Cs-137 activity concentration of water, herbivorous fish, and carnivorous fish was measured in two different lakes. It increases in that order due to accumulation in the trophic chain. In the same lake environment, the effective half-times describing the decrease of the activity concentrations in time are roughly the same. In lake Vorsee, where the Cs-137 is mainly bound to organic material in sediment and watershed, the effective half-times are between 2.7 years (pike) and 4.8 years (water). In Lake Constance, where illitic clay mineral particles are mainly responsible for the fixation of Cs-137, the effective half-times for the longer-term transfers from the catchment are between 2.3 years (pike and eel) and 2.5 years (water).

In the spruce forest Altdorfer Wald, the aggregated transfer factor soil - plant was measured for fern, bilberry, clover, and blackberry and it has been shown that its magnitude decreases in that order, partially due to the different root systems overlapping more or less with the vertical Cs-137 distribution in the forest soil. The ecological half-times are between 2.3 years (blackberry) and 5.9 years (clover). This means that in spruce forest soils with thick organic humus deposits, the transfer factor is not only high (T_{ag} between 0.1 and 1 m^2/kg for fern), but it decreases also very slowly in time.

The aggregated transfer factor soil - roe deer was measured for the spruce forest of Ochsenhausen. Taking only data from January to June in order to eliminate the influence of mushroom grazing, an ecological half-time of 3.8 years was observed. This corresponds to the values measured for the grazing plants in the Altdorfer Wald.

In conclusion, we find from our data that in different semi-natural environments, the observed half-times of Cs-137 contamination and transfer factors are in the order of 2 years and are determined by irreversible fixation of Cs-137 at illitic clay minerals. Longer half-times – up to 6 years – are observed if relatively large inventories of Cs-137 are reversibly bound to organic matter and remain bioavailable due to an insufficient fixation capacity by clay minerals.

5. References

1. Kaminski, S.; Richter, T.; Walser, M.; Lindner, G. Redissolution of cesium radionuclides from sediments of freshwater lakes due to biological degradation of organic matter. Radiochimica Acta 66/67 (1994), 433-436.
2. Sansone, U.; Voitsekhovitch, O.; Eds. Modeling and study of the mechanisms of the transfer of radioactive material from terrestrial ecosystems to and in water bodies around Chernobyl. Final report EUR 16529 EN. Brussels, Luxembourg (1996), p.93
3. Robbins, J. A.; Lindner, G.; Pfeiffer, W.; Kleiner, J.; Stabel, H. H.; Frenzel, P. Epilimnetic scavenging and fate of Chernobyl radionuclides in Lake Constance. Geochim. Cosmochim. Acta 56 (1992), 2339.
4. Mangini, A.; Christian, U.; Barth, M.; Schmitz, W.; Stabel, H. H. Pathways and residence times of radiotracers in Lake Constance, in M. M. Tilzer and C. serruya (Eds.), Large Lakes, Ecological Structure and Function, Springer Verlag, Berlin (1990), 245-264
5. Santschi, P. H.; Bollhalder, S.; Zingy, S.; Luck, A.; Farrenkother, K. The self cleaning capacity of surface waters after radionuclide fallout. Evidence from European lakes after Chernobyl 1986-88. Environ. Sci.Technol. 24 (1990), 519-527.
6. Smith, J. T.; Leonard, D. R. P.; Hilton, J.; Appleby, P. G. Towards a generalized model for the primary and secondary contamination of lakes by Chernobyl-derived radiocesium. Health Phys. 72(6) (1997), 880-892.
7. De Haan, F.A.M.; Bolt, G. H.; Pieters, B. G. M. Diffusion of Potassium-40 into an illite during prolonged shaking. Soil Science Society Proceedings 29 (1965), 528-530
8. Smith, J. T.; Fesenko, S. V.; Howard, B. J.; Horrill, A. D.; Sanzharova, N. I.; Alexakhin, R. M.; Elder, D. G.; Naylor, C. Temporal change in fallout ^{137}Cs in terrestrial and aquatic systems: A whole ecosystem approach and
Zibold, G.; Förschner, A.; Drissner, J.; Klemt, E.; Miller, R.; Walser, M. Time- dependency of the Cs-137 concentration factor in pike from lake Vorsee. UIR Topical Meeting Mol, Belgium, 1.-5. June 1998, to be published in proceedings
9. Drissner, J.; Bürmann, W.; Enslin, F.; Heider, R.; Klemt, E.; Miller, R.; Schick, G.; Zibold, G. Availability of Cesium Radionuclides for Plants - Classification of Soils and Role of Mycorrhiza. J. of Environ. Radioecology (1998), in press.
10. Kiefer, P.; Pröhl, G.; Müller, H.; Lindner, G.; Drissner, J.; Zibold, G. Factors affecting the transfer of radiocaesium from soil to roe deer in forest ecosystems of Southern Germany. Sci. Total. Environ. 192 (1996), 49-61.
11. Klemt, E.; Drissner, J.; Flügel, V.; Kaminski, S.; Lindner, G.; Walser, M.; Zibold, G. Bioavailability of Cesium Radionuclides in Prealpine Forests and Lakes. Mitt. der Österr. Bodenkundl. Ges. 54 (1996), 267-274.

EXPERIMENTAL STUDY AND PREDICTION OF DISSOLVED RADIONUCLIDE WASH-OFF BY SURFACE RUNOFF FROM NON-AGRICULTURAL WATERSHEDS

A.A. BULGAKOV and A.V. KONOPLEV
Institute of Experimental Meteorology, SPA "Typhoon", Obninsk, RUSSIA

Yu. V. SHVEIKIN and A.V. SCHERBAK
Ukrainian Hydrometeorological Research Institute, Kiev, UKRAINE

1. Introduction

The surface wash-off from watersheds is an important mechanism of radionuclide redistribution between terrestrial (agricultural, meadow and forest) ecosystems and water bodies. Within the present work using artificial irrigation of runoff plots, an experimental study was conducted of wash-off of dissolved forms of ^{137}Cs and ^{90}Sr by surface rainfall runoff from non-agricultural watersheds in the proximity of the Chernobyl NPP. A method is proposed for calculating radionuclide concentrations in the runoff as a function of characteristics of rainfall, runoff, soil and radionuclide distribution in the soil profile.

2. Materials and methods

The study of radionuclide wash-off by the surface rainfall runoff was carried out in 1986-1991 using artificial irrigation of the runoff plots located on typical (soddy-podsolic and alluvial) soils of watershed areas of the rivers in the 30-km zone of the Chernobyl NPP. For this purpose a rainfall simulator developed in Ukrainian hydrometeorological research institute (Kiev) was used which permits uniform irrigation of a plot of 5.4 m^2 with intensity of 0.2 to 2 mm/min. Two symmetric plots (1m^2) were arranged in the irrigated area. One of them had edges and bottom of sheet aluminum and a trough connected with a measuring device for automatic recording of rainfall intensity. The second plot is a ground plot fenced with steel edges 10 cm high from three sides and a trough connected with a runoff intensity measuring device from the fourth (lower) side.

The water flowing out of the runoff measuring device after filtration through the paper filter was transferred to plastic bottles on which sampling time was indicated. The sample volume was 0.2-1 L which corresponds to 0.2-1 mm of surface runoff. On

I. Linkov and W.R. Schell (eds.), Contaminated Forests, 103–112.
© 1999 *Kluwer Academic Publishers. Printed in the Netherlands.*

completion of rainfall 5 soil samples were collected at the plot (in the corners and in the middle) using a steel ring 5 cm high and 14 cm in diameter. The soil samples were divided into layers 0-0.5, 0.5-1, 2-3 and 3-5 cm. Then the water soluble form of the radionuclide was extracted from soil layers with distilled water and the exchangeable form - by 1M CH_3COONH_4 [1]. In each the content of ^{137}Cs was determined by gamma-spectrometry [2] and the content of [90]Sr - by the radiochemical method [3]. In the runoff samples, water and acetate extracts the concentrations of calcium, magnesium, potassium and sodium were determined by the method of atomic adsorption.

3. Results and discussion

3.1. DEPENDENCE OF RADIONUCLIDE CONCENTRATION IN RUNOFF ON HYDROPHYSICAL AND HYDROCHEMICAL CHARACTERISTICS OF RAINFALL AND RUNOFF

Contaminants pass into the surface runoff either from the pore solution or as a result of desorption from the soil surface. The concentration of non-sorbed substances in the runoff decreases sharply with time. For example, after 10 min rainfall with intensity 1 mm/min, the content of bromide-anion in the runoff was three orders of magnitude lower than in the first runoff samples[4]. The concentration of sorbed substances decreases slower in keeping with the value of the adsorption retardation factor. In our experiments, with the duration of irrigation of 1 hour and rainfall intensity about 1 mm/min there was no significant decrease in ^{137}Cs concentration in the runoff and ^{90}Sr concentration decreased by less than 30%. This suggests that the key mechanism of ^{90}Sr and ^{137}Cs supply to the runoff is desorption from the soil surface. The absence of effect of rainfall and runoff intensity on radionuclide runoff concentration suggests that the processes of sorption and desorption in the upper soil layer/runoff system proceeds rather quickly and the interphase distribution of radionuclides in this system can be considered to be at equilibrium. The examples of dependence of the ^{90}Sr concentration on rainfall and runoff intensity for one of the plots are shown in Fig. 1 and 2.

The sorption of radiocesium and radiostrontium by soils occurs by the mechanism of cation exchange and, consequently, the equilibrium concentration depends on the cation composition of the runoff. The dependence of radionuclide concentration on the content of cations in the runoff was investigated by irrigation of the plots with rain water with different concentrations of Ca^{2+} and K^+. It has been shown that the activity of ^{90}Sr in the runoff is directly proportional to Ca^{2+} concentration (see Fig. 3). For ^{137}Cs the dependence of activity on potassium concentration in the runoff is also close to the linear (see Fig. 4).

Thus, the radionuclides in the system upper soil layer/runoff are at cation-exchangeable equilibrium which for homovalent exchange is described by the Kerr isotherm equation:

$$[R]_w = [R]_{ex}[M]_w / K_c^R [M]_{ex} \qquad (1)$$

where $[R]_w$ is the radionuclide concentration in the runoff, Bq/L; $[M]_w$ is the

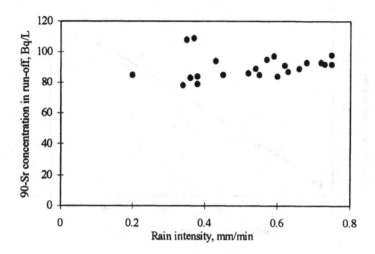

Fig 1. Dependence of ^{90}Sr concentration in the runoff on rainfall intensity at the plot in the vicinity of Chernobyl in 1990

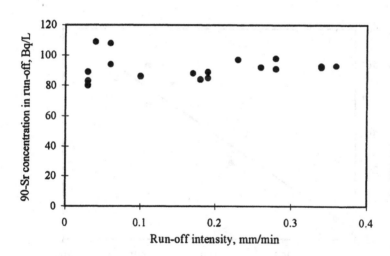

Fig 2. Dependence of ^{90}Sr concentration in the runoff on runoff intensity at the plot in the vicinity of Chernobyl in 1990

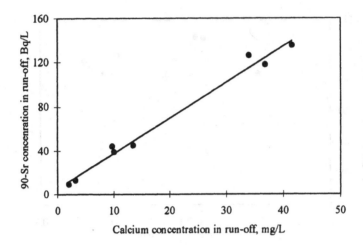

Fig 3. Dependence of ^{90}Sr concentration on calcium concentration in the rainfall surface runoff at the plot in the vicinity of Chernobyl in the spring-summer 1988.

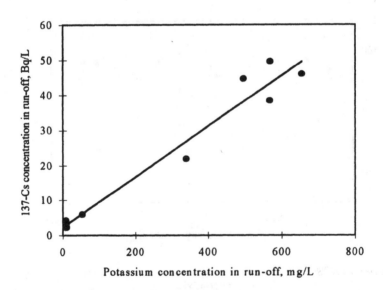

Fig 4. Dependence of ^{137}Cs on potassium concentration in the surface rainfall runoff from the plot in the vicinity of Kopachi in 1991.

concentration in the runoff of the competing cation with the same charge as the radionuclide cation (Ca^{2+} for ^{90}Sr and K^+ for ^{137}Cs), mg/L; $[R]_{ex}$ and $[M]_{ex}$ are the concentration of the exchangeable form of the radionuclide and competing cation in the upper soil layer, Bq/kg and mg/kg, respectively; K_c^R is the effective selectivity coefficients in the soil/runoff system.

3.2. EXPERIMENTAL DETERMINATION OF PARAMETERS REQUIRED FOR CALCULATION OF RADIONUCLIDE CONCENTRATION IN THE SURFACE RUNOFF

Using equation (1) one can calculate the expected concentrations of ^{137}Cs and ^{90}Sr in the runoff, if the cation composition, the content of exchangeable ^{137}Cs, ^{90}Sr and the cations in the soil and selectivity coefficients are known.

The cation concentration in the runoff is determined by its concentration in the rainfall water and soil solution and processes of cation exchange in the soil/runoff system. The study of the cation composition of the runoff has shown that it is very similar to rain water composition (see Table 1). The content of cations in the runoff is not dependent on the rainfall and runoff intensity and duration. In most cases, the standard deviation of the mean cation concentrations in the runoff is about 10% which can be almost completely explained by the measurement error.

TABLE 1. Cation concentrations in water extract from the upper half-centimeter soil layer, rain water and runoff from some experimental plots

Site	Cation	Concentration, mg/l		
		Water extract (1:1)	Rain water	Runoff
Benevka 1988	Ca^{2+}	1300±50	18±2	15±5
	Mg^{2+}	63±21	7.4±0.9	8±2
	K^+	78±17	5.6±0.2	6.2±0.5
	Na^+	52±3	36±1	38±4
Chernobyl 1989	Ca^{2+}	95±30	58±2	51±6
	Mg^{2+}	14±2	11±1	9±1
	K^+	180±40	9.5±0.3	16±6
	Na^+	10±3	35±2	35±3
Kliviny 1989	Ca^{2+}	10±3	21±3	20±2
	Mg^{2+}	1.8±0.2	3.7±0.2	2.7±0.1
	K^+	26±7	1.7±0.2	2.7±1.0
	Na^+	14±5	5.4±0.3	6.7±0.3

The mean concentration of Ca^{2+} in the runoff was not different from its concentration in the rainfall by more than 20% and the mean concentration of K^+ - by more than 40%. However, the initial concentration of these cations in the pore solution of the upper soil layer in some cases was two orders of magnitude higher than in the rain water. This suggests that prior to the start of the runoff, dissolved cations from the pore solution in the upper soil layer are washed out into the lower layers. Therefore, their contribution to formation of the runoff composition is not significant.

For determination of the content of the exchangeable form of radionuclides and cations in soil a standard method is available [1]. Given their non-uniform distribution in the soil profile the result of determination depends on the depth of soil layer sampled for analysis. In such cases the approximation of the complete mixing layer (CML) is used, which is the assumption that a contaminant sorbed in the surface layer of depth h is in equilibrium with the runoff. The depth h for our plots was estimated from the dependence of the radionuclide concentration in the runoff with time. If the approximation of the complete mixing layer accounts for the wash-out process accurately enough, the logarithm of the concentration of the washed-out material is a linear function of time [4] :

$$\ln[R^+]_W = \ln[R^+]_0 - [I/h(\rho K_d^R + \theta)]t \qquad (2)$$

where $[R^+]_0$ is the radionuclide concentration in the runoff at the runoff start, Bq/L; I is the rain intensity, mm/min; K_d^R is the effective distribution coefficient in the soil/runoff system; ρ and θ are soil volumetric density, kg/L and porosity respectively; and, t is time after the start of the runoff, min.

The change in the ^{90}Sr concentration logarithm in the course of the runoff in our experiments was well approximated by the linear function. An example of such a dependence for one of the experimental plots is shown in Fig 5.

Table 2 contains the coefficient values of equation (2) derived by the least square method for experiments with rainfall duration of not less than 2 hours and the CML depths calculated with them.

In calculating the distribution coefficient value of ^{90}Sr in the soil/runoff system, we used data on the concentration of the exchangeable radionuclide content in the upper half-centimeter soil layer, as it is practically impossible to separate the soil samples into the layers of millimeter depth. As shown by the data of Table 2 , 0.5 cm is a rather good estimate of the CML depth. The derived values of h are in complete agreement with those for other contaminants. The calibration of the ARM model using experimental data on the wash-off of agricultural chemicals gave the values of the mixing layer depth from 0.2 to 0.6 cm [5]. The values of h determined in [6] for the wash-off of phosphate-anion labeled with P-32 isotope (0.2-0.3 cm) fall in the same range.

The experimental concentrations of exchangeable forms of radionuclides and cations in the upper half-centimeter soil layer and their concentrations in the surface runoff were used for calculation of effective selectivity coefficients in the soil/runoff system. The calculation results are presented in Table 3. For comparison this table also presents the values of selectivity coefficients in the soil/water extract system obtained for the 0-5 mm layer by the method of sequential extraction.

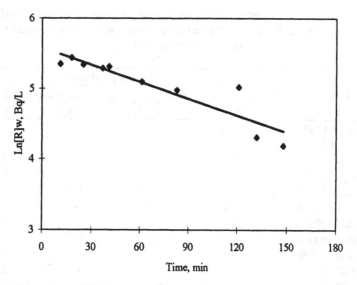

Fig 5. Dependence of the logarithm of ^{90}Sr concentration in the runoff on time, Benevka 1988.

TABLE 2. Equation (2) regression coefficients, estimated CML depths and values of the soil parameters used for its calculation

Plot	$\ln[R^+]_r^0$, $\ln(Bq/L)$	$I/h(\rho K_d^R+\theta)$	I, mm/min	K_d^R, L/kg	ρ, kg/L	h, mm
Benevka 1988	5.4±0.2	$(6.3\pm1.8)\times10^{-3}$	0.88	240	0.8	0.7
Chernobyl 1988	5.0±0.1	$(2.0\pm0.7)\times10^{-3}$	0.52	43	1.6	3.7
Chernobyl 1989	4.8±0.1	$(4.9\pm1.3)\times10^{-3}$	0.68	42	1.6	2.0
Korogod 1990	4.3±0.1	$(5.0\pm1.1)\times10^{-3}$	0.58	37	1.5	2.1

The values of K_c^R (Sr-Ca) and K_c^R (Cs-K) averaged over all experiments are 2.1 ± 0.9 and 30 ±18, respectively. They appeared for the Sr-Ca exchange by 2.3± 1.1 times and for the Cs-K exchange by 6.1±3.8 times higher corresponding selectivity coefficients in the soil/water extract system. The reason for this may be differences in concentrations of the exchangeable form of cations and radionuclides in the complete mixing layer and in the 0-5 mm layer. Also, the concentrations of the exchangeable form measured with the method used may be different from the real content of the radionuclides and the cations capable of fast exchange with the runoff.

TABLE 3. Effective selectivity coefficients for radionuclides in the soil/runoff system and selectivity coefficients in the soil/water extract system determined in the experiments at runoff plots

| Plot | Selectivity coefficient | | | |
| | soil/runoff | | soil/water extract | |
	^{90}Sr-Ca	^{137}Cs-K	^{90}Sr-Ca	^{137}Cs-K
Benevka 1988	2.2	-	-	-
Kopachi 1988	1.1	-	-	-
Korogod 1988	3.5	35	-	4.9
Korogod 1989	4.6	67	1.1	8.3
Korogod 1990	2.6	-	1.2	-
Chernobyl 1988	1.2	25	-	7.4
Chernobyl 1989 (1)	2.0	25	1.1	2.0
Chernobyl 1989 (2)	1.6	19	0.9	-
Chernobyl 1990	1.4	-	0.6	6.1
Rudnya Il`inetskaya 1989 (1)	23	20	0.6	6.1
Rudnya Il`inetskaya 1989 (1)	2.0	-	-	-
Kliviny	1.1	16	0.7	4.6

A variety of factors influencing the value of K_c^R and the complexity of experimental determination of parameters required for its calculation make it difficult to calculate K_c^R by soil and radionuclide properties. For prediction purposes the values of K_c^R given in Table 3 may be used. In those cases when it is sufficient to derive the upper estimate of the radionuclide concentration in the rainfall surface runoff, K_c^R in equation (2) can be replaced by the selectivity coefficient in the soil/water extract system. The recommended values of which for different soils can be found in reference [7].

3.3. EXAMPLE OF PREDICTING THE CONCENTRATION OF LONG-LIVED RADIONUCLIDES IN RAINFALL INDUCED SURFACE RUNOFF FROM WATERSHEDS CONTAMINATED AFTER THE CHERNOBYL ACCIDENT

In order to verify the method developed for estimating radionuclides concentrations in the surface rainfall runoff, we compared estimated and experimental values of normalized wash-off coefficients. The selectivity coefficients K_c^R were taken from Table 2. The concentrations of exchangeable forms of cations and radionuclides in the CML were determined by 1N ammonium acetate extraction. The data on the composition of rain water in the northern areas of Kiev region were provided by the Voeikov main geophysical observatory (Leningrad).

TABLE 4. Comparison of calculated and experimental normalized wash-off
coefficients

Field plot location	Normalized wash-off coefficients, 10^{-6} mm^{-1}			
	Calculated		Measured	
	^{90}Sr	^{137}Cs	^{90}Sr	^{137}Cs
Pripyat watershed				
Chernobyl	22	1.6	24±13*	2.5±1.2*
Kopachi	12	-	18±8*	
Il'ya watershed				
Staraya Rudnya	-	3.4		
Rudnya	60	5.2	36.9±12.2**	5.0±2.2**
Il'yanetskaya				
Kliviny	17	2.4		

*Wash-off coefficients were calculated using the radionuclides activity measured in the natural runoff samples
taken on the plots during the rainfall of 06.08.88.
**1988 averaged wash-off coefficients calculated for the Il'ya river watershed by Nikitin at al. [8]

The normalized wash-off coefficients for ^{90}Sr and ^{137}Cs were calculated as follows:

$$K_L = [R]_w / C_s \qquad (3)$$

where C_s is the runoff plot contamination density, Bq/m^2.

For the plots located in the Pripyat watershed, calculation results were
compared with experimentally determined normalized coefficients for natural runoff
from the two plots during the rainfall of 8 July 1988. For the plots in the watershed of
the river Il'ya, the comparison was conducted with the values K_L calculated in [8] by
radionuclides concentrations in the river outlet in 1988. The calculation was done
using concentrations of exchangeable radionuclides in the 0-5 mm layer measured in
the Pripyat watershed in June 1988 and in the Il'ya watershed in June 1989. Therefore,
the values calculated for the Il'ya river are somewhat underestimated, though the
magnitude of underestimation is not more than 20 %. As can be seen from data of
Table 4 there is a good agreement between the calculated and measured values.

Since for the plots in the Pripyat watershed we did nothing but recalculate the
radionuclides concentrations with the change in potassium and calcium concentration
in the runoff, the results seem trivial. They confirm the possibility of making, using
the proposed method, a rather reliable prediction of radionuclide concentrations in the
runoff from small plots, given that the required parameters are measured and
estimated fairly accurately. The consistency of calculated and measured values for the
Il'ya river seem to be of more interest. Though the radionuclide concentration in the
river is determined not only by their surface runoff from the watershed, but also by
river regime and radionuclides exchange between bottom sediments and water, the
calculated and measured wash-off coefficients are in good agreement. This indicates
that the proposed method can, probably, be used for prediction of radionuclides wash-
off from real watershed areas.

Applicability of the parameters in Table 3 for prediction of radionuclide concentrations in the runoff from real watersheds is also suggested by the fact that the concentrations of ^{90}Sr and ^{137}Cs in the runoff are not significantly dependent on the size and geometry of the runoff plot [9]. The normalized wash-off coefficients of these radionuclides appeared to be practically the same for rainfall runoff from the closely-spaced plots, both large (about 600 m^2) and small (1 m^2) [10].

The conditions and limits of the method, and therefore its applicability to real watersheds calls for additional study.

4. References

1. 1.Pavlotskaya F. I. (1974). Migration of radioactive materials of global fall-out in soils. Atomizdat, Moscow, (In Russian).
2. Silant'ev A.N. (1969). Spectrometric analysis of radioactive environmental samples. Gidrometeoizdat, Leningrad (In Russian).
3. 3.Sereda G.A., Shulepko Z.S. (1966). Collection of methods for determination of environmental radioactivity. Methods of radiochemical analysis, Moscow (In Russian).
4. Ahuja L.R. (1986). Characterisation and modelling of chemical transfer to runoff. *Adv. Soil Sci.*, v.4, pp.149-188.
5. 5.Donigan A.S., Jr., Beyerlein D.C., Davis H.H., Crawford N.H.(1977). Agricultural runoff management (ARM) model, version II: refinement and testing. EPA 600/3-77-098, Environ, res. lab USEPA Athens, GA, US Government Printing Office, Washington, DC.
6. Ahuja L.R., Sharpley A.N., Yamamoto M., Menzel R.G.(1981). The depth of rainfall-runoff-soil interaction as determined by P-32. *Water Resou.Res*, v.17, pp. 969-174.
7. Konoplev A.V., Bulgakov A.A. (1997). Prediction of ^{90}Sr and ^{137}Cs distribution in natural soil-water system In: Chernobyl Lessons Learned (Environment and Countermeasures). Sandia National Laboratory, Albuquerque, NM, USA, in press.
8. Nikitin A.I., Bovkun L.A., Chumichev V.B., Khersonsky E.S, Martynenko V.P. (1992). Field studies of cesium-137 and strontium-90 removal from the watershed affected by the releases from the damaged unit at the Chernobyl NPP. In: Ecological and geophysical aspects of nuclear accidents. Hydrometeoizdat, Moscow, p50-56 (In Russian).
9. Bulgakov A.A, Konoplev A.V., Popov V.E., Shcherbak A.V.(1990) Dynamics of wash-off from soil by the surface runoff of long-lived radionuclides in the area of the Chernobyl NPP. Pochvovedenie (Soil Science), N4, pp. 47-54 (In Russian).
10. Borzilov V.A., Konoplev A.V. et al. (1988). Experimental study of radionuclides wash-off of radionuclides deposited on soil as a result of the Chernobyl accident. Meteorologiya i Gidrologiya (Meteorology and Hydrology), N11, pp. 43-53 (In Russian).

BEHAVIOUR OF [137]CS IN SLOPING SEMI-NATURAL ECOSYSTEMS IN GREECE

G. ARAPIS, D. DASKALAKIS AND A. GODORA
Laboratory of Ecology and Environmental Sciences, Agricultural University of Athens, Iera odos 75, 118 55 Athens, GREECE.

1. Introduction

Among the radioactive fission products ^{137}Cs is considered of great importance due to its long residence half time and its low mobility in the majority of all soil types. The regions of Greece, where heavy ^{137}Cs deposition from the Chernobyl accident was observed, were those of Northwest Thessalia and West Macedonia. Radiocaesium's slow rates of migration in the soil, both horizontally and vertically (downward migration), is due to its irreversible sorption onto various soil particles, its fixation by mineral and organic components, and the low content of ^{137}Cs mobile forms, i.e. forms that are ionic, exchangeable and water-soluble [1,2,3,4,5].

Radiocaesium's horizontal migration mainly depends on water runoff. This leads to the runoff of radioactive material in its dissolved state, and in the solid state caesium where absorbed on microparticles can migrate. Surface runoff depends on the quantity of radionuclides in the upper soil's surface layer, and their water solubility. As a result of the radionuclide's mechanical transportation by water runoff, accumulation at local geochemical barriers, and in forest soils at the lower borders of slopes was reported.

In order to examine the above mentioned fact, we carried out a survey which included a) soil sampling from hill slopes at four radiopolluted locations of northern Greece, b) ^{137}Cs measurements using gamma-spectrometry, and c) primary soil analysis. Each sampling site was divided into four or five levels (from the upper to the lower part of the slope) with all sampling points of a level being at the same height, in order to detect easily any difference in radiocaesium concentration, between the various levels of the slope.

2. Materials and methods

The chosen sampling sites were part of a large area with an important ^{137}Cs deposition ranging from 20 to 40 kBq/m^2 (see Map 1) The majority of these sites were semi-natural slopes of North, North-Western exposure with the main vegetation cover consisting of tree species such as beech (*Fagus orientalis* and *Fagus silvatica*) and oak (*Quercus secciflora* and *Quercus coccifera*).

I. Linkov and W.R. Schell (eds.), Contaminated Forests, 113–119.

114

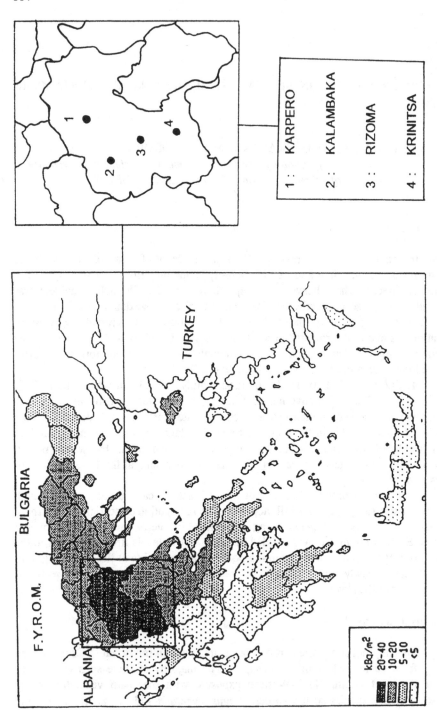

1 : KARPERO

2 : KALAMBAKA

3 : RIZOMA

4 : KRINITSA

Map 1. ^{137}Cs deposition in Greece.

In considering possible ^{137}Cs mechanical transportation by water runoff, resulting in its accumulation at local geochemical barriers as well as at the lower parts of slopes, we planed the soil sampling in a way that any important redistribution of radiocaesium, would be detected. The detection of horizontal displacement of the radionuclides demands a number of specific conditions such as small variation in the radioactive deposition for the soil sampling plots, as well as the low variation in environmental factors such as the plant cover, soil properties and/or the direction of run-off during the material (organic and inorganic) transport.

Four sampling areas (Karpero, Kalambaka, Rizoma and Krinitsa) were selected (see Map 1) where hill-slopes (height 200 - 250 m and declination 15-30 degree) located between the regions of Grevena and Kalambaka, in the North of Greece. The lower level of the slopes end at a relatively flat cultivated field, while the slope itself is uncultivated.

Each sampling site was divided into four or five levels (from the upper to the lower part of the slope with a distance of almost 100 meter between each level). The upper level is near the top of the hill while the lower level is near the cultivated field at the bottom of the hill. On each level we consider a representative sampling of at least five "sampling points" (from which we take the soil samples) at equal distances between them (approximately 10 m).

The soil sampling took place on April 1995. The best period to collect soil samples in Greece is considered to be from the end of autumn until the beginning of winter as well as at the beginning of Spring. At these times the soil moisture is sufficient enough for the easier extraction of the samples. The soil-sampler consisted of a metal cube of 1000 cm^3 volume (10x10x10cm) which was inserted into the ground at the constant depth of 5 cm. The bottom edge of the tube was sharpened in order to minimize the disturbance of the soil and to push it down into the ground easier. In this way, the soil-samples obtained had the standard geometry of 10x10x5cm.

In the laboratory, stones and roots were removed and the soil was homogenized, dried at 70°C for about 24 hours and weighed. Part of the sample was destined for granulometric analysis in order to classify the soils. The homogenized samples were put into pre-weighed plastic tubes of 5 cm height and 7,5 cm diameter and analyzed for ^{137}Cs using gamma -spectrometry, with a Ge(Li) detector (efficiency 18% and resolution 1,9 KeV (FWHM) at 1332 KeV, shielded by 10 cm thick lead). The detector was calibrated with IAEA standard reference-sources using the same geometry presented above.

3. Results

The content of ^{137}Cs measured in the soil (per Kg d.w.) was calculated and the results are shown in Figures 1 and 2. Statistical analysis of data was performed using

Sampling site of KARPERO

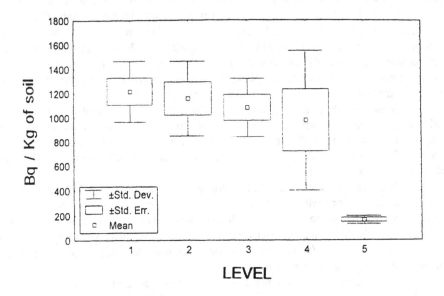

Sampling site of RIZOMA

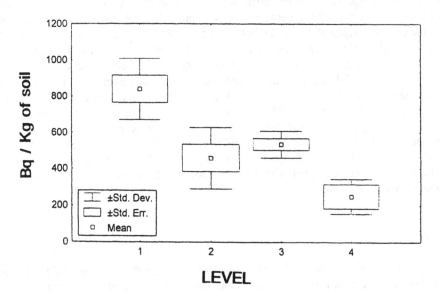

Fig. 1. Content of ^{137}Cs in the top 5cm of soil in Karpero and Rizoma.

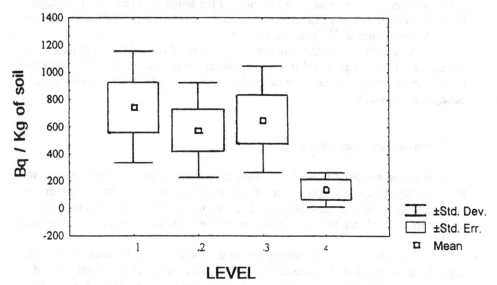

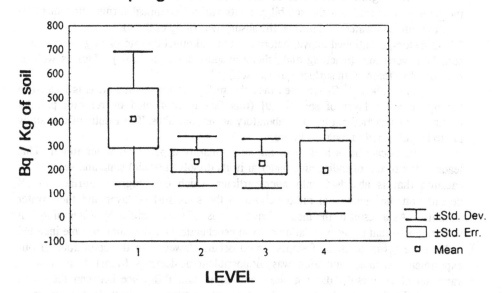

Fig. 2. Content of ^{137}Cs in the top 5cm of soil in Kalambaka and Krinitsa.

the t-test. From the analysis of variation it seems that there are no significant differences between the mean values for most of the sampling levels for all the studied areas. At the present time, the uncertainty analysis is being carried out in order to verify the occurrence of ^{137}Cs surface transfer.

In connection with the granulometric analysis of the soils we determined that the samples from Karpero and Rizoma can be characterized as Clay (C), while the samples from Krinitsa can be characterized as Sandy-Loam (SL) and from Kalambaka Sandy-Clay-Loam (SCL).

4. Discussion and conclusions

At most of the sampling sites, despite their differences in basic soil parameters and their slightly different vegetation cover, we ended up being unable to detect any redistribution or trend to movement of radiocaesium superficially from the upper to the lower part of the slopes. However, further uncertainty analysis of the present results is needed.

The behaviour of any radionuclide in the soil as well as its migration velocity depends on the physical and chemical properties of the soil which is strongly related to soils characteristics such as acidity, humic or mineral component content, surface-horizon structure, granulometric composition, adsorption ability and overall water content.

The migration behaviour of radionuclides (controlled by different migration mechanisms) depends on the solubility of its various chemical forms (movement by solution into soil water) and also on the absorptive ability of the soil.

These factors, mentioned above, determine the radionuclide rate of migration in the soil, both vertically (leaching and other transport down the soil profile) as well as horizontally (removal in surface water runoff) [6].

The bulk of ^{137}Cs scattered over the surface of the studied areas is located in the upper 5 cm layer of soil [7,8,9] (this fact is confirmed by relevant parallel investigations carried out by our Laboratory at the four sites. The results of this study are to be published in the near future).

Radiocaesium's horizontal migration mainly depends on water runoff. This leads to the runoff of radioactive material in its dissolved (ionic) state and solid state-caesium that is absorbed onto microparticles which can migrate. Surface runoff depends on the quantity of radionuclides in the soils surface layer and their water solubility. As a result of the mechanical transport of radionuclides by water runoff, it was supposed that the accumulation at local geochemical barriers and in some inclined soils at the lower borders of slopes would occur. However, in the conditions of our experiments, such accumulation was not possible to be detected. In fact the analysis of variation of our results did not show any significant difference between the mean values of the four sampling areas. At all sampling sites, despite their differences in basic soil parameters such as granulometric composition, pH, humic and other organic content and their slightly different plant cover, there was no evident redistribution of the radiocaesium between the upper and the lower part of the slope [10,11,12].

It is important to note that despite the original hypothesis, according to which we expected accumulation of ^{137}Cs at the lower parts of slopes, the lower levels of each sampling site seem to present lower ^{137}Cs content than the upper ones. This inconsistency could be due to the fact that those lower parts border regions of cultivated soils witch, generally, are supposed to be ploughed.

The observed non-homogeneity of the radioactivity values among some levels of the same sampling site can also be attributed to channel runoff of radioactive material via occasional water ways.

Finally, from the results obtained in this study we were not able to detect any obvious accumulation of [137]Cs. This seems to be in accordance with the information reported [6] that five to ten years following a nuclear accident, the mean value of the runoff coefficient for the main radioisotopes is approximately 0.2% per year. These measurements show that in our case, during the period of approximately ten years following the Chernobyl accident, only 5% of the deposited ^{137}Cs would be horizontally displaced.

5. References

1. Arapis G., Petrayev E., Shagalova E., Zhukova O., Sokolik G. & Ivanova T. Effective Migration Velocity of ^{137}Cs and ^{90}Sr as a Function of the Type of Soils in Belarus. Journal of Environmental Radioactivity, Vol. 34, No 2, pp. 171-186, 1997.
2. Cremers A., Elsen A., De Preter P. & Maes A.Quantitative analysis of radiocaesium retention in soils. Nature, Vol. 335, pp. 247-249, 1988.
3. Fawaris B. & Johanson K. A Comparative Study on Radiocaesium (^{137}Cs) Uptake from Coniferous Forest Soil. Journal of Environmental Radioactivity, Vol. 28, No 3, pp. 313-326, 1995.
4. Isaksson M. & Erlandsson B. Experimental Determination of the Vertical and Horizontal Distribution of ^{137}Cs in the Ground. Journal of Environmental Radioactivity, Vol. 27, No 2, pp. 141-160, 1995.
5. Rosen K. Transfer of radiocaesium in sensitive agricultural environments after the Chernobyl fallout in Sweden. II. Marginal and semi-natural areas in the country of Jamtland. The Science of the Total Environment, Vol. 182, No 1-3, pp. 135-146, 1996.
6. G. N. Romanov, D. A. Stukin & R. M. Aleksakhin. Peculiarities of ^{90}Sr migration in the environment. CEC-Radiation Protection-53. EUR 13574, pp. 421-435, 1990.
7. G. C. Bonanzzola, R. Ropolo & A. Facchinelli. Profiles and downward migration of [134]Cs and [106]Ru deposited on Italian soils after the Chernobyl accident. Health Physics, Vol. 64, No 5. pp. 479-484, 1993.
8. K. G. Anderson & J. Roed. The behaviour of Chernobyl ^{137}Cs, [134]Cs and [106]Ru in undisturbed soil: Implications for external radiation. Journal of Environmental Radioactivity, Vol. 22, pp. 183-196, 1994.
9. E. D. Stukin. Characteristics of primary and secondary Caesium-radionuclide contamination of the countryside following the Chernobyl NPP accident. CEC-Radiation Protection-53. EUR 13574, pp. 255-300, 1990.
10. E. M. Korobova & P. A. Korovaykov. Landscape and geochemical approach to drawing up a soil distribution profile for Chernobyl radionuclides in distant areas. CEC-Radiation Protection-53. EUR 13574, pp. 309-326, 1990.
11. A. V. Konoplev & T. I. Bobovnikova. Comparative analysis of chemical forms of long-lived radionuclides and their migration and transformation in the environment following the Kyshthym and Chernobyl accidents. CEC-Radiation Protection-53. EUR 13574, pp. 371-396, 1990.
12. J. T. Smith, J. Hilton & R. N. J. Comans. Application of two simple models to the transport of ^{137}Cs in an upland organic catchment. The Science of the Total Environment, Vol. 168, pp. 57-61, 1995.

STUDY OF BIODIVERSITY IN NATURAL ECOSYSTEMS OF SOUTH TURKEY

A. EVEREST, A. OZCIMEN, O. BALIKCI, L. SEYHAN, and A. UCAR
Mersin University, Faculty of Science, Department of Biology, Çiftlikköy, Mersin-TURKEY

The Turkish Electrical Generation and Transmission Company has plans for its Nuclear Program in 1996 where the first unit of the Nuclear Power Plant is

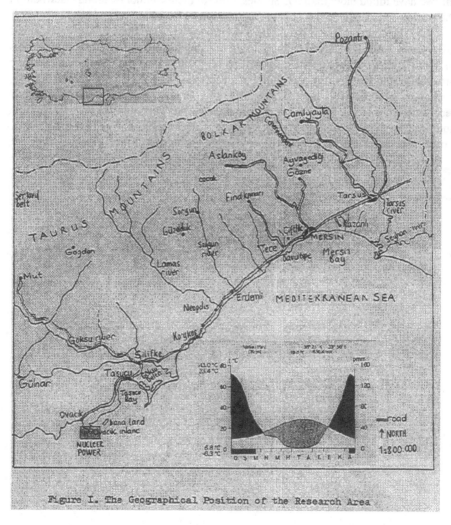

Figure I. The Geographical Position of the Research Area

I. Linkov and W.R. Schell (eds.), Contaminated Forests, 121–122.

expected to be commissioned in 2005. In this paper we have studied the biodiversity of natural ecosystems around the site proposed for the Nuclear Plant.

We have investigated natural resources of the Mediterranean Sea coastal area between the city of Mersin to Akkuyu in the south Turkey (Camlıyayla, Ayvagedıgı, Aslankoy and Kazanlı, Goksu Delta wetland, Neopolis, Korikos maquis hills and Mersin bay, Tece, Davultepe, Erdemlı, Tasucu bay, Fig. 1). These include Mersin high plateaus, Mersin wetlands and Mersin-Akkuyu seasides. We found many types of vegetation along the Mediterranean coast icluding a total of 8 tree, 17 shrub and 276 grass species in the wetland area. Pinus halepensis, Scrophularia trichopoda, Alkanna aucherana, Phlomis monocephala (for Korykos and Neopolis), Stachys rupestris, Thymus cilicica, Salvia hypargenia (for highplateaus) are unique for Turkey. The biodiversity is even higher for the highplateaus areas in the Bolkar Mountains. So far, we have found more than 600 species, some of them can not be found elsewhere in the world. The detailed report of our findings can be found in [1-2].

The construction of the Akkuyu nuclear power plant may affect these natural ecosystems. Moreover, it can cause damage to many historical and archeological sites in South Turkey. The site for the nuclear power plant is near the Ecemis fault seismic area that would increase the probability of an accident. The Adana earthquike in 1998, destroyed a settlement located close to this area. Therefore, we recommend reconsideration of the nuclear energy program by the Turkish Electrical Generation and Transmission Company. Alternative energy sources such as wind and solar energy can be of potential use for the climatic and natural resources of the Turkey.

References

1. Everest, A., et all. 1998. The Floristical Research of Mersin-Kazanlı, Adana-Seyhan Coast, 20-22.May.1998, II.International Science Congress of Kırıkkale Press, (252-258).
2. Everest, A., 1993. The Floristic Research of Göksu Delta, Nature and Human, June 93, issue 2, Rekmay press,(24-28).

AIRBORNE HEAVY METALS OVER EUROPE: EMISSIONS, LONG-RANGE TRANSPORT AND DEPOSITION FLUXES TO NATURAL ECOSYSTEMS

G.PETERSEN

GKSS Research Centre,Institute of Hydrophysics
Max-Planck-Strasse, D-21502 Geesthacht, GERMANY

Abstract

This paper presents a brief review of the processes by which airborne heavy metals are transported from the main emission areas in Europe and become subject to deposition and absorption into terrestrial and aquatic ecosystems with subsequent transport and transformation within the biotic and abiotic media that comprise these ecosystems. Results from numerical simulation models capable of simulating long-range transport of heavy metals over Europe together with measurement data of heavy metal concentrations in air and precipitation and the corresponding dry and wet deposition fluxes are reported. European wide inventories of anthropogenic heavy metal emissions based on location and capacity of their dominating source categories such as fossil fuel burning in power plants, industrial and residential combustion, waste incineration and road traffic are briefly described. Emission reduction scenarios with respect to introduction of lead free gasoline are outlined.

The critical gaps of knowledge on heavy metals in the atmosphere are identified focusing on uncertainties associated with emission fluxes in Eastern Europe and the scarcity of measurement data in that area. Future research is needed to estimate the effects of emission reductions on deposition fluxes of heavy metals to sensitive ecosystems such as forested areas in Europe is recommended. Special emphasis is placed on mercury, lead and cadmium which have been defined within the European convention on long-range transboundary air pollution of the United Nations-Economic Commission Europe (UN-ECE) to be the priority heavy metals of concern.

1. Introduction

Heavy metals emitted into the atmosphere by anthropogenic activities are deposited partly in the vicinity of emission sources but mostly 'en route' during their long-range transport over spatial scales from about 100 km to continental. Therefore, environmental policy strategies aiming at the reduction of emissions and deposition fluxes of heavy metals require international agreements. In Europe, such agreements

I. Linkov and W.R. Schell (eds.), Contaminated Forests, 123–132.

124

are now underway within the United Nations Economic Commission for Europe, Convention on Long-Range Transboundary Air Pollution (UN-ECE-LRTAP).

In view of the protocol for heavy metals within the UN-ECE-LRTAP and the already existing Arctic Monitoring and Assessment Program (AMAP), the Oslo and Paris Commission (OSPAR) and Helsinki Commisssion (HELCOM) programmes on inputs of atmospheric heavy metals to the Arctic, the North Sea and the Baltic Sea, respectively, significant efforts have been devoted in Europe to the monitoring of heavy metals to European marginal seas eg. [1-3], inland lakes and forest areas [4-6], and the Norwegian Arctic [7]. Recently, a data base has been established by the EMEP Co-operative Programme for Monitoring and Evaluation of the Long-range Transmission of Air Pollutants in Europe [8]. This data base contains concentrations of heavy metals in ambient air and precipitation for the time period 1987 to 1996 from 69 measurement sites in 20 European countries.

Increasing attention is now given to numerical modeling of the long-range transport and depositon of heavy metals over Europe. In recent years it has been widely recognized that the use of numerical models is important in formulating effective control strategies for the reduction of atmospheric deposition fluxes of heavy metals to the above mentioned ecosystems. The only way of delineating their atmospheric transport pathways and hence the emitter-receiver relationship is through numerical modeling. The major sources of information for this review are those listed in the reference section. In addition, information was obtained from internal reports and personal communication with scientists from European and North American research institutes and environmental protection agencies.

2. Atmospheric cycling of heavy metals

Once they are released into the atmosphere, heavy metals are subjected to various physical and chemical processes that determine their ultimate environmental fate. Fig.1 shows the fundamental features of the atmospheric cycle for heavy metals, beginning with emissions to the atmosphere and ending with deposition on the Earth's surface. The physical and chemical properties of a particular heavy metal and the prevailing environmental conditions determine the atmospheric pathways that it will follow during its residence time in the atmosphere. The term 'atmospheric pathways' refers to the multitude of possible processes and interactions in which heavy metals may participate from the time it enters the atmosphere until it leaves this environmental compartment.

2.1. ANTHROPOGENIC EMISSIONS

High temperature processes, such as coal and oil combustion in electric power generating stations and in district heating and industrial plants, roasting and smelting of

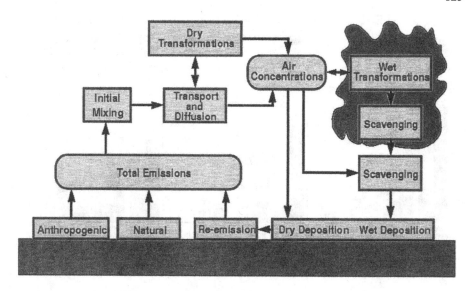

FIGURE 1. Airborne heavy metals: emission-to-deposition cycle

ores in non-ferrous metal smelters, melting operations in ferrous foundries, and waste incinerators, emit various heavy metals.

The quantity of atmospheric emissions from these sources depends upon:

- the contamination of fossil fuels and other raw materials
- physico-chemical properties of heavy metals, affecting their behavior during the industrial processes
- the technology of industrial processes
- the type and efficiency of control equipment

A detailed discussion as to what extent the above parameters affect the emission of heavy metals is presented in [9]. The first quantitative worldwide estimate of the annual industrial input of 16 heavy metals into the air, soil, and water has been published in [10]. This study illustrated that human activities are significantly altering the cycling of many heavy metals in the environment on a global scale. One of the first attempts to estimate atmospheric emissions of heavy metals from anthropogenic sources was completed at the beginning of the 1980s [11]. This work included information on emissions of 16 compounds. Previous works have dealt with either a single metal or a given source category. Most of the effort was spent on making inventories for emissions of lead, cadmium and mercury[12, 13].

In Central and Eastern Europe, the political changes of the 1990s initially brought a sharp decline of industrial activities, followed by significant restructuring of the economy and industry. The resulting emission reductions for lead, cadmium and mercury together with the relative changes of the main source categories are shown in Fig. 2.

126

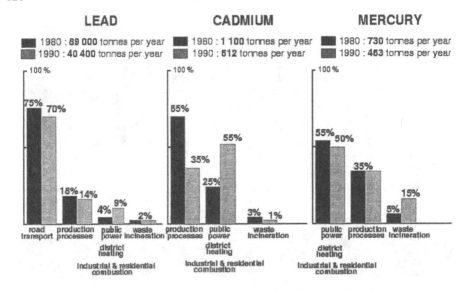

FIGURE 2. Main emission categories for heavy metals in Europe 1980 [12 , 13] and
1990 [14].

2.2. INITIAL MIXING, TRANSPORT AND TRANSFORMATIONS

Initial mixing refers to the physical processes that act on heavy metals immediately
after their release from an emission source into the mixed layer, which is the lower
region of the troposphere with typical extensions of 1-2 km during daytime and a few
hundred meters at night. In this layer, heavy metals are relatively free to circulate and
disperse vertically as well as horizontally due to small-scale turbulence, which
promotes intimate contact between vapor phase and aerosol associated heavy metals.
Such direct contact is important for physico-chemical transformations of heavy metals
near the source before extensive dilution has occurred and while air concentrations are
still relatively high.

Diffusion and transport processes occur simultaneously in the atmosphere.
Diffusion is caused by turbulent motion or eddies which develop in air that is unstable
or influenced by strong wind shear. Pollutant transport results from air mass
circulations driven by local or global forces. The actual distance travelled by pollutants
depends strongly on the amount of time a specific heavy metal resides in the
atmosphere. As a result of dispersive and removal processes such as precipitation
scavenging, some heavy metals associated with relativily large particles are deposited
from the mixing layer quickly. Metals on small particles or in gaseous and water
unsoluble form, which are removed more gradually, can be transported over severel
hundred or even thousands of kilometers. During transport and diffusion through the
atmosphere some of the metals such as selenium and mercury can participate in
complex chemical reactions. These processes can transform a metal from its primary

physical and chemical state to another state that may have similar or very different characteristics. For example, mercury is mainly present in the atmosphere in its gaseous elemental form but undergoes a series of physico-chemical reactions. These reactions include oxidation by ozone in the gas and aqueous phase, complex formation of oxidized mercury with subsequent back-reduction and adsortion on particulate matter in the aqueous phase (Fig.3).

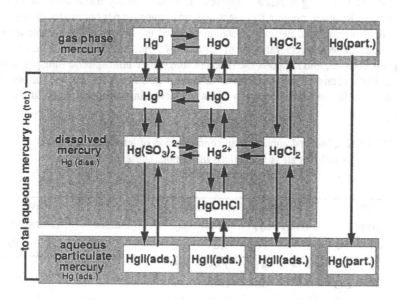

FIGURE 3. Chemical transformations of atmospheric mercury species,
(adopted from [15])

2.3. DEPOSITION PROCESSES

Removal processes of heavy metals can be conveniently grouped into two categories: dry deposition and wet deposition. Dry deposition denotes the direct transfer of gaseous and particulate heavy metals to the Earth's surface. Wet deposition encompasses all processes by which heavy metals are transferred to the Earth's surface in aqueous form (i.e. rain, snow, fog). The current understanding of wet deposition far exceeds the knowledge of dry deposition. Wet deposition is relatively simple to measure, even though considerable uncertainty exists if one attemps rigorous conceptual or mathematical description. By comparison, dry deposition is difficult to measure; therefore, the existing data base on this process is relatively small and contains many uncertainties. For both, dry and wet deposition the atmospheric pathways are much better described and understood for heavy metals associated with particles than for metals in gaseous form. Current knowledge of dry and wet deposition processes of heavy metals are extensively reviewed in [16] and [17].

2.4. OBSERVED ATMOSPHERIC CONCENTRATIONS

Observations of heavy metals in the atmosphere have been reported over the last 15 to 20 years for a number of locations throughout the European continent. However, the scarcity of measurements and uncertainties in the quality of data in parts of Eastern and Southern Europe is such that no generalized conclusions for Europe can be drawn. Nevertheless, the following trends are apparent: individual heavy metal concentrations vary widely with location, ranging from less than one nanogam per cubimeter in remote areas to tens of micrograms per cubicmeter in polluted urban areas. The remote areas recorded measurable concentrations of some of the elements, indicative of anthropogenic sources, and supports the thesis of long-range atmospheric transport into these areas. Data obtained for urban areas in central and Northern Europe, although having considerable variation, generally show a decreasing temporal trend in most elements.

TABLE 1. Observed annual mean concentration of lead and cadmium
in ambient air at 5 locations in Europe. Units: ng m^{-3}
(adopted from [8])

	Ispra (Italy)		Deuselbach (Germany)		Liesek (Slovakia)		Preila (Lithuania)		Lista (Norway)	
	Cd	Pb	Cd	Pb	Cd	Pb	Cd	Pb	Cd	Pb
1988			0.74	41	0.83	38				
1989		243	0.68	36	1.09	29				
1990		162	0.50	30	0.89	31				
1991		123	0.52	27	0.80	19			0.06	2
1992		120	0.46	21	1.18	39	0.17	13	0.05	2
1993		105	0.36	19	1.07	11	0.17	9	0.07	3
1994		104	0.34	15			0.41	12	0.08	3

Heavy metals have not yet been officially included in the UN-ECE convention on long-range transboundary air pollution in Europe, but in the framework of the preparatory phase of the heavy metal protocol the EMEP Chemical Coordinating Centre has established a data base containing observations from 1987 and onwards from different national and international European programmes such as HELCOM, AMAP, and OSPARCOM [8]. Twenty European countries have contributed data from 69 measurement sites. In Table 1, observations at 5 selected locations ranging from a densely populated area with considerable road traffic in Italy to a remote station in Norway are shown to reflect typical concentrations and temporal trends for cadmium and lead in Europe.

2.5 NUMERICAL SIMULATION MODELS

Because processes involved in atmospheric transport of heavy metals are so complex, numerical simulation models are invaluable in understanding the cause-and-effect relationship between the emissions of heavy metals from particular sources in Europe and the spatial pattern of atmospheric deposition fluxes to various ecosystems. These models aid in predicting which proposed control strategy can succeed in reducing these fluxes to acceptable levels. An example of numerical model capabilities is illustrated for a model predicting exposure of the Baltic Sea to lead emissions in Europe in 1980 is shown in Fig. 4. The black bars indicate the predicted contribution of 6 European countries to the atmospheric deposition fluxes of lead to the Baltic Sea. Comparison between the black and grey bars reveals the substantial flux reduction if only lead free gasoline would be used in road traffic of these countries.

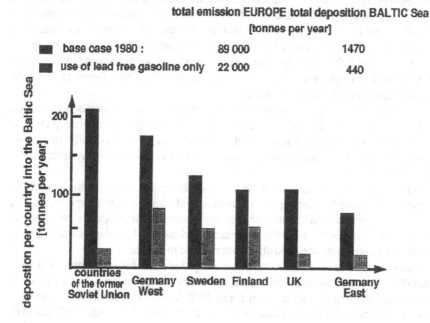

FIGURE 4. Emission reduction scenario 'lead free gasoline in Europe' 1980
(adopted from [18])

Despite intense interest in using models for heavy metals only a few operational models, which are validated through comparison with experimental data, are currently available in Europe. These models have been designed to simulate long-range and long-term transport of heavy metals assumed to be chemically inert and in association with particulate matter using a combined trajectory-climatologic approach [19], namely statistical approach using a Gaussian plume model and a trajectory model [3] and an EMEP type Lagrangian trajectory model [2], [18], [20]. More recently, three-dimensional models using an Eulerian reference frame have been introduced simulating various inert species on regional geographic scales in Europe [21 - 23] . Moreover, the development and application of dispersion models for heavy metals has progressed from formulations for inert species to models accounting for the physico-chemical transformations of atmospheric mercury species [15] , [24 - 26].

3. Conclusions

The current knowledge of airborne heavy metals over Europe has been briefly reviewed. There is strong evidence that recent progress in understanding atmospheric physico-chemical processes has permitted investigation of atmospheric heavy metal transport, transformation and deposition by means of field measurements and numerical simulation models. However, growing international concerns about heavy metals in Europe's atmosphere, as demonstrated in the recent agreement of the Heavy Metals Protocol of the UN-ECE, have led to a realization that the underlying scientific knowledge is incomplete and requires additional interdisciplinary analysis of heavy metals in the atmosphere. Listed below are the three main areas of uncertainty and the required activities needed to improve assessments of heavy metal deposition fluxes to terrestrial and aquatic ecosystems in Eastern Europe:

- A coordinated atmospheric monitoring programme for heavy metals does not yet exist in Europe. Hence, it is not possible at present to get a complete picture of deposition fluxes and their spatial and temporal variability especially in Eastern Europe, where air and precipitation measurements are few and highly uncertain in some areas.

- Field measurements close to the main industrial sources in Eastern Europe would be necessary to identify existing heavy metals and their chemical speciation, e.g mercury and organic and inorganic mercury species. These species are hard to identify with currently available analytical techniques but they could have a significant influence on the deposition pattern close to sources.

- Results from a model intercomparison study for lead have demonstrated their capability to predict monthly and annual deposition fluxes within a factor of two of observations over spatial scales from about 1000 km to continental [27]. However, model calculations for other metals such as cadmium and zinc greatly underestimate observations, even though calculations are correlated to measurements[18] , [19]. This shows that emissions for some heavy metals are underestimated.

4. References

1. HELCOM (1991) Airborne Pollution Load to the Baltic Sea 1986-1990. Baltic SeaEnvironment Proceedings No. 39. Baltic Marine Environment Protection Commission - Helsinki Commission -.

2. Petersen, G.,Weber, H., and Grassl, H. (1989) Modelling the Atmospheric Transport of Trace Metals from Europe to the North Sea and the Baltic Sea, in J.M. Pacyna and B.W. Ottar (eds.), *Control and Fate of Atmospheric Trace Metals*, NATO-ASI Series,Series C: Mathematical and Physical Sciences -

Vol.268,pp 57-83,Kluwer Academic Publishers, Dordrecht.

3. van Jaarsveld, J.A., van Aalst ,R.M., and Onderdelinden, D. (1986) Deposition of metals fromthe Atmosphere into the North Sea: model calculations. Report No. 84 20 15 002, National Institute of Public Health and Environmental Protection (RIVM), Bilthoven, The Netherlands.

4. Michaelis, W., Schönburg, M., and Stößel, R.-P. (1989) Deposition of Atmospheric Pollutants into a North German Forest Ecosystem, in H.-W. Georgii (ed.), *Mechanisms and Effects of Pollutant-Transfer into Forests,* Kluwer Academic Publishers, pp 3-12.

5. Rühling, A., Rasmussen, L.,Pilegaard, K.,Mäkinen, A. and Steinnes, E. (1987) Survey of Atmospheric Heavy Metal Deposition in the Nordic Countries in 1985.Report NORD 1987: 21, prepared for 'The Steering Body for Environmental Monitoring', The Nordic Council of Ministers.

6. Iverfeldt,A ., Munthe, J., and Hultberg,H. (1996) Terrestrial Mercury and Methylmercury Budgets for Scandinavia, in W. Baeyens, R. Ebinghaus, and O.F.Vassiliev (eds.), *Regional and Global Mercury Cycles,Fluxes and Mass Balances,*NATO-ASI Series,Series C: Mathematical and Physical Sciences,Kluwer Academic Publishers,Dordrecht, pp.381-401.

7. Ottar, B. and Pacyna, J.M. (1984) Sources of Ni, Pb and Zn during the Arctic Episode in March 1983,*Geophysical Research Letters* 11,441-444.

8. Berg, T.,Hjellbrekke,A.G. and Skjelmoen, J.E. (1996) Heavy Metals and POPs within the ECE Region,EMEP/CCC-Report 8/96,Norwegian Institute for Air Research,P.O.Box 100, N-2007 Kjeller,Norway.

9. Pacyna, J.M. (1989) Technological Parameters Affecting Atmospheric Emissions of Trace Elements from Major Anthropogenic Sources, in J.M. Pacyna and B.W. Ottar (eds.),*Control and Fate of Atmospheric Trace Metals*, NATO-ASI Series,Series C: Mathematical and Physical Sciences - Vol.268,pp 15-39,Kluwer Academic Publishers,Dordrecht.

10. Nriagu, J.O. and Pacyna, J.M. (1988) Quantitative Assessment of Worldwide Contamination of Air,Water and Soils with Trace Metals, *Nature* 333,134-139.

11. Pacyna, J.M. (1984) Estimation of the atmospheric emissions of trace elements from anthropogenic sources in Europe,*Atmospheric Environment* 18,41-50.

12. Axenfeld, F., Münch, J., and Pacyna, J.M. (1991) Europäische Test-Emissionsdatenbasis von Quecksilberverbindungen für Modellrechnungen. In: G. Petersen (1992). Belastung von Nord- und Ostsee durch ökologisch gefährliche Stoffe am Beispiel atmosphärischer Quecksilberverbindungen. Abschlußbericht des Forschungsvorhabens 104 02 726 des Umweltforschungsplanes des Bundesministers für Umwelt, Naturschutz und Reaktorsicherheit. Im Auftrag des Umweltbundesamtes, Berlin.

13. Axenfeld, F., Münch, J., Pacyna, J.M., Duiser,J.A. and Veldt,C. (1992) Test-Emissionsdaten-basis der Spurenelemente As, Cd, Hg, Pb, Zn und der speziellen organischen Verbindungen Lindan, HCB, PCB und PAK für Modellrechnungen in Europa. Abschlußbericht des Forschungsvorhabens 104 02 588 des Umweltforschungsplanes des Bundesministers für Umwelt, Naturschutz und Reaktorsicherheit. Im Auftrag des Umweltbundesamtes, Berlin.

14. Umweltbundesamt (1997) The European Atmospheric Emission Inventory of Heavy Metals and Persistent Organic Pollutants.Umweltbundesamt (Federal Environmental Agency) P.O. Box 330022, D-14191 Berlin,Germany.

15. Petersen, G.,Munthe,J.,Bloxam,R.,and Vinod Kumar,A. (1998) A Comprehensive Eulerian Modelling Framework for Airborne Mercury Species: Development and Testing of the Tropospheric Chemistry Module (TCM).*Atmospheric Environment - Special Issue on Atmospheric Transport,Chemistry,and Deposition of Mercury* (edited by S.E.Lindberg,G.Petersen and G.Keeler),Vol.32,No.5,pp.829-843.

16. Davidson, C. and Yee-Lin, W. (1989) Dry Deposition of Trace Elements, in J.M. Pacyna, B.W. Ottar (eds.),*Control and Fate of Atmospheric Trace Metals*, NATO-ASI Series, Series C: Mathematical and Physical Sciences - Vol.268,pp 147-202,Kluwer Academic Publishers,Dordrecht.

17. Barrie, L.A. and Schemenauer, R.S. (1989) Wet Deposition of Heavy Metals, in J.M. Pacyna and B.W. Ottar (eds.),*Control and Fate of Atmospheric Trace Metals*, NATO-ASI Series,Series C: Mathematical and Physical Sciences - Vol.268,pp 203-231,Kluwer Academic Publishers,Dordrecht.

18. Petersen, G. and Krüger, O. (1993) Untersuchung und Bewertung des Schadstoffeintrags über die Atmosphäre im Rahmen von PARCOM (Nordsee) und HELCOM (Ostsee) Teilvorhaben: Modellierung des großräumigen Transports von Spurenmetallen.Abschlußbericht des For- schungsvorhabens 104 02

    583 des Bundesministers für Umwelt, Naturschutz und Reaktor sicherheit. Im Auftrag des Umweltbundesamtes, Berlin.Externer Bericht GKSS 93/E/28.

19. Alcamo, J. ,Bartnicki, J., and Olendrzynski, K. (1991) Modeling Heavy Metals in Europe's Atmosphere:A Combined Trajectory - Climatologic Approach, in H.van Dop and D.G. Steyn (eds.), *Air Pollution Modeling and its Application VIII*,Plenum Press, New York,pp.389-398..

20. Krüger O. (1996) Atmospheric deposition of heavy metals to North European marginal seas: Scenarios and trend for lead,*GeoJournal* 39.2,117-131.

21. Bartnicki, J. (1994) An Eulerian model for atmospheric transport of heavy metals over Europe:Model description and preliminary results,*Water,Air and Soil Pollution* 75, 227-263.

22. Galperin, M.,Sofiev, M.,Gusev,A.,and Afinogenova, O. (1995) The Approaches to Modeling of Heavy Metals Transboundary and Long-Range Airborne Transport and Deposition in Europe,EMEP/MSC-E Technical Report 7/95, EMEP Meteorological Synthesizing Center-East,Kedrova str.8-1,Moscow, Russia.

23. Pekar, M. (1996) Regional Models LPMOD and ASIMD.Algorithms,Parameterization and Results of Application to Pb and Cd in Europe Scale for 1990, EMEP / MSC-E Report 9/96, EMEP Meteorological Synthesizing Center-East, Kedrova str.8-1,Moscow, Russia.

24. Petersen, G. (1992) Atmospheric Input of Mercury to the North Sea.Paris Convention for the Prevention of Marine Pollution. 9th Meeting of the Working Group on the Atmospheric Input of Pollutants to Convention Waters. London: 5-8 Nov. 1991 (PARCOM-ATMOS 9/13/1).

25. Petersen, G. (1992) Atmospheric Input of Mercury to the Baltic Sea.Baltic Marine Environment Protection Commission - Helsinki Commission -. Ninth Meeting of the Group of Experts on Airborne Pollution of the Baltic Sea Area (EGAP). Solna, Sweden: 19-22 Mai 1992 (HELCOM-EGAP 9/2/5).

26. Petersen, G.,Iverfeldt, A. and Munthe, J. (1995) Atmospheric Mercury Species over Central and Northern Europe.Model Calculations and Comparison with Observations from the Nordic Air and Precipitation Network for 1987 and 1988,*Atmospheric Environment* 29, 47-67.

27. Sofiev, M.,Maslyaev, A., and Gusev, A. (1996) Heavy metal model intercomparison Methodology and results for Pb in 1990.EMEP/MSC-E Technical Report 2/96, EMEP Meteorological Synthesizing Center-East,Kedrova str.8-1, Moscow, Russia.

PERSPECTIVES IN FOREST RADIOECOLOGY

Report of the Working Group on Measurement and Data[1]

T. K. RIESEN
Paul Scherrer Institut,
Division for Radiation Protection and Waste Management
CH-5232 Villigen PSI, SWITZERLAND

S. FESENKO
Russian Institute of Agricultural Radiology,
RU-249020 Obninsk, RUSSIA

K. HIGLEY
Oregon State University,
Dept. Of Nuclear Engineering, Corvallis, OR 97330, USA

1. Introduction

During the NATO Advanced Research Workshop on "Contaminated Forests: Recent Developments in Risk Identification and Future Perspectives" a Working Group (WG) entitled "Measurements and Data" was conducted. At the end of the first session the participants of the Working Group selected three topics from a list to be discussed in more detail. The Working Group decided:

- to work out recommendations for environmental sampling,
- to work out recommendations on how to share data, and
- to rank radioecological processes in forest ecosystems by using a matrix method and applying the expert knowledge of the WG participants.

The limited time available allowed only a very general consideration of the three topics.

---

[1] Members of Working Group: T.K. Riesen (Chairman), S. Fesenko (Co-Chairman), K. Higley (Rapporteur), G. Arapis, R. Avila, A. Bulgakov, V. Davydchuk, P. Kiefer, E. Klemt, A. Kliashtorin, A. Konoplev, V. Krasnov, Y. Kutlakhmedov, A. Orlov, A. Özgimen, O. Povetko, B. Rafferty, N. Sanzharova, A. Shcheglov, G.Sokolik, E. Steinnes, S. Yoshida

I. Linkov and W.R. Schell (eds.), Contaminated Forests, 133–140.
© *1999 Kluwer Academic Publishers. Printed in the Netherlands.*

2. Recommendations for environmental sampling

The result of the working group discussion clearly showed that in the future a "**new philosophy**" has to be followed for field sampling in view of the fact that radioecological research in Western countries is cut back and financial support is decreasing. Three measures are recommended:

1. **Conduct environmental sampling under a more general view.** This means that sampling should not be restricted to the act of simply taking the sample for radioecological purposes, but should include a description of the environment around the sample. The forest or landscape around the sample should be characterized and environmental conditions described in as much detail as possible. When planning a sampling campaign other fields of science should be considered, e.g. ecotoxicology, geobotany, air pollution research, etc. because samples could be shared. By this measure
   - the value of samples will improve and
   - the spectrum of radioecology becomes broader which will help this discipline to survive.

2. **Report your data** in a way which makes them accessible and applicable across disciplines and to other research programs. This measure helps to combine research programs and avoids double track sampling.

3. **Don't throw samples away.** As a consequence of points one and two, it is recommended to keep samples (or portions) after they have been collected. "Historical" samples can be used by other research groups for other purposes at a future date. New analytical techniques might be developed which then are applicable to these samples.

The Working Group was aware of certain constraints which can limit compliance with its recommendations:
1. The Working Group has no authority to require compliance
2. Sampling time and finances to follow the above recommendations are restricted
3. New samples have to be comparable to old ones (backward compatibility).

For more detailed information concerning sampling we refer to a not yet published report about a recent workshop on "Measuring Radionuclides in the Environment: Radiological Quantities and Sampling Designs" jointly organized by ICRU, EC and GSF in November 1997.

3. Recommendations how to share present and future data

The Working Group was convinced that a huge number of samples and data exist which could be exchanged and used in other projects. By communicating the existence of these samples and data we would improve their value and initiate new collaboration and joint publications within the field of radioecology and other disciplines. Four possibilities were proposed to improve the present situation:

1. To indicate the existence of samples and data sets in the web on the home page of institutes and scientists with cross links to other sites.
2. To set up a database within an international program or organization which supports the idea (EC, IAEA, NATO). Especially the EC could be interested to have a database of the data collected in the frame of their programs. It is suggested to submit a proposal for the 5th programme.
3. To strive for collaboration with UIR (International Union of Radioecologists) which already runs a database in one of their Task Forces.
4. To establish a News Group to share activities.

4. Ranking of radioecological processes in forest ecosystems

During the presentations of this NATO-ARW it became very obvious that most investigations are restricted to more or less one specific forest type. It was also shown that it is very unlikely that a single model can describe all radionuclide fluxes and a tendency was seen to develop a framework of sub-models. Therefore it is important to identify key processes and get an insight into them. This knowledge would help to define which data are needed in the future.

In a first step, the Working Group members were asked to rank 11 factors (given by one participant), in order of their importance. Members were asked to use a scale from 1 (not important) to 20 (very important). Table 1 shows the importance of forest type, soil type and type of landscape on radionuclide migration which primarily depends on the type of the radionuclide and its mobility. This result also reflects the fact that most radioecological forest models are very site specific and confirms the findings of the Working Group in section one to do environmental sampling under a more general view.

In a second step, the importance of radioecological processes in forest ecosystems was ranked by using an interaction matrix previously applied by Avila and Moberg [1] for the development of a conceptual model of the migration of ^{137}Cs in forest ecosystems. The heterogeneous composition of the Working Group with chemists, physicists, soil scientists, biologists, geochemists, ecotoxicologists, health physicists and ecologists gave a sound basis for ranking the processes given in the matrix (Figure 1). The starting point of the evaluation was a pulse contamination with ^{137}Cs.

TABLE 1. Ranking of factors influencing radionuclide migration in forest eco-system on a landscape level. Scale: 1 (not important) - 20 (very important).

Factor	Mean ranking (n=7)	Standard deviation
Mobility of radionuclides	15.6	3.5
Type of forest	15.7	4.6
Type of soil	15.4	3.7
Type of landscape	15.6	4.7
Climatic conditions	12.1	5.6
Humidity of forests	10.7	4.0
Forest Fires	9.9	4.1
Geobotanical variety in forest	7.7	4.7
Biological effects of dose, factors on migration process	7.3	4.8
Removing of forest products	5.7	5.0
Pastures in forest	4.4	3.4

Participants were asked to rank the importance of single processes which are specified in the off diagonal boxes of the matrix. Rankings were from 4 to 0 (critical- 4, strong- 3, medium- 2, weak- 1, no interaction 0) for ^{137}Cs cycling and to specify the forest type for which they were doing the evaluation. The ranking had to be done process-related and independent of the question if such a flux is measurable in practice. Figure 2 gives the results of the evaluation by indicating the mean value, standard deviation and relative deviation for each process.

Based on Figure 2 the following 10 interactions between compartments were found to be most important in forest ecosystems in decreasing order (mean value = 3.0):

1. Root uptake (soil organic-fungi)
2. Root uptake (soil organic-understorey)
3. Root uptake (Soil organic-tree)
4. Decomposition (litter-soil organic)
5. Leaf fall (tree-litter)
6. Root uptake (fungi-tree)
7. Ingestion (fungi-wild animals)
8. Ingestion (understorey-wild animals)
9. Diffusion (soil organic- mineral)
10. Root uptake (fungi-understorey)

The above list clearly shows that processes at the soil - plant interface are rated to be most important for radiocaesium cycling in forests followed by ingestion of fungi and understorey vegetation by animals. Decomposition of litter, leaf fall and fixation of ^{137}Cs to mineral soil are other important processes. By adding the process ranked 11th, the translocation from tree to leaves, the cycle of radiocaesium in forests can be closed. It is astonishing that the ^{137}Cs transfer from understorey to litter was not ranked higher compared to litter fall from trees. Another very obvious result is that fungi and especially mycorrhizae keep a most important position in the whole cycle of radiocaesium which is exactly the opposite of the knowledge of understanding we have.

By adding the numbers in each row we obtain an indication of the way a certain component affects ^{137}Cs cycling. Table 2 gives the corresponding values.

Atmosph. (Air)	Inter-ception, Rainfall Snowfall	Inter-ception	Inter-ception, Rainfall, Snowfall			Inter-ception,	Inter-ception, Rainfall, Snowfall	Inter-ception, Inhalation
Transpir. Burning	**Tree leaves**	Translo-cation	Leaf fall, weather-ing			Weather-ing, Inter-ception	Weather-ing, Inter-ception	Ingestion
Burning	Translo-cation	**Tree other**	Weather-ing, Inter-ception	Fertili-sation	Fertili-sation	My-corrhizae	Weather-ing, Inter-ception	Ingestion
Resus-pension		Rain splash	**Litter**	Decom-position, Percola-tion	Percola-tion	Root upake	Rain splash	Ingestion
		Root Upake	upward transport, Meso-fauna	**Soil oganic**	Percola-tion, Diffusion, Advect.	Root upake	Root uptake	Soil Ingestion
		Root Uptake			**Soil mineral**	Uptake	Uptake	Soil Ingestion
		Root uptake (My-corrhizae)	Fertili-sation	Fertili-sation	Fertili-sation	**Fungi**	Root uptake, My-corrhizae	Ingestion
Transpir. Burning			Leaf fall, Weather. Intercept.	Fertili-sation	Fertili-sation		**Under-storey**	Ingestion
			Fertili-sation			Con-sumption	Con-sumption	**Wild animals**

Fig. 1. Interaction matrix with 9 diagonal elements to describe migration of ^{137}Cs in a forest ecosystem. The matrix is read clockwise. After Avila and Moberg [1].

Atmosph. (Air)	*1.2* **2** 1.5 1.0	*1.3* **1** 1.1 1.0	*1.4* **1** 1.4 1.0	*1.5* **0** 0.3 3.9	*1.6* **0** 0.3 3.9	*1.7* **0** 0.5 1.5	*1.8* **1** 1.0 1.1	*1.9* **0** 0.5 1.5
2.1 **2** 1.2 0.8	**Tree leaves**	*2.3* **2** 1.2 0.6	*2.4* **3** 0.8 0.2	*2.5* **0**	*2.6* **0**	*2.7* **1** 0.8 1.6	*2.8* **1** 1.1 0.9	*2.9* **2** 1.0 0.5
3.1 **2** 1.1 0.7	*3.2* **3** 1.0 0.4	**Tree other**	*3.4* **1** 0.7 0.5	*3.5* **1** 1.2 1.0	*3.6* **1** 0.7 1.4	*3.7* **2** 1.4 0.9	*3.8* **1** 0.9 1.2	*3.9* **1** 1.1 1.4
4.1 **0** 0.5 1.3	*4.2* **0** 0.8 3.9	*4.3* **0** 0.8 2.1	**Litter**	*4.5* **4** 1.1 0.3	*4.6* **1** 1.3 1.1	*4.7* **3** 1.2 0.4	*4.8* **1** 0.9 1.7	*4.9* **1** 1.1 0.9
5.1 **0**	*5.2* **1** 1.4 2.6	*5.3* **4** 0.6 0.2	*5.4* **1** 1.0 1.5	**Soil oganic**	*5.6* **3** 1.3 0.4	*5.7* **4** 0.8 0.2	*5.8* **4** 0.9 0.3	*5.9* **1** 1.1 1.7
6.1 **0**	*6.2* **0** 0.4 2.6	*6.3* **2** 1.1 0.6	*6.4* **0**	*6.5* **0** 0.7 1.8	**Soil Mineral**	*6.7* **1** 1.1 1.1	*6.8* **1** 1.1 0.7	*6.9* **0** 0.6 1.9
7.1 **0**	*7.2* **0** 1.2 2.7	*7.3* **3** 1.3 0.4	*7.4* **2** 0.9 0.4	*7.5* **3** 0.7 0.3	*7.6* **1** 0.9 1.1	**Fungi**	*7.8* **3** 0.8 0.3	*7.9* **3** 0.8 0.2
8.1 **1** 0.7 0.5	*8.2* **0**	*8.3* **0**	*8.4* **2** 0.9 0.4	*8.5* **1** 1.0 0.8	*8.6* **1** 0.7 1.1	*8.7* **0** 0.6 2.2	**Under-storey**	*8.9* **3** 0.7 0.2
9.1 **0**	*9.2* **0**	*9.3* **0**	*9.4* **1** 1.0 0.8	*9.5* **0** 1.0 3.9	*9.6* **0**	*9.7* **1** 1.2 1.1	*9.8* **1** 1.1 1.3	**Wild animals**

Fig. 2. Coded matrix of the long-term migration of ^{137}Cs in a forest ecosystem. Meaning of the numbers within fields from top: numeric identification of the process; interaction index of process (critical- 4, strong- 3, medium- 2, weak- 1, no interaction 0); standard deviation; relative deviation. n=15, values were rounded.

TABLE 2. Interaction matrix from Figure 2. The sum of columns indicates how high the effect of the system onto a component is, the sum of rows indicates how high the effect of a component onto the system is

4.9	5.7	12.3	12.5	9.3	6.5	11.3	12.4	11.6	
at-	*1.2*	*1.3*	*1.4*	*1.5*	*1.6*	*1.7*	*1.8*	*1.9*	5.9
mosph	1.5	1.2	1.5	0.1	0.1	0.3	0.9	0.3	
2.1	leaves	1.9	3.4	0.0	0.0	0.5	1.2	2.0	10.5
1.6									
3.1	2.9	Tree	1.3	1.2	0.5	1.6	0.8	0.8	10.6
1.5									
4.1	0.2	0.4	litter	3.5	1.3	2.9	0.5	1.1	10.4
0.4									
5.1	0.5	3.6	0.7	soil org.	3.1	3.6	3.6	0.7	15.7
0.0									
6.1	0.1	1.9	0.0	0.4	soil min.	1.0	1.5	0.3	5.3
0.0									
7.1	0.5	3.3	2.1	2.5	0.9	fungi	3.0	3.3	15.6
0.0									
8.1	0.0	0.0	2.3	1.3	0.7	0.3	under-storey	3.1	9.0
1.4									
9.1	0.0	0.0	1.3	0.3	0.0	1.1	0.9	animals	3.5
0.0									

The highest effects on ^{137}Cs cycling have the soil organic compartment and the fungi compartment. The effect of the tree and litter compartment are intermediate while the role of the mineral soil compartment is very small.

During the evaluation additional processes have been identified by single participants which were not contained in the original matrix of Avila and Moberg [1], e.g., bioturbation (compartment 9.4 and 9.5, Figure 1), deposition and percolation (compartment 1.5 and 1.6, Figure 1) and the transfer of ^{137}Cs from understorey to fungi by mycorrhizae (compartment 8.7, Figure 1).

5. Conclusions

Discussions within the Working Group on "Data and Measurement" are concluded as follows:

- On an ecosystem level future data collection and measurements have to be done under a more general view. Collaboration with projects of other disciplines of science have to be considered. By this measure the value of samples will be improved and costs reduced. The extension of the activities will also help this discipline to survive.
- On a process level, key processes were identified by a matrix method. Future research in forest ecosystems has to be directed towards a better understanding of processes where fungi and mycorrhiza are involved in, such as root uptake, especially in the organic soil horizon, and decomposition of litter.

- It is essential that we find a method to share data by initiating the creation of a widely accessible database.

6. References

1. Avila, R. and Moberg, L. A systematic approach to the migration of Cs-137 in forest ecosystems using interaction matrices, accepted for publication in J. Environ. Radioactivity.

Part 2

Radionuclide Fate and Transport Modeling

REDUCING UNCERTAINTY IN THE RADIONUCLIDE TRANSPORT MODELING FOR THE CHERNOBYL FORESTS USING BAYESIAN UPDATING

I. LINKOV
Department of Physics, Harvard University, Cambridge, MA 02138, USA.

D. BURMISTROV
Urals Research Center for Radiation Medicine, RUSSIA

M. KANDLIKAR
Department of Engineering and Public Policy, Carnegie Mellon University, Pittsburgh, PA 15213, USA.

W.R. SCHELL
Department of Environmental and Occupational Health, University of Pittsburgh, Pittsburgh, PA 15261, USA.

Abstract

Our current understanding of fundamental processes that influence radionuclide behavior in contaminated natural forests is very limited. There is: (i) uncertainty in our knowledge, and (ii) natural variability of the ecosystems. Nevertheless, remediation of contaminated areas requires immediate policy decisions. Bayesian updating has been shown to be a useful tool for making decisions. It allows incorporation of the available experimental data to reconcile and reduce uncertainty both in the input parameters of a model and the future model predictions. This paper illustrates an application of the Bayesian updating for radioecology in general and specifically for contaminated forests of the Chernobyl Exclusion Zone, Belarus. The results of uncertainty reduction in a dynamic model FORESTPATH [1] for the fate and transport of radionuclides in the Chernobyl forests are presented. The implementation of the Bayesian updating is discussed: uncertainty in the input data determines the model complexity and the selection of the Bayesian updating technique.

1. Model Complexity and Uncertainty in the Current Knowledge on the Ecosystem

Developing a usable model for predicting doses from radionuclide release and interaction of radionuclides within the environment requires an evaluation of the many *uncertain* parameters and variables. These parameters are uncertain because: 1) ecosystem characteristics vary significantly from one ecosystem to another depending

I. Linkov and W.R. Schell (eds.), Contaminated Forests, 143–150.

144

on climate, soil types, plant diversity, etc.; and 2) available experimental data are unreliable due to the scarcity of measurements. Moreover, many model parameters are represented by aggregated values averaged over time, space and species. In contrast, the actual measurements *are site*-specific. Therefore, in developing an ecological model we should consider uncertainty as an integral part of the modeling process, and account for it by incorporating available information about the specific ecosystem under question, and about ecological processes in general.

Excessive model complexity will also result in a limited utility of the model because parameter uncertainty decreases the reliability of the model predictions. On the other hand, using overly simplistic models in situations when a better understanding of the ecosystem has already been achieved could lead to simplified model results that could possibly contradict with the available experimental data. Therefore, an optimal model complexity should be selected that depends on the availability of environmental data and study objectives (Figure 1).

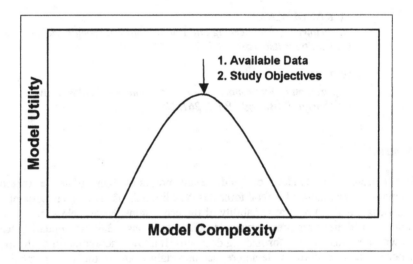

Fig. 1. Optimal Model Complexity

Our experience suggests that the dynamic compartmental multimedia modeling is an approach that is flexible enough to achieve a balanced combination of model complexity and utility for many ecosystems. It allows inclusion and consideration of the key ecological processes but does not require any detailed description of their specific mechanisms. Another advantage is that it places only modest requirements of computer resources. In addition it allows application of advanced uncertainty and sensitivity analyses.

Bayesian updating has been found to be a useful methodology that gives the analyst the ability to incorporate a limited knowledge of a specific ecosystem in order to improve model predictions and reduce or reconcile associated uncertainties. The input parameters of the model are treated as random variables with known subjective probability distributions (prior distributions) based on existing knowledge of this or other similar ecosystems. A representative sample of model parameters is generated from their probability distribution. The model is then run iteratively for each vector of sampled inputs. The resulting outputs characterize the probability distributions for each output variable. In Bayesian updating, these output distributions are compared with those observed experimentally and new input parameter distributions (posterior distributions) are constructed to provide the maximum likelihood of measured data. These posterior distributions incorporate additional knowledge about the ecosystem. Uncertainty is often reduced by the use of Bayesian updating, though in principle, it can also increase. If updated outputs are still far from the observed data, the procedure can be repeated again using distributions received after the first updating as inputs.

Even though Bayesian updating is conceptually easy, its implementation is associated with extensive computations. As a result, application of Bayesian updating has been limited to very simple models. We are developing a new approach to the Bayesian updating methodology: a stepwise Bayesian procedure that implements an efficient random search algorithm and takes into account the information about the utility function obtained in the previous runs. This approach can increase the efficiency of the current Bayesian updating procedures especially in its applications for environmental models.

2. Bayesian Updating in FORESTPATH Model for Chernobyl forests

Linkov [1] and Schell et al. [2] developed the generic model for radionuclide transport in forests, FORESTPATH, that calculates time series of inventories for a specific radionuclide distributed within the following eight compartments: Understory, Tree, Organic Layer, Labile Soil, Fixed Soil and Deep Soil. To incorporate details of radionuclide migration in the Organic Layer and fungi, the model was developed further (Fig. 2). The Organic Layer was represented by three horizons: Ol (litter), Of and Oh [1, 3].

In our previous studies [1-2] uncertain model parameters were estimated for the generic model application from the literature and satisfactory model predictions provide a general view on radionuclide fate and transport. For site-specific applications, the available literature data were limited to the ecosystems close to the site under consideration and site-specific parameters were thus estimated. Nevertheless, this deterministic approach has a limited site-specific application because it does not provide uncertainty estimates for the radionuclide concentrations in the compartments. Therefore, it cannot be used to estimate the confidence intervals for radiation doses required in risk assessment.

In this paper the model uncertainty is treated probabilistically. Results of a literature review show that values for model parameters are very uncertain and can be presented only by broad probability distributions (Table 1). A triangular shape for the distributions is assumed and characterized by three parameters: minimal and maximal values and mode. In the Bayesian updating procedure, model predictions for the radionuclide concentration in forest compartments at a specific date with the actual measurements are compared.

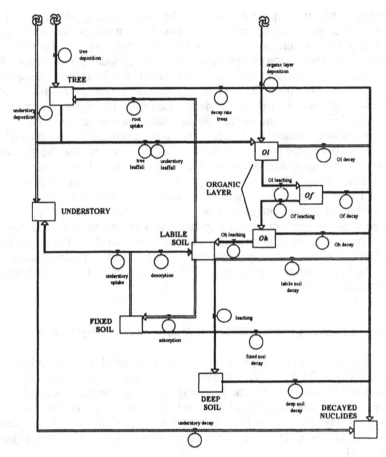

Fig. 2. FORESTPATH Model

TABLE 1. Uncertain FORESTPATH parameters for ^{137}Cs in the Chernobyl forests: Prior distributions and results of the Bayesian updating using 1993-1994 field data.

Parameter (yrs)	Prior Distributions			Updated Distributions		
	Min	Max	Mode	95% confidence interval		Mode
				Low	High	
1) Adsorption half-time	0.2	1	0.6	0.254	0.867	0.575
2) Desorption half-time	0.5	10	1.1	0.812	7.89	3.37
3) Leaching half-time	350	3000	800	592	2462	1299
4) Ol removal half-time	0.1	60	0.6	1.46	6.89	2.54
4) Of removal half-time	0.1	60	1.3	3.65	44.7	15.3
5) Oh removal half-time	0.1	60	1.7	5.20	47.4	21.0
6) Tree removal half-time long	0.5	10	3	0.643	9.58	5.42
7) Understory removal half-time long	0.1	2.8	0.2	0.22	2.2	0.904
8) Tree uptake half-time	0.1	100	8	7.32	78.2	33.0
9) Understory uptake half-time	0.1	100	8	5.42	95.2	49.8

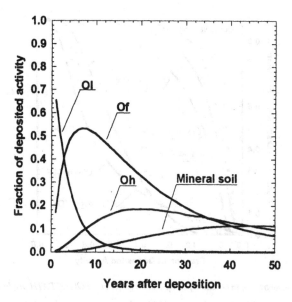

Fig 3. [137]*Cs concentration in soil compartments of a coniferous forest in Chernobyl over 20 years from an initial acute deposition (kBg/kg f.w.).*

Figure 3 shows radionuclide accumulation by the Organic Layer (Ol and Of) compartments, in a coniferous forest in Chernobyl described in [1,4]. Reported residence time for Cs in these horizons (see [1] for review) along with values for other FORESTPATH parameters for the Chernobyl forests [1-2] were used for the prior distribution construction (Table I). Organic Layer compartments exhibit complex time dynamics. The litter compartment is significantly contaminated immediately after the deposition, but looses much of its activity during the first year due to wash-off and leaching towards deeper layers. Of and Oh horizons show accumulation peaks at about two and four years respectively following the initial deposition.

At the first step, we reduce the uncertainty in the model input parameters using distributions of the radionuclide concentration in Ol and Of compartments measured in 1994 (Dvornik et al, unpublished). Figure 4 presents model prediction for radionuclide accumulation in the Ol and Of compartments along with measured accumulation in Belarussian forests in 1994 as cumulative distribution functions (CDF). The prior model distributions are significantly broader than the observed data. Moreover, the median values of the distributions are shifted towards the higher accumulation, i.e. the model predicts higher accumulation in these compartments as compared to the field observations. These site-specific measurements along with the uncertain generic distributions obtained from the literature were used for Bayesian updating. The updated distributions for the input model parameters (Table 1) are significantly narrower than the prior distributions.

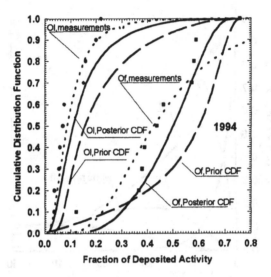

Fig 4. [137]Cs concentration in Ol and Of compartments: generic FORESTPATH predictions, experimental data for the Chernobyl forests in 1994 and Bayesian updated predictions for 1994

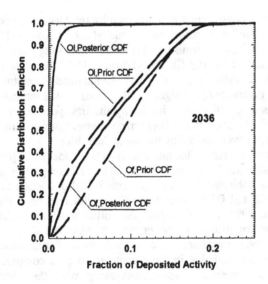

Fig 5. [137]Cs concentration in Ol and Of compartments: generic FORESTPATH predictions and updated predictions for 2036

The updated probability distribution for the model parameters with reduced uncertainties can be used in the model to predict future radionuclide concentrations. Figure 5 shows generic and updated predictions for the radionuclide accumulation in forest compartments for 2036.

4. Conclusions

Uncertainty and variability are inherent in the very nature of radioecology. Therefore we need to use appropriate tools to define, represent and analyze this uncertainty. The analyst should weigh current knowledge and associated uncertainties to develop a model that is appropriate for a specific ecosystem. Model oversimplification may ignore available data and knowledge on specific issues while excessive model complexity may lead to limited predictability and applicability of the model because the high uncertainty in model inputs results in even higher uncertainty in model outputs. The appropriate model complexity also depends on the specific objectives of the researcher: if one is to describe a limited dataset, a global comprehensive model may not be necessary.

A good example illustrating these points is found in fate and transport models developed for forest radioecology. Attempts to use comprehensive fate and transport models developed for relatively well-controlled agricultural ecosystems have not been fruitfully applied to contaminated forests primarily because of the high natural variability of forests in element uptake, soil type and conditions, plant species, etc. On the other hand, a simplistic description using transfer factors does not represent the entire complexity of dynamic processes in contaminated forests. The necessity to use dynamic parameters and thus compartmental modeling for forest ecosystems is now clearly realized [5].

The uncertainty and variability inherent in radioecology requires probabilistic treatment of model inputs and outputs since no single value can describe the complexity of the processes involved in radionuclide fate and transport in the environment. The probabilistic treatment not only provides a more realistic description of the ecosystem but also allows application of one of the most powerful tools developed to deal with the model and parameter uncertainty. Bayesian updating allows reduction and reconciliation in the uncertainty of the model input using information about model output and observations of the same quantity in the natural. We have demonstrated how a wide range of variations in the generic values of input model parameters can be reduced based on the site-specific data collected in the Chernobyl Forests.

In many cases Bayesian updating requires extensive measurement and experimental data that are difficult to collect or unavailable at all. If measured data are limited, the uncertainty in model predictions can be reduced by limiting the ranges of variation in input parameters based on expert judgments. The expert judgment can also be used to reduce the uncertainty in the stepwise Bayesian procedure that is described in this paper.

Bayesian updating procedure requires intensive computer calculations. The present study required several days of calculations on an IBM Pentium-166 PC, while one model run for the given values of parameters takes less than one minute. Therefore, an efficient reduction of the uncertainty in complex environmental applications requires: 1) increase computational speed and efficiency, 2) improve model structure to decrease the number of uncertain model parameters, i.e. decrease model complexity, and 3) improve efficiency of Bayesian updating procedure itself. The main reason for the low efficiency of the Bayesian updating procedure is the lack of memory in the algorithm of random sampling: random sampling does not take into account all the information obtained in the previous calculations. In addition, methodology for incorporation of the available experimental data in the updating scheme should be further developed. Use of the all array of the available information is desired to obtain the maximal decrease in the model uncertainties. But excessive volume of experimental data could lead to a very complex model likelihood function and may result in considerable fluctuations in the posterior distributions. We are currently developing a sampling technique for Bayesian updating that should address these issues.

5. Acknowledgements

The authors like to thank Dr. A. Dvornik and T. Zhuchenko for the field data. We are very grateful to Dr. M. Steiner for the most fruitful discussion and critical review of the manuscript This work was supported by the NATO Collaborative Research Grant and by the US NAS/NRC fellowship.

6. References:

1. Linkov, I. Radionuclide Transport in Forest Ecosystems: Modeling Approaches and Safety Evaluation. PhD Dissertation. University of Pittsburgh; 1995.
2. Schell, W.R., Linkov, I., Myttenaere, C., and Morel, B. (1996) "A dynamic model for evaluating radionuclide distribution in forests from nuclear accidents." *Health Physics*. 70, 318
3. Schell, W.R.; Linkov, I., Belinkaia, E., Morel, B. (1996) "Application of a dynamic model for evaluating radionuclide concentration in fungi." In: Proc. of 1996 Int. Congress on Radiation Protection. Vienna; p.2-752.
4. Schell, W.R., Linkov, I., Rimkevich, et al.. (1996) "Model-directed sampling in Chernobyl forests: general methodology and 1994 sampling program" *Science of the Total Environment* 180, 229.
5. Linkov, I., Schell, W.R., eds "Contaminated Forests: Recent Developments In Risk Identification and Future Perspectives," Kluewer, Amsterdam 1999 (in preparation)

REVIEW OF FOREST MODELS DEVELOPED AFTER THE CHERNOBYL NPP ACCIDENT

T.K. RIESEN
Paul Scherrer Institut,
Division for Radiation Protection and Waste Management,
CH-5232 Villigen PSI, SWITZERLAND

R. AVILA, L. MOBERG, L. HUBBARD
Swedish Radiation Protection Institute
SE-171 16 Stockholm, SWEDEN

1. Introduction

The investigation of transfer processes and the description of radionuclide circulation in forest ecosystems by appropriate models is a difficult and challenging task. The Chernobyl accident showed the importance of understanding such fluxes in these complex ecosystems. Due to the long ecological half-lives of released radionuclides it became necessary to develop models for dose and risk assessment which also could be used to assess the effect of countermeasures and to develop strategies to remediate such forests. The necessity of having such models is underlined by national and international organisations such as the IUR with a subgroup for forest ecosystems, the IAEA with a Forest Working Group within the BIOMASS program (BIOspheric Modelling and ASSessment) and the EC with the SEMINAT [1] and LANDSCAPE [2] projects.

The aims of the present review are:
- to compile a list of forest models,
- to compare the models developed after the Chernobyl accident in relation to the compartments they include, the transfer processes they describe and the conditions which they have been developed for and
- to look at the endpoint predictions of the models.

2. Models considered and their general features

Myttenaere *et al.* [3], Schell *et al.* [4] and Avila *et al.* [5] have previously published reviews of forest models. In this review emphasis is put on models developed after the Chernobyl accident. Not included are forest models developed before the Chernobyl

I. Linkov and W.R. Schell (eds.), Contaminated Forests, 151–160.

accident (TABLE 1), models describing radionuclide dynamics in fruit trees or models describing only processes in selected parts of a forest ecosystem. Also not included in this review are models which have been presented the first time during this NATO workshop.

TABLE 1. Overview of forest models developed before 1986 (adapted from Schell *et al.* [4])

References	Forest	Source	Compartments used
Olson [6]	Liriodendron trees	caesium inoculation	source, leaves, bark, roots, undercover, littermat, soil
Prohorov and Ginzburg [7]	generic forest	radionuclide enter via leaves, twigs and litter	litter, soil, roots, trunk, twigs, leaves
Aleksakhin *et al.* [8]	deciduous, coniferous	Sr-90 enter via leaves, twigs and litter	litter, soil, branches, wood, bark, herbs, needles/leaves
Mednik *et al.* [9]	deciduous (birch), coniferous (pine)	Sr-90 enter via leaves, twigs and litter	litter+soil, branches, wood, bark, leaves/needles
Jordan *et al.* [10], Jordan and Kline [11]	Tropical rain forest	strontium and manganese from atmospheric weapon testing	canopy, litter, soil, wood
Garten et al. [12]	deciduous (oak)	plutonium and fission products	consumer, ground vegetation, soil, litter, soil fauna, leaves, wood, roots
Croom and Ragsdale [13]	deciduous (oak)	caesium injection from atmospheric bomb testing into the litter	tree, litter, lower soil, upper soil; available for uptake, upper soil

TABLE 2. Overview of forest models developed after 1986 in chronological order

Acronym/reference	*	Nuclide	Forest type	Soil type	No. of sites	
RADFORET [14]	A	Cs	decidous	loamy clay, sandy	2	USA
Prohorov and Ginzburg [7], applied by Alexakhin *et al.* [15]	B	Sr	birch pine	chernozem soddy podzol	2	RUS
FORESTPATH [1] [4]	D	Cs, Pu	decidous coniferous	generic	na	USA, BEL
FOA [2, 5]	C	Cs	boreal, coniferous	not specified	1	SWE
RIFE I [16]	E	Cs	different types	not specified	2 2	RUS, UKR IRL, GER
FORESTLIFE [1][16, 17]	F	Cs	pine	not specified	13	BEL
FORM [18]	G	Cs	generic	generic	na	
ECORAD [19]	H	Cs, ^{14}C, [129]I	generic	soddy podzol	4/4	RUS/UKR
RIFE2 [1] [20]	I	Cs	pine	not specified	5	Europe
LOGNAT [2] [2, 5]	K	Cs	needle, deciduous	not specified	1	ITA
FORESTLAND [1] [21]	This model was presented during the NATO-ARW					
Seymour [1] [22]	This model was presented during the NATO-ARW					

*	internal reference (cf *Figures 1* and *2*, TABLE 3)
[1]	model described in this volume
[2]	model under development
na	not applicable

From TABLE 2 which shows 10 post-Chernobyl models it can be seen that
- the majority of the models describe caesium fluxes.
- the majority of the models have been calibrated for one forest type with the exception of RIFE 1 and 2 [1, 16] which offer the possibility to select between different types of sites. Generic models are designed to be applicable for several sites and are calibrated using a wide range of available experimental data.
- the models were parameterized with site specific experimental data (between 1 and 13 sites).

3. Model structure

The migration of radionuclides in forest ecosystems involves multiple components and interactions. As the different models have been designed for special purposes and selected sites a comparison is only partly valid. However, there are three major predictive compartments which can be reasonably well compared: the tree, the understorey and the soil compartment.

TABLE 3 gives an overview of the compartments included in the 10 studied models. Considering transfer processes in trees, the whole range of differentiation can be found. FORESTPATH [4] and the FOA model [2, 5] which are focused on generic applications on an ecosystem and landscape level have one compartment for the whole tree or the perennial vegetation, respectively. In contrast, Prohorov and Ginzburg's model [7, 15] and ECORAD [19] have a needle, branch, bark and wood compartment. The most common to all models is an additional model-specific compartment.

Considering transfer processes in soil, it is found that all models have a litter and a more or less differentiated soil compartment. In the original version of FORESTPATH [4] the organic soil compartment includes litter. Some models including a further developed version of FORESTPATH [23] additionally distinguish between litter and one or two organic horizons. In FORESTPATH [4], FORM [18], ECORAD [19] and RIFE 2 [20] mineral soil is further divided in several compartments. In RADFORET [14] and FORESTLIFE [16] corrections of transfer rates are made to allow a description of vertical migration and sorption processes in soil. So far FORESTPATH [4] is the only model which distinguishes between a labile/available soil compartment and a fixed/unavailable compartment permitting a better description of sorption-desorption processes in soil. ECORAD [19] considers three layers of surface organic horizons and divides mineral and mineral organic layers into 1-cm layers down to 15 cm. Such a detailed description creates difficulties in applying this model to other sites. RIFE 2 [20] offers 6 soil layers which can be defined to suit the need of the corresponding modeler.

All models have varying descriptions of an understorey compartment except LOGNAT [2, 5]. Another specific compartment is the fungi/mushroom compartment which is included in 4 models. The FOA model [2, 5] appears to be the only model at present that identifies retention of radionuclides in a moss or lichen carpet, potentially constituting a significant secondary source to radioactive caesium transferred into

154

circulation. The only other model considering animals is FORM [18], which has a game and roe deer compartment. Another endpoint in this model is the ingestion dose from forest products calculated from activity assessments in mushrooms, berries, honey and milk.

TABLE 3. Overview of compartments considered in 10 post-Chernobyl forest models and their assignment (X) to single models

compartment	Model*									
	A [14]	B [15]	C [2,5]	D [4,23]	E [16]	F [16]	G [18]	H [19]	I [20]	K [2,5]
leaf/needle	X	X					X	X		X
leaf internal									X	
leaf external									X	
tree external				X	X					
tree internal					X					
external bark								X		
internal bark										
bark		X					X			
branches	X	X						X		
living tree						X				X
bole	X									
wood		X				X	X	X	X	
xylem										
perenn. veg.			X							
fresh litter						X				
Litter/Ol	X	X	X	X	X	X	X	X	X	X
Of				X	X		X	X	X	X
Oh				X			X	X		
soil 1	X	X	X	labile	X	X	X	X	X	X
soil 2				fixed			X	X	X	
soil 3				deep			X	X	X	
roots	X					X		X		
understorey	X	herbs		X	herbs	X	X	X		
perenn. veg.				X						
ground veg.										
soil fauna										
mushrooms					(X)		X		X	
game							X			
forest product							X			
mammals										
distributive								roots,		
pool								fungi		
moss/lichen			X	(X)						
competitors			X							

* letters from TABLE 2.

From TABLE 3 it can be concluded that all long-term models have at least 5 compartments:

1. a needle or leaf compartment, which is reasonable considering the important role played by leaf fall in caesium recycling within the system.
2. a wood compartment which is more or less differentiated
3. a litter or organic soil compartment
4. a mineral soil compartment
5. an understorey compartment

4. Transfer processes in forest ecosystems

A summary of the migration of radionuclides described in the 10 forest models is shown in a matrix diagram (*Figure 1*) which is described more detailed elsewhere [24]. In this case the matrix is only used to visualise the various transfer processes and their inclusion frequency in the 10 models. The leading diagonal elements display the main compartments of radionuclide cycling as identified before. The off-diagonal elements correspond to the transfer processes between these compartments. The diagram is read clockwise. The shading of the single boxes corresponds to the frequency the process is included in the 10 models. The diagram does not indicate the importance of the process in relation to the whole ecosystem. However, the frequency numbers give an indication about the availability of data, about the state of the understanding of a process and about the consensus among modelers to include the process in their model. This may be illustrated by the example of litterfall. Litterfall is relatively easy to measure and is described in nearly all the models. It is a necessity to include this pathway which returns radionuclides from trees and understorey vegetation to the ground in a research model. However, the importance given to this pathway by modelers does not necessarily correspond to the importance of it when the whole forest ecosystem is considered. The soil compartment is represented by a single box in this matrix although it was shown that most models have several soil compartments. To describe the processes occurring in soil a second matrix was drawn (*Figure 2*). Taking into account that soil can contribute to external radiation in a dose assessment model and following the tendency to work with aggregated transfer factors, e.g., in the RESTORE project of the EC, the representation of soil by one box seems justified. The example of the soil-plant transfer also demonstrates, how differently this pathway is modelled in contrast to the previously mentioned litterfall.

leaf needle	1	3		8		1			
2	branch			3		1			
4	3	Tree (bole, bark)		2		2			
			under-storey	5		2			
		1	1	litter A$_f$, A$_h$	1	7			1
1	1	3		1	Roots	1			
2	1	6	5	2	3	soil	1	1	3
							game		
								berries and others	
									fungi

process considered in 1-2 models
process considered in 3-5 models
process considered in > 5 models

Figure 1. Matrix description of caesium fluxes in a forest ecosystem and frequency of the transfer processes applied in 10 forest models (cf TABLE 2)

5. Endpoints from a radiation protection perspective

Ultimate endpoints in forest models can be concentrations in forest components, or from a radiation protection point of view, doses due to external radiation or internal radiation caused by contaminated forestry products for human consumption. Other possible endpoints are external radiation from forest products of industrial use or radiation effects from the forest itself. According to Table 2 the soil and wood compartment could be used as a basis for the corresponding dose calculations in all models. However, the food pathway is only included in the FORM model [18]. With few exceptions the connection of the models under consideration with existing food-chain models is not yet realised. Dose calculations are not addressed, probably because methods for calculating doses from activity concentrations already exist. What is missing at the moment therefore is an extension of present forest models to dose assessments or the connection with dose assessment models as it is done between FORESTLIFE [16] and FORESTDOSE [25]. Also FORESTPATH has been applied for dose and risk assessment [26].

plant			2	4			
3	root		1	1			
2	1	litter	5	3			
2	1		A$_h$, A$_f$ (organic layer)	5	2		
0	2	2		soil general			3
1	1				soil labile	1	2
1					1	soil fixed	
1	1						deep soil

░ process considered in 1-2 models
▨ process considered in 3-4 models
■ process considered in > 4 models

Figure 2. Matrix description of caesium fluxes in the soil compartment and frequency of the transfer processes applied in 10 forest models (cf TABLE 2)

6. Factors influencing caesium migration

Table 4 shows how the factors which influence caesium migration are considered in the 10 models. A factor is considered in the model if it is possible to use the model for evaluating the degree of influence of this factor on caesium migration. A model considers a factor if the following two conditions are fulfilled:
(i) it describes the transfer processes that are significantly influenced by the "factor" and (ii) it provides relationships between transfer rates and some qualitative and quantitative measure of the "factor". The factors influencing caesium migration are in general poorly considered by the models and most models include only a few or none of the parameters listed in TABLE 4.

TABLE 4. Factors influencing caesium migration considered by 10 models (cf Table 2).

Factor	Model
Type of deposition [1]	
Season of the deposition	
Type of forest (deciduous, coniferous)	FORESTPATH, FORM, Prohorov and Ginzburg
Biomass growth	RADFORET, FORESTLIFE
Age of the trees	RADFORET, FORESTLIFE, FORM
Soil characteristics	RADFORET

[1] dry, wet, chronic, pulse deposition

7. Conclusions

- Most of the transfer processes in the reviewed models are described with rate constants. These are often highly variable and include several processes.
- The models can hardly be applied for explaining and predicting differences in the behaviour and distribution of radionuclides in different forest types. All models except FORESTPATH [4] and FORM [18] are site specific and the characteristics of forests used for calibrations are usually not provided, which basically makes it difficult to apply them for other sites. The parameters in FORESTPATH and FORM are given for generic forest ecosystems but some of them vary within two orders of magnitude.
- The improvement of models is constrained by the availability of experimental data and by lack of sufficient knowledge of migration mechanisms. Future effort should be put into describing some important transfer rates at a process level.

It seems impossible to have one model describing all relevant processes and types of forest ecosystems. Generic models can be used to show general trends. A simple basic model describing the main processes with attached sub-models for complex compartments, e.g., a soil compartment, and selected forest types is a possible format for assessing radionuclide distribution in different forest environments.

8. References

1. Belli, M. (1998) SEMINAT, Long-term dynamics of radionuclides in semi-natural environments: derivation of parameters and modelling. Mid-Term Report 1996-1997, EC contract F14P-CT95-0022, ANPA, Roma.

2. Moberg, L., Hubbard, L., Avila, R., Wallberg, L., Feoli, E., Scimone, M., Milesi, C., Mayes, B., Iason, G., Rantavaara, A., Vetikko, V., Moring, M., Bergman, R., Nylén, T., Palo, T., White, N., Raitio, H., Aro, L., Kaunisto, S., and Guillitte, O. (1998) LANDSCAPE, An integrated approach to radionuclide flow in semi-natural ecosystems underlying exposure pathways to man, Mid-Term Report 1996-1997, EC contract F14P-CT96-0039b, SSI, Stockholm.

3. Myttenaere, C., Schell, W.R., Thiry, Y., Sombre, L., Ronneau, C., and Van der Stegen de Schrieck, J. (1993) Modelling of the Cs-137 cycling in forest: Recent developments and research needed, Sci. Total Environ. 136, 77-91.

4. Schell, W.R., Linkov, I., Myttenaere, C. and Morel, B. (1996) A dynamic model for evaluating radionuclide distribution in forest from nuclear accidents, Health Phys 70 (3), 318-335.

5. Avila, R., Moberg, L. and Hubbard, L. (1998) Modelling of radionuclide migration in forest ecosystems. A literature review, SSI-report 98:07 (Swedish Radiation Protection Institute, ISSN 0282-4434).

6. Olson, J.S. (1965) Equation for cesium transfer in a Liriodendron forest, Health Phys. 11, 1385-92.

7. Prohorov, V.M. and Ginzburg, L.R. (1972) Modeling the process of migration of radionuclides in forest ecosystem and description of the model, Soviet J. Ecology 2, 396-402.

8. Aleksakhin, R.M., Ginsburg, L.R., Mednik, I.G. and Prohorov, V.M. (1976) Model of Sr-90 cycling in a forest biogeocenose, Soviet J. Ecol. 7, 195-202.

9. Mednik, I.G, Tikhomirov, F.A., Prohorov, V.M. and Karaban, R.T. (1981) Model of Sr-90 migration in young birch and pine forests, Soviet J. Ecology 12, 40-45.

10. Jordan, C.F., Kline, J.R. and Sassger, D.S.A. (1973) A simple model of strontium and manganese dynamics in a tropical rain forest, Health Phys. 24, 477-89.

11. Jordan, C.F. and Kline, J.R. (1976) Strontium-90 in a tropical rain forest: 12th-yr validation of a 32-yr prediction, Health Phys. 24, 477-89.

12. Garten, C.T. Jr., Gardner, R.H. and Dahlman, R.C. (1978) A compartment model of plutonium dynamics in a deciduous forest ecosystem, Health Phys. 34, 611-19.

13. Croom, J.M. and Ragsdale, H.L.A. (1980) A model of radiocesium cycling in a sand hills-turkey oak (Quercus Laevis) ecosystem, Ecological modeling 11, 55-65.

14. Van Voris, P., Cowan, C.E., Cataldo, D.A., Wildung, R.E., and Shugart, H.H. (1990) Chernobyl case Study: Modelling the dynamics of long-term cycling and storage of Cs-137 in forested ecosystems, in G. Desmet, P. Nassimbeni and M. Belli (eds.), *Transfer of radionuclides in natural and semi-natural environments*, Elsevier, London, pp. 61-73

15. Alexakhin, R.M., Ginsburg, L.R., Mednik, I.G. and Prohorov, V.M. (1994) Model of Sr-90 cycling in a forest biogeocenosis, Sci. Total Environ. 157, 83-91.

16. Shaw, G., Mamikhin, S., Dvornik, A., Zhuchenko, T. (1996) Forest model descriptions. In *Behaviour of radionuclides in natural and semi-natural environments*, Experimental collaboration project No 5., Final report, European Commission EUR 16531, Luxembourg, pp. 26-31.

17. Dvornik, A. and Zhuchenko, T. (1999) Phenomenologic model FORESTLIFE and prediction of radioactive contamination of forest in Belarus, in I. Linkov (ed.), *Contaminated forests: Recent developments in risk identification and future perspectives*, NATO ASI Series 2-, Kluwer Academic Publishers, Dordrecht, this volume.

18. Frissel, M.J., Shaw, G., Robinson, C., Holm, E. and Crick, M. (1996) Model for the evaluation of long term countermeasures in forests, in F.F. Luykx and M.F. Frissel (eds.), *Radioecology and the restoration of radioactive-contaminated sites*, NATO ASI Series 2-13, Kluwer Academic Publishers, Dordrecht, 137-154.

19. Mamikhin, S.V., Tikhomirov, F.A., and Shcheglov, A.I. (1997) Dynamics of Cs-137 in the forests of the 30-km zone around the Chernobyl nuclear power plant, Sci. Total Environ. 193, 169-177.

20. Shaw, G., Belli, M. (1999) The RIFE models of radionuclide fluxes in European forests, in I. Linkov (ed.), *Contaminated forests: Recent developments in risk identification and future perspectives*, NATO ASI Series 2-, Kluwer Academic Publishers, Dordrecht, this volume.

21. Avila, R., Moberg, L., Hubbard, L., Fesenko, S., Spiridonov, S. and Alexakhin, R. (1999). Conceptual overview of FORESTLAND – A model to interpret and predict temporal and spatial patterns of

radioactively contaminated forest landscapes, in I. Linkov (ed.), *Contaminated forests: Recent developments in risk identification and future perspectives*, NATO ASI Series 2- , Kluwer Academic Publishers, Dordrecht, this volume.

22. Seymour, E.M., Mitchell, P.I., Léon Vintró, L. and Little, D.J. (1999) A model for the transfer and recycling of Cs-137 within a deciduous forest ecosystem, in I. Linkov (ed.), *Contaminated forests: Recent developments in risk identification and future perspectives*, NATO ASI Series 2- , Kluwer Academic Publishers, Dordrecht, this volume.

23. Schell, W.R., Linkov, I., Belinkaia, E. and Morel, B. (1996) Application of a dynamic model for evaluating radionuclide concentration in fungi, in *IRPA9, 1996 International Congress on Radiation Protection*, Proceedings Vol. 2, International Radiation Protection Association, Seibersdorf, 752-754.

24. Avila, R. and Moberg, L. (accepted for publication) A systematic approach to the migration of Cs-137 in forest ecosystems using interaction matrices, J. Environ. Radioactivity.

25. Zhuchenko, T. and Dvornik, A. (1999) Model FORESTDOSE and evaluation of exposure doses of population from forest food products, in I. Linkov (ed.), *Contaminated forests: Recent developments in risk identification and future perspectives*, NATO ASI Series 2- , Kluwer Academic Publishers, Dordrecht, this volume.

26. Linkov, I., Morel, B. and Schell, W.R. (1997) Remedial policies in radiologically- contaminated forests: environmental consequences and risk assessment, Risk Analysis **17**, 67-75.

THE RIFE MODELS OF RADIONUCLIDE FLUXES IN EUROPEAN FORESTS

G. SHAW
Imperial College Centre for Environmental Technology,
Silwood Park, Ascot, Berkshire, SL5 7PY, UK

M. BELLI
ANPA, Via Vitaliano Brancate 48, 00144, Rome, ITALY

1. Introduction

In another paper [1] a collaborative research project, SEMINAT, designed to collect data on radionuclide fluxes in forests across Europe, is described. As part of the SEMINAT project a suite of computer models is being compiled, based on the data collected from field sites in Austria, Germany, Ireland, Italy and the United Kingdom. This suite of models is called RIFE and is intended to provide a computational framework within which the behaviour of Radionuclides in Forest Ecosystems can be evaluated for the purposes of nuclear accident consequence assessment and management. This paper gives an overview of each of the three component models which together make up RIFE, as well as the underlying rationale and capabilities of each model.

2. Background to the RIFE Models

Until recently, few models of forest radioecology existed within the literature. Those that did exist were not freely available for use by the radioecological community and were generally not applicable to contemporary problems in forest radioecology. One of the major concerns of radioecologists in Europe over the last few years has been the impact of the Chernobyl pulse within forests across the continent. The radionuclide of main current concern is ^{137}Cs, but as research has begun to demonstrate the lengthy ecological half life of this radionuclide it has also become clear that the potential for persistence of other radionuclides within forest ecosystems is a cause for concern. In developing our understanding of forest radioecology, therefore, it is desirable to be able to predict both acute and longer-term behaviour of radiocaesium and other radionuclides within forest ecosystems. There is certainly a relative lack of data on radionuclides other than radiocaesium in forests, but any predictive model should at

161

I. Linkov and W.R. Schell (eds.), Contaminated Forests, 161–171.

least have the capacity to facilitate the use of such data as and when they become available.

It is against this background that the RIFE suite of models has been developed. The key aim of RIFE is to allow the incorporation and use of data derived from forest sites contaminated with radionuclides to provide a structured means of organising and using that data. From reviews of existing forest models (eg [2-3]) it is clear that the approach of different authors to structuring models of radionuclide behaviour in forests can be very diverse. The final model structure adopted will, to a large extent, be controlled by the quantity and reliability of data available to the modeller. However, any model must always be suitable for its intended purpose and this consideration should also be influential in determining the final stucture of a model. The list of possible end-uses of a forest radioecology model is probably as diverse as the range of possible model structures and all end-uses are unlikely to be realised by a single model. With this in mind, three individual models have been created within the RIFE suite, as follows.

i.   RIFEQ - an equilibrium 'screening' model allowing rapid calculations of radionuclide distributions in major components of the forest ecosystem, using minimal data inputs.

ii. RIFE1 - a relatively simple (5 compartment) dynamic model of radionuclide flux following either pulse or chronic inputs from the atmosphere.

iii. RIFE2 - a more complex (10 compartment) dynamic model, allowing time-dependent computations of radionuclides in forests following pulse or chronic inputs from the atmosphere.

3. The RIFEQ Model

As a preliminary step in SEMINAT's development of dynamic models appropriate to forest ecosystems a screening model was developed which allowed preliminary data from each of the research groups' forest sites to be used in a probabilistic uncertainty analysis of radiocaesium distributions in major components of each ecosystem. The screening model developed is the RIFEQ model which uses a combination of aggregated transfer coefficients (T_{agg} values) and biomass estimates in a mass balance calculation, as described below.

Dynamic modelling of semi-natural ecosystems, such as forests and upland pastures, presents a challenge because of the complexity of the ecosystems but also because of uncertainty associated with transfer parameters within the ecosystems. The inherent variability in transfer rates in these ecosystems makes it particularly difficult to estimate reliable rate coefficients or fluxes with which to calibrate dynamic models. One answer to this problem has been the adoption of 'aggregated transfer factors' (T_{agg} values) for animals and their products, as well as for fungi, herbaceous vegetation and trees. T_{agg} values have been criticised for their simplicity, their site-specificity and their apparent lack of rigour in facilitating understanding of transfer processes within the

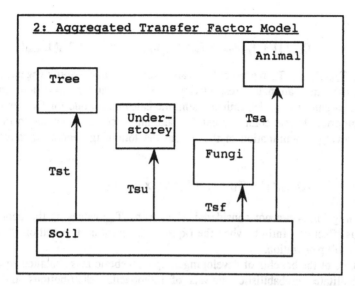

Fig. 1. Structure of the RIFEQ screening model

ecosystem of concern. One of the key requirements of any ecosystem-level model is that it should take into account the mass balance of radionuclides introduced into it. In the normal usage of T_{agg} values such considerations are usually forgotten, and this detracts from the usefulness of this parameter. As part of the RIFE modelling exercise a simple methodology has been developed which uses T_{agg} values in a rigorous way from the point of view of accounting for the mass balance of a radionuclide within a forest.

The structure of this simple 'screening model' is shown in Figure 1. In the example shown the model consists of five compartments but, in principle, any number of compartments can be accommodated in the RIFEQ approach. For each compartment an estimate of aggregated transfer coefficient (T_{agg} value) and biomass is required. T_{agg} values are defined as follows:

$$T_{agg} = \frac{Bq\ kg^{-1}\ in\ product\ (eg.\ wood\ or\ mushrooms)}{Bq\ m^{-2}\ within\ the\ soil} \quad (unit\ m^2\ kg^{-1})$$

The basic mass balance equation is simply the sum of activity inventories (Q, Bq m^{-2}) in each compartment. Where this inventory is not known it can calculated by multiplying the activity concentration within the compartment (C, Bq kg^{-1}) by the mass of the compartment (M, kg m^{-2}). Hence:

$$
\begin{aligned}
Q_{tot} &= Q_{soi} + (C_{tree}M_{tree}) + (C_{understorey}M_{undrestorey}) + (C_{fungi}M_{fungi}) + (C_{animal}M_{animal}) \\
&= Q_{soil} + (T_{st}Q_{soil})M_{tree} + (T_{su}Q_{soil})M_{understorey} + (T_{sf}Q_{soil})M_{fungi} + (T_{sa}Q_{soil})M_{animal} \\
&= Q_{soil}[1 + T_{st}M_{tree} + T_{su}M_{understorey} + T_{sf}M_{fungi} + T_{sa}M_{animal}]
\end{aligned}
$$

and

$$Q_{soil} = Q_{tot} / [1 + T_{st}M_{tree} + T_{su}M_{understorey} + T_{sf}M_{fungi} + T_{sa}M_{animal}]$$

where T_{st}, T_{su}, T_{sf} and T_{sa} represent T_{agg} values for trees, understorey vegetation, fungi (mushrooms) and animals, respectively. Concentrations in each individual compartment can then be determined using the calculated value for Q_{soil} and the T_{agg} value appropriate to that compartment. To take into account losses due to radioactive decay following contamination of the system the following general equation can be used.

$$Q_{soil(t)} = (Q_{tot(0)} / [1 + T_1M_1 ... + T_nM_n]) \times exp\text{-}\lambda_R t$$

Soil leaching losses are not considered when using T_{agg} values as the depth of soil assumed is effectively infinite when the Bq m^{-2} soil inventory is based on the depth of maximum soil penetration.

One of the benefits of developing simple algebraic mass balance equations is that it facilitates probabilistic analysis of radionuclide distributions and activity concentrations in forest ecosystems. Figure 2 shows the results of a probabilistic uncertainty analysis of radiocaesium activity distributions in a generic forest ecosystem using uniformly distributed ranges of T_{agg} values and compartmental masses. This information is based on data from SEMINAT forest sites in Austria, Ireland, Italy and the UK and for the purposes of the analysis an initial total deposition of 1kBq kg^{-1} was assumed. Figure 2 indicates that the majority of radiocaesium is expected to reside within the soil compartment (up to almost 100% in extreme cases) with a probable maximum percentage distribution of approximately 25% of radiocaesium in the standing biomass of trees. The percentage distribution of radiocaesium in mushrooms is centred around 0.5%, although possible maximum distributions are higher than this (approximately 2%). The results of such calculations allow an instant 'snapshot' of the expected ranges of radiocaesium activity concentrations and distributions with forests to be generated, which gives important insights into the relative importance of individual compartments with respect to the development of dynamic models (site-specific calculations of radiocaesium distributions using RIFEQ are intended to assist with the calibration of dynamic models). These results also begin to provide answers to straightforward management questions such as the potential feasibility of suggested countermeasures within the forest [4]. For instance, with the percentage distributions of radiocaesium in mushrooms centred around 0.5% the suggestion that systematic removal of mushrooms could be used to decontaminate a forest soil does not appear to hold true (removal of 0.5% of the total ecosystem radioactivity in successive years results in a decontamination half time of nearly 140 years).

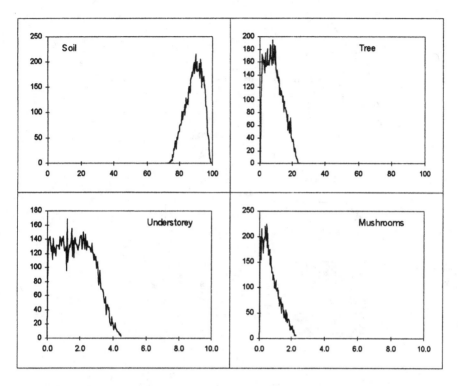

Fig. 2. Frequency plots of % distributions of radiocaesium calculated probabilistically using the RIFEQ Model. The Y axes represent frequencies while the X axes represent % distribution of the total inventory of radiocaesium within the forest.

4. The RIFE1 Model

While the RIFEQ screening model can provide simple answers to forest management questions and is useful in assisting with calibration of dynamic forest models, it does not itself provide a tool for interpreting and forecasting radionuclide behaviour in forests on a time-dependent basis. For this reason SEMINAT is developing a dynamic modelling capability using data from forest sites within five EU countries. The RIFE1 model was the first dynamic model to be developed within SEMINAT and was initially based on data obtained from experimental studies in the Cherrnobyl 30 km zone and the Irish republic [5]. The data requirements for RIFE1 are reasonably modest and it is this model which can be considered the 'workhorse' of the three RIFE models. The level of complexity represented by RIFE1 is intended to be sufficient to provide information which may be required for making basic decisions on the management of contaminated forests. Uncertainties surrounding such decision-making are of obvious importance and RIFE1 can thus be operated in either deterministic or probabilistic modes.

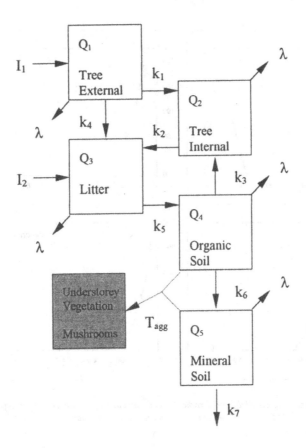

Fig. 3. Conceptual Structure of the RIFE2 Model

RIFE.1 is a five compartment dynamic model (see Figure 3) which was originally calibrated for ^{137}Cs fluxes using post-Chernobyl data derived for the far-field (Waterford, Ireland) and the near-field (Kopachi, 5 km due south of the Chernobyl Nuclear Power Plant). These sites differ in the quantity of radioactivity deposited in 1986, but more importantly in the nature of the deposition. The Irish site received ^{137}Cs in aerosol form which proved to be easily transported down the soil profile and readily incorporated into the tissues of pine trees growing on the site. By contrast, Kopachi received fuel fragments ('hot particles') from the Chernobyl reactor in which the physical mobility and bioavailability of ^{137}Cs was much reduced. After calibrating the RIFE.1 model for both sites the rate coefficients obtained were treated as extremes of a uniform distribution of values. Steady state solutions have been obtained for the

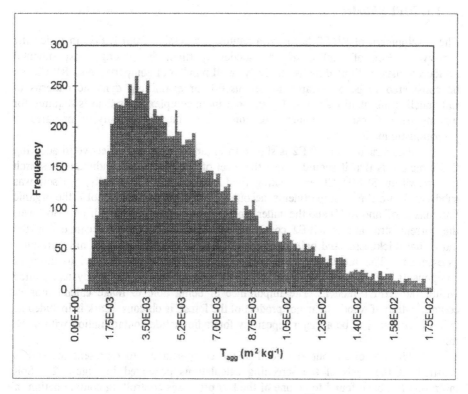

Fig. 4. Frequency distributions of T_{agg} values predicted for ^{137}Cs in Pinus sylvestris wood using steady-state solutions for the RIFE1 model.

RIFE.1 differential equations and, using the distributions of rate values obtained from the near-field and far-field model calibrations, a Monte Carlo analysis was carried out to determine the range of likely T_{agg} values for ^{137}Cs in trunk wood which would result. The results of this exercise are shown in Figure 4. The overall range of T_{agg} values calculated probabilistically using RIFE1 is 0.00032 - 0.041 m^2 kg^{-1}, which compares favourably with a range of measured T_{agg} values from Belarus and the Russian Federation of 0.0001 - 0.003 m^2 kg^{-1}. It is interesting that, compared to the measured data, RIFE1 tends to overpredict the T_{agg} values slightly, possibly as a result of undue weight being given to the relatively high tree root uptake rate coefficients from Ireland by adopting uniform frequency distributions for this parameter. However, the paucity of data on rate coefficients applicable to radiocaesium migration in forest ecosystems makes the establishment of more realistic frequency distributions for use in Monte Carlo simulations extremely difficult. Nevertheless, work is currently under way to allow probabilistic, time-dependent solutions of the RIFE1 equations to be obtained using numerical techniques.

5. The RIFE2 Model

The development of RIFE2 has been a natural progression from RIFE1 in which the perceived level of reality of the model (particularly amongst experimental radioecologists) is limited by its relatively small number of compartments. RIFE2 can be considered to be a research model, useful for examining dynamic patterns of radionuclide distributions within forests on a more complex basis than is required for making basic forest management decisions. This model can only be operated in deterministic mode.

The structure of RIFE2 is shown in Figure 5. The key objective in adopting this structure is that it should reflect the level of detail at which individual research groups within SEMINAT are making field measurements. Initially, the soil was subdivided into three compartments, accounting for the litter layer (AoL), the organic horizons (AoF and AoH) and the underlying mineral horizons, as in RIFE1. However, the current structure of RIFE2 comprises six soil layers: the user can define the individual thicknesses and bulk densities of these to suit the needs of any particular assessment. The remaining four dynamic compartments are devoted to different components within trees. Of these, bark and wood were not previously represented within the RIFE1 model. The importance of being able to make calculations of contamination of wood, the major product of the forest, is obvious. Bark is included as it has been found to be a major repository for radionuclide contamination within the tree.

The use of six out of a total of ten compartments to represent the soil is justified on the basis of the screening calculations presented in Figure 2. Soil migration has been found to be one of the key processes controlling contamination of mushrooms and understorey vegetation [6]. The mushroom and understorey compartments are not included as part of the dynamic structure of the RIFE2 model and therefore are not taken into account as part of the system mass balance. This is also justified on the basis of the screening calculations shown in Figure 2. Contamination of mushrooms and understorey vegetation is calculated using transfer factors or T_{agg} values, which are modified according to the distribution of the fungal mycelium or the understorey root system within the six soil layers. These distributions can be specified by the user, as can the distribution of tree roots.

Fluxes of radionuclides through the ten dynamic compartments are represented mathematically as a series of coupled first order differential equations (rate coefficients which are considered to be constant with time). RIFE2 allows the user to solve these equations for a variety of initial conditions (distributions of radionuclides within the system can be specified before running the model) and for the case of a single pulse input or for chronic deposition from the atmosphere over any number of years. For the latter, specific input files must be used. Currently, time-dependent input files are available for Lady Wood near Sellafield in the UK and Hoglwald in Bavaria, Germany. Solution of the model's differential equations is carried out by numerical integration using a Runge-Kutta method and mass balance of radionuclides introduced into the system is automatically accounted for.

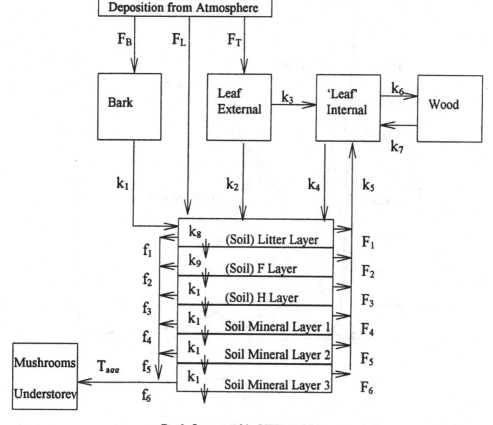

Fig. 5. Structure of the RIFE2 Model

Calibration of the RIFE2 model is more difficult than for the previous two models because of the requirement for more detailed information on radionuclide fluxes in the forest ecosystem. Ideally, detailed time course data should be used for such calibrations, but these data are especially hard to acquire, particularly over the long term. The best medium to long term data available for radiocaesium fluxes in forests is undoubtedly from the geographical areas within and close to the Chernobyl 30 km zone. Dvornik and Zhuchenko [7] have based their own model, FORESTLIFE, on data from sites in these areas including Kruki, a Pine forest situated in Belorus, some 25 km to the north west of the Chernobyl Nuclear Power Plant. Figure 6 shows data from Kruki superimposed on time curves calculated using RIFE2. These predictions are somewhat simplified, with the six possible soil compartments within RIFE2 reduced to three curves (litter, organic and mineral soil) and the four possible tree compartments reduced to just a single curve for wood. However, this figure demonstrates that, with

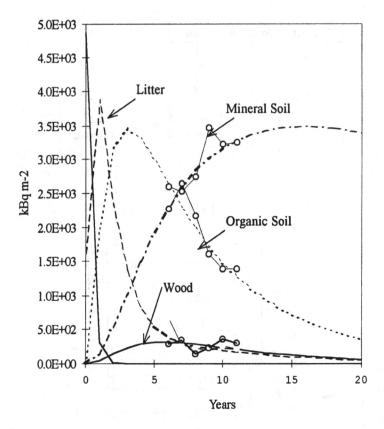

Fig.6. Dynamics of ^{137}Cs within a Pine forest at Kruki, Belorus, calculated using RIFE2 (smooth curves) and determined by field measurements (circles joined by solid lines)

appropriate calibration, first-order compartment models can be used to represent the fluxes of radionuclides within forest ecosystems with reasonable reliability, at least over the time scale of 6 to 11 years after a pulse deposition event. Clearly, continued measurements are still required to ensure that model projections beyond the present day are also reliable.

6. Conclusions

This paper has summarised the state of development of three models of radionuclide fluxes in forest ecosystems under the common title Radionuclides in Forest Ecosystems (RIFE). This suite of models is being constructed under the central philosophy that models should be fit for their intended purpose. Accordingly, each of the component models described here is designed for somewhat different purposes. The ability to address uncertainty is seen as a crucial challenge both in estimating suitable values for model parameters and in developing models capable of undertaking probabilistic analyses of radionuclide behaviour in forests. The RIFEQ and RIFE1 models have the latter capability built-in. Finally, time course results from the vicinity of the Chernobyl 30 km zone have shown that dynamic compartmental models such as RIFE2 are capable of producing reliable interpretations of radionuclide dynamics in forests. Longer term predictions can be made, but further validation of these predictions will be required over the forthcoming years.

7. Acknowledgements

The RIFE models are being developed as part of the SEMINAT project, funded under the Radiation Protection programme of the European Commission, whose financial backing is gratefully acknowledged. GS would also like to thank A. Dvornik and T. Zhuchenko for the provision of data and the Royal Society (London) for providing funds for collaboration between Imperial College, London, and the Belorussian Forestry Institute, Gomel.

8. References

1. Belli, M., et al. (1996) Dynamics of radionuclides in forest environments. pp. 69 - 80 in the proceedings of the First International Conference of the European Commission, Belarus, Russian Federation & Ukraine on the Radiological Consequences of the Chernobyl Accident (Minsk, 18-22 March, 1996). EUR 16544 EN, ISBN 92-827-5248-8.
2. Schell, W., I. Linkov, B. Morel and C. Myttenaere (1996) A dynamic model for evaluating radionuclide distribution in forests from nuclear accidents. Health Physics 70:318-335.
3. Riesen, T.K., Avila, R., Moberg, L. and Hubbard, L. (1999) Review of forest models developed after the Chernobyl NPP accident, in I. Linkov (ed.), Contaminated forests: Recent developments in risk identification and future perspectives, NATO ASI Series 2- , Kluwer Academic Publishers, Dordrecht, this volume
4. Guillitte, O., F. A. Tikhomirov, G. Shaw and V. Vetrov (1994) Principles and practices of countermeasures to be carried out following radioactive contamination of forest areas. *Science of the Total Environment*, 157, 399 - 406
5. Belli, M., Bunzl, K., Delvaux, B., Gerzabeck, M., Rafferty, B., Riesen, T., Shaw, G., Wirth, E. (1999). Dynamics Of Radionuclides In Semi-Natural Environments. These proceedings.
6. Ruehm, W., M. Steiner, E. Wirth, A. Dvornik, T. Zhuchenko, A. Kliashtorin, B. Rafferty, G. Shaw & N. Kuchma (1996) Dynamics of radionuclide behaviour in forest soils. pp. 225 - 228 in the proceedings of the First International Conference of the European Commission, Belarus, Russian Federation & Ukraine on the Radiological Consequences of the Chernobyl Accident (Minsk, 18-22 March, 1996). EUR 16544 EN, ISBN 92-827-5248-8.
7. Dvornik, A. and T. Zhuchenko (1999). Model FORESTLIFE And Prediction Of Radioactive Contamination Of Forests In Belarus in I. Linkov and W. R. Schell (eds.), Contaminated forests: Recent developments in risk identification and future perspectives, NATO ASI Series 2- , Kluwer Academic Publishers, Dordrecht, this volume.

CONCEPTUAL OVERVIEW OF FORESTLAND - A MODEL TO INTERPRET AND PREDICT TEMPORAL AND SPATIAL PATTERNS OF RADIOACTIVELY CONTAMINATED FOREST LANDSCAPES

R. AVILA, L. MOBERG, L. HUBBARD
Swedish Radiation Protection Institute
SE-171 16 Stockholm, SWEDEN

S. FESENKO, S. SPIRIDONOV, R. ALEXAKHIN
Russian Institute of Agricultural Radiology and Agroecology
249020, Obninsk, RUSSIA

1. Introduction

Forested areas contaminated with radioactive substances are characterized by an extremely high spatial variability of radionuclide levels. The horizontal and vertical patterns of the radioactive contamination may also show substantial changes with time, especially during the first days and months after the deposition. Furthermore, the effective ecological half-lives tend to be very long in some forest components. The contamination level of a specific forest component at a specific time and location is due to multiple factors and processes interacting in a complex way. The extrapolation of measured activity levels in space and time can, therefore, often not be done by common statistical methods or by using simple heuristic relationships. A conceivable approach to solve this problem is to combine the capacity of Geographical Information Systems (GIS) for management and representation of large amounts of data with the generalization possibilities of dynamic ecosystem models. Here we give a conceptual overview of a forest ecosystem model (FORESTLAND) that could be used for this purpose. In particular, we present some features important for a future coupling with GIS.

2. General Description of FORESTLAND

FORESTLAND is a dynamic ecosystem model to interpret and predict temporal and spatial patterns of the radioactive contamination of forest ecosystems. The model is focused on ^{137}Cs migration pathways leading to internal and external radiation doses to man. FORESTLAND can be applied to both the acute and long-term phases after an aerial deposition of radioactive substances. The present version of the model consists of five individual models:

- FORBIO: A model of the biomass dynamics of trees and the understorey vegetation,
- FORACUTE: A dynamic model of the migration of radionuclides in forest

I. Linkov and W.R. Schell (eds.), Contaminated Forests, 173–184.

ecosystems during the acute phase of the contamination,
- FORGAME: A dynamic model of the long-term migration of radionuclides in forest food chains, in particular wild animals,
- FORTREE: A model of the long-term migration of radionuclides in forest trees,
- FOREXT: A dosimetric model for calculation of gamma dose rates in the forest.
- FORDOSE: A model for calculation of the internal and external doses to the population (presently under development).

The main characteristics and possibilities of the individual models are discussed below. These models can be used stand-alone or as a sequence within FORESTLAND, where the outputs of one model are inputs to other models (Fig. 1).

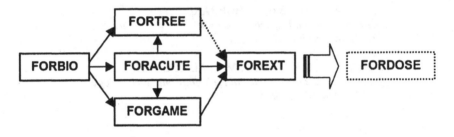

Fig. 1. Connection between the individual models of FORESTLAND

The connections between the individual models of FORESTLAND can be described as follows:
- The biomass densities calculated with FORBIO are used in the migration models (FORACUTE, FORGAME and FORTREE) for calculation of several transfer rates. For example, the uptake rates by trees in FORTREE are expressed as a function of the biomass of the leaves (needles). The values of biomass density (FORBIO outputs) are used with the values of the radionuclide content in different forest components (in Bq m^{-2}) calculated with the migration models to obtain the radionuclide concentration (in Bq kg^{-1}) in these components.
- The initial distribution of radionuclides in different forest components needed in FORTREE and FORGAME (initial conditions) are calculated with FORACUTE. Alternatively the user can define directly the initial conditions.
- FORACUTE and FORGAME provide the values of activity levels in different forest components needed for calculation of the dose rates with FOREXT. A similar connection between FORTREE and FOREXT (dashed arrow in Fig. 1) is being implemented. The endpoints of the 3 migration models are concentrations in different components (plants, animals, etc.).
- The radionuclide concentrations calculated with the migration models and the dose rates calculated with FOREXT will be used as input data for FORDOSE.

The forest ecosystem has been classified into four different categories. Each category corresponds to a different type of tree (coniferous or deciduous) and landscape (automorphic or semi-hydromorphic). A set of model parameters, consisting of best estimate values and intervals of variation, is given for each forest category.

3. FORBIO - Model of biomass dynamics

In FORBIO, a simple approach for describing seasonal and long-term biomass dynamics of trees and understorey vegetation has been applied. The input parameters of FORBIO are the growth rate constants, the biomass decrease rate constants, the maximum biomass that can be reached by different types of vegetation, the tree mortality and the times marking the beginning and end of different stages of the seasonal dynamics (start of the vegetative period, time of maximal biomass, end of the vegetative period). Default values (generic values) of these parameters are presently incorporated in the model. These values are derived from long-term statistics/data of biomass dynamics in several types of forests of the Bryansk region of the Russian Federation.

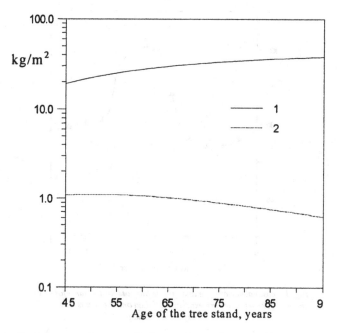

Fig. 2. Long-term dynamics of the biomass density of trees (1) and needles (2) in a coniferous forest in the Bryansk region (automorphic soil)

176

For the understorey vegetation and mushrooms, biomass growth is simulated with a logistic model, while an exponential decrease is assumed during senescence. Differentiation is made between summer and autumn mushrooms and between fruits of berries and the whole plant (animal feeds).

The biomass growth of an individual tree is also described with a logistic model while an exponential equation is used for calculation of tree mortality. A linear differential equation, obtained by combining the equations for growth and mortality, is used for simulating the long-term changes of tree biomass density (kg m^{-2}).

For tree leaves (needles) distinction is made between seasonal and long-term biomass dynamics. The yearly values of leaf (needle) biomass depend on the age of the tree. It is assumed that the contribution of leaves (needles) to the total tree biomass decreases with the age of the trees. The seasonal variation of the leaf (needle) biomass is described with a logistic model during the periods of growth and senescence.

The endpoints of this model are the biomass density of different types of vegetation as a function of time. An example of calculations with FORBIO of the long-term changes of tree biomass and needle biomass is shown in Fig. 2. The seasonal variation of needle biomass for a coniferous forest of the Bryansk region are presented in Figure 3.

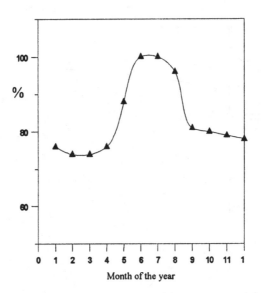

Fig. 3. Seasonal variation of total needle biomass for the same forest as in Fig.2 (The triangles are experimental data and the line is a model fit to the experimental data)

4. FORACUTE – Model for the acute phase

FORACUTE is a dynamic model of the migration of radionuclides during the acute phase, lasting up to a few years after an aerial deposition. The model describes the primary interception of the radionuclides by the above ground vegetation and their subsequent redistribution in the system. Transfer processes like weathering, translocation in the tree and the understorey vegetation, and root uptake from the upper soil-litter layer are included in the model. The model also permits evaluation of the dynamics of the radionuclide levels in forest products consumed by man, including forest game.

To model the interception of the radionuclide by the above-ground phytomass, the phytomass is viewed as a set of four successive filters: the tree leaves (needles), the tree bark, the understorey vegetation and the upper soil-litter layer (Fig. 4).

The interception by the understorey vegetation is calculated with an exponential function of the biomass density (Chamberlain's equation). A method, similar to the one commonly used for evaluating the passage of light through tree crowns, is used to simulate the initial retention of radionuclides by trees. It is assumed that the initial retention by trees is proportional to the "projective cover" of the tree crowns. This cover can be calculated from the crown closure (relative area of crowns) and the crown tracery coefficient, which depends on the tree species. The values of these parameters can be mapped for different areas, which facilitates using the model in combination with GIS. Calculations with the model that illustrate the extent of the influence of the crown closure on the initial retention of radionuclides by trees are presented in Fig. 5. The initial retention depends on the season of the deposition, reaching a maximum value when the leaf biomass is maximal.

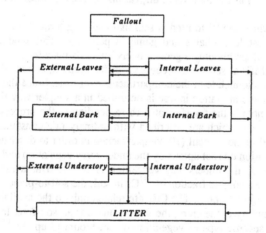

Fig. 4. Overview of the model for the acute phase FORACUTE

The present version of the model has been calibrated for dry deposition in both coniferous and deciduous forests of the Bryansk region. The calibration for a wet deposition event has not yet been carried out.

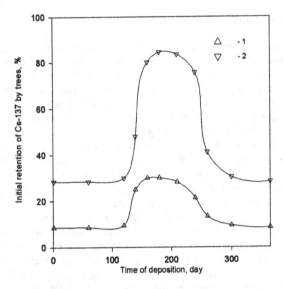

Fig. 5. Initial retention of the deposition as a function of the season and the crown closure for a deciduous forest (1- the crown closure is 0.3; 2- the crown closure is 1.0)

5. FORGAME – Model of the long-term migration of radionuclides in forest food chains

FORGAME is a dynamic model to predict seasonal and long-term changes of ^{137}Cs activity concentrations in forest food chains and game in particular. This model is described by Fesenko et al. [1] and therefore only a general overview of its main features is given here.

In FORGAME a set of 20 coupled differential equations describes the net accumulation of the radionuclide in the compartments over time. Since the model is focused on forest food chains, the migration in tree is described in a simpler way than in FORTREE (see below). An example is the transfer rates corresponding to the processes of root uptake and translocation in trees, which are described with ordinary rate constants. The soil on the other hand is modeled in more detail (10 compartments) in order to describe the influence of roots and mycelia location on root uptake by the understorey vegetation and mushrooms.

Root uptake by the understorey vegetation is described as a function of the root distribution in soil, the available fraction of ^{137}Cs in soil, the soil-to-plant concentration ratio (CR) and the biomass growth rates. The CRs are related only to the available fraction of the radionuclide in soil and have, therefore, the same values for all soil-litter layers. They reflect the capacity of each specific type of vegetation or mushroom to uptake ^{137}Cs from the soil. The CRs commonly reported in the literature are, however, related to the total content of the radionuclide in soil or in a specific soil layer. They depend on the soil properties and vary with the soil depth and time.

The intake rate of ^{137}Cs by game (roe deer and moose) is described by a function of the total feed intake, the share of different feeds in the daily animal diet and the activity concentrations in different feeds. It is assumed that the radionuclides incorporated by the animal via ingestion are instantly distributed in the animal body and that the elimination rate from muscles (edible meat) is proportional to the activity levels in this part of the animal body.

6. FORTREE – Model of migration of radionuclides in trees

FORTREE is a dynamic compartment model of the long-term kinetics of ^{137}Cs activity concentrations in the tree. One endpoint of the model is the seasonal change of activity concentrations in leaves (needles). The conceptual models of FORTREE for deciduous and coniferous trees are presented in Figures 6 and 7 respectively. Five compartments are used to describe the time and space variation of ^{137}Cs levels in soil. The first soil layer (0-10 cm) contains most active parts of the root and is responsible for root uptake during the first decades after the radioactive contamination. Each soil layer is divided into an available and an unavailable fraction of ^{137}Cs in order to consider sorption/desorption processes leading to fixation and remobilization of the radionuclides. It is assumed that the available fraction of the radionuclide for root uptake is equal to the fraction that is available for vertical migration in the soil.

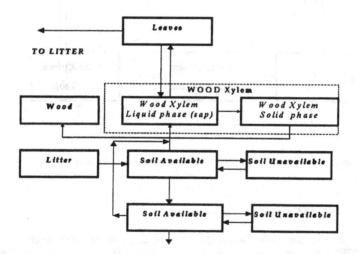

Fig. 6. Conceptual model of FORTREE for a deciduous forest

The tree is described by the following compartments: two compartments for the living part of the wood (liquid and solid phases of the wood xylem), one compartment for the dead wood (structural wood), one compartment for leaves (deciduous trees) and four compartments for needles of different age class (coniferous trees).

Root uptake by trees is described by a function of the water flux (transpiration flow) through the xylem during the vegetative period. The main part of this flux is due to evaporation from leaves, which is proportional to the leaves (needles) biomass. The root uptake rates are therefore expressed as a function of the leaf (needle) biomass, which depends on tree biomass and age. Thus the model can be used to evaluate the influence of these factors on the accumulation of ^{137}Cs by trees. An example of calculations in given in Fig. 8. This permits a reduction of the uncertainties of the prediction of the root uptake by trees, especially if detailed information of biomass and tree age is made available through a coupling with GIS.

The values of the parameters corresponding to the processes of translocation in the tree were estimated from experimental data of stable potassium contents in different parts of the tree. The experimental data was collected during a program of long-term observations carried out in the Bryansk region. A comparison of the model predictions using the parameter values obtained this way with independent experimental observations of ^{137}Cs concentrations in different parts of the tree shows a good agreement.

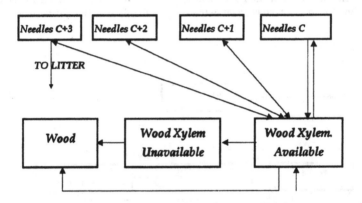

Fig. 7. Conceptual model of FORTREE for a coniferous forest (only the modifications of the model in Fig. 6 are presented)

7. FOREXT – Model for evaluation of external doses

FOREXT is a dosimetric model for calculation of the dose rates in radioactively contaminated forests. FOREXT calculates the dose rate from ^{137}Cs at 1 m from the soil surface. The vertical column of the forest is divided into 7 successive layers (Fig. 9) with different average densities. The soil is described with four layers, one for each soil horizon (L, Of, Oh and A). The fifth layer is from the soil surface to the average height of the understorey vegetation. The frontier between the sixth and seventh layer is set at the average height of the bottom of the tree canopy.

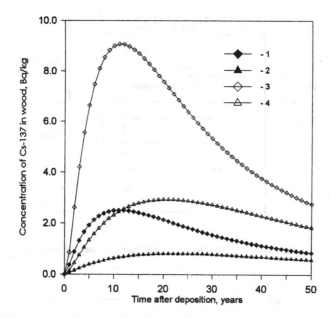

Fig. 8. Activity levels in wood for different types of forest soil and ages of trees - deposition 1 kBq/m² of ¹³⁷Cs
(1, 2 - automorphic soil, 3, 4 - semi-hydromorphic soil; 1, 3 - initial age 20 years, 2, 4 - initial age 80 years)

Each layer is considered to be a plane source of finite thickness. The activity of each source is calculated from the values of ^{137}Cs activity concentrations in different components of the layer provided either by FORGAME or FORACUTE. The variation of these values with time leads to time variation of the estimated dose rates. The attenuation and scattering of the photons in different layers is calculated with the equations commonly used for shielding calculations in radiation physics. The attenuation and scattering depend on the density of the layers, which varies with time due to biomass changes. As a result, the contribution of different layers to the calculated dose rates depends on the time elapsed after a deposition event, especially during the acute phase of the contamination (Fig. 10).

8. Concluding comments

We conclude this paper by indicating the main directions for further development of FORESTLAND and by making some comments on how the coupling of the model with GIS is presently viewed by us. Some improvements that will be introduced in the individual models have been mentioned above. Other directions for the improvement of FORESTLAND are the following:

Testing of the model. The main approaches to be used for this purpose are comparison of the model predictions with independent experimental data, comparison with predictions of :

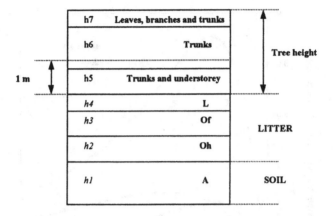

h7	Leaves, branches and trunks	Tree height
h6	Trunks	
h5	Trunks and understorey	
h4	L	LITTER
h3	Of	
h2	Oh	
h1	A	SOIL

1 m

Fig. 9. Geometry for calculations of the external dose rates (arbitrary scale)

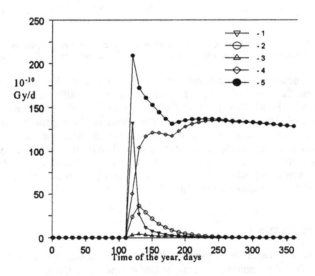

*Fig. 10. Calculated ccontributions from different layers to the dose rate as
a function of time after the deposition (deposition 1 kBq/m² of ¹³⁷Cs)
(1- tree canopy, 2 - tree trunks between canopy and understorey,
3 - understorey, 4 - litter, 5-total)*

- other models and evaluation of the uncertainties of the model predictions by propagating the estimated uncertainties in input parameters with Monte Carlo simulations.
- Development of a model for calculation of internal and external doses to man (FORDOSE). This model will take into account factors that can influence horizontal and time variations of the doses such as the land use, the distribution of the population in the contaminated areas, intake variations, etc.

Optimization of the interconnection between the individual models. FORESTLAND is implemented in a "model maker" (Stella software) and other possible computer implementations, such as writing codes in C++ for the individual models, are being considered.

Coupling of FORESTLAND with GIS

In addition to creating a forest ecosystem model, a coupling with GIS has been a goal from the beginning of FORESTLAND development. The way this will be done in practice has, however, not been decided. A direct implementation of the model in a GIS environment seems to be unrealistic because of the complexity of the model. We are presently considering the following alternatives:

- FORESTLAND is used to produce a multidimensional matrix of desired endpoints for the contaminated forest. This matrix is then combined with maps of forest characteristics in a GIS environment.
- FORESTLAND can be used to investigate how different forest characteristics influence the endpoints. The results of these investigations and expert opinions are used to design a neural network (NN). The NN is applied, instead of FORESTLAND, to carry out generalizations from the available experimental data.

9. Acknowledgements

This work has been partially financed by the European Commission through contract No F14P-CT96-0039 - The Landscape project (the Swedish group), and by SSI contract RYS 6.15 (the Russian group).

10. References

1. Fesenko, S., Spiridonov, S., Avila, R. "Modeling of ^{137}Cs behavior in forest game food chains". (In this volume).

MODEL FORESTLIFE AND PREDICTION OF RADIOACTIVE CONTAMINATION OF FORESTS IN BELARUS

A. DVORNIK AND T. ZHUCHENKO
Forest Institute of the National Academy of Sciences of Belarus, 71 Proletarskaya Street, 246654 Gomel, BELARUS

1. Introduction

Fallout from the accident at the Chernobyl NPP in 1986 revealed the potential of forests in Belarus as an accumulator of radioactivity. The forest ecosystem proved to be a natural barrier to the flux of radionuclides, preventing this redistribution. Of the Belorussian woodland (each one-quarter hectare) is radioactive-contaminated. A ^{137}Cs contamination density of soil exceeding 1 Ci/km^2 is typical for a million hectares.

Unlike other natural ecosystems, forests are capable of retaining traces of nuclear accidents over a long period of time. The forest is the most closed and conservative natural ecosystem, where radionuclides are now included in the long-term ecological nutrient cycle. After ten years the problems of growing low contaminated agricultural products were solved successfully. However, the problems of radioactive-contaminated forests still remain [1]. These problems are as follows:

- a high level of contamination in forest products,
- annual losses of timber are greater than 2.0 millions m^3. By 2015 the losses will have increased by a factor of 1.6,
- the radiation dose of forestry workers is on the average 2.5 times higher than that of the rural population,
- the contribution of the forest component (through the consumption of mushrooms, berries, game meat and the use of forest pastures) to the internal dose to the population of forested areas is currently 30 to 80%. This contribution to the internal dose tends to increase.

To estimate the radioactive situation in forests, to provide for the safe use of woodlands, to rehabilitate radioactive-contaminated areas, and to take measures to decrease radiation doses from contaminated forest lands and forest products, it is necessary to have the prediction capability for the near and long term periods of time. It is possible to make such predictions only on the basis of mathematical models of radionuclide migration in forest ecosystems. The parameters for these models have

I. Linkov and W.R. Schell (eds.), Contaminated Forests, 185–194.
© 1999 *Kluwer Academic Publishers. Printed in the Netherlands.*

been determined using the data obtained by full-scale observations of the radionuclide redistribution among the compartments of the forest biogeocoenosis.

The current paper reports some experimental data on the ^{137}Cs behaviour in forest stands of Belarus and the prediction of the dynamics of ecosystem contamination. The predictions were calculated using the FORESTLIFE model.

2. Methodology

In 1992-1997 experimental data were collected from the 10 plots established in 20 to 40 year old pure moss pine forests of the Belorussian Polesye where soddy-podsolic sandy soils exist under uniform growth conditions [2]. Such forests are typical in Belarus, including the radioactive-contaminated areas, and occupy about 65% of the total forested area. Nine plots were established in deciduous forests where the following main species occur: birch, oak, aspen and alder. The forested areas where these plots were established received a ^{137}Cs deposition varying from 700 to 20000 kBq/m^2.

The age dependence of the ^{137}Cs transfer factor from soil into wood and other components of pine trees was studied at three plots established in areas where pine stands of different ages exist (from 10 to 80 years old).

3. Experimental data

Forest soils and timber stands are known to be in close association. The soil determines the forest type, while the stand influences the soil development and nutrient biological cycle of the forest-soil system.

Since one of the objectives of our study was to investigate contamination levels of stands in relation to the factors responsible for the input and accumulation of radionuclides in woody plants, we shall view the features of the radionuclide migration in the soil and phytocoenosis in the light of forest types and stand conditions. In accordance with the methodological principles, main features were studied with pine stands that are the most typical for Belorussian Polesye. The data on these stands are statistically representative. The systematic approach to the selection of plots and carrying out regular sampling of soil and vegetation over several years made it possible to accumulate a large quantity of experimental data on radionuclide contents in different components of pine biogeocoenoses.

On the basis of the data obtained, we can infer that the transfer factor of ^{137}Cs into pine trees increases in the following sequence: wood - bark - shoots - needles, which demonstrates the variability of physiological activity of different vegetative organs of the tree. For the plots with the different ^{137}Cs depositions, this ratio of specific activities of the components of trees of different species to the specific activity of their wood was calculated. These data are given in Tabl.1.

TABLE 1. Ratios of contamination levels in components of trees of different species compared to wood.

Species	Components of tree			
	Wood	Bark	Shoots	Leaves, new needles
Pine	1	4.7 ± 0.7	10.0 ± 1.0	12.9 ± 0.8
Birch	1	3.8 ± 0.2	10.7 ± 0.9	9.8 ± 0.8
Oak	1	5.7 ± 1.2	19.8 ± 2.0	-
Aspen	1	6.0 ± 1.8	11.1 ± 1.8	13.5 ± 2.7
Alder	1	5.3 ± 1.2	6.0 ± 1.2	-

According to the levels of contamination the following three groups were determined: wood, bark and shoots with leaves (or needles). The highest levels of contamination are found in vegetative organs of trees, the least in wood. Bark apparently retains traces of initial superficial contamination. Thus, the trees of different species show a single tendency for contamination with radionuclides which is governed by physiological peculiarities of their nutrition.

The dynamics of the behaviour of ^{137}Cs in components of a tree is a question of particular interest. The data from 6 years observations indicate the tendency for ^{137}Cs to redistribute in the phytomass. These data are also valid for confirmation and calibration of mathematical models. For example, Table 2 presents the ^{137}Cs distribution in components of pine from 1992 to 1997.

TABLE 2. Dynamics of ratios of the ^{137}Cs transfer factors in the components of pine trees compared to wood as 1.

Years	Wood	Bark	Needles of current year	Needles of last years	Shoots
1992	1	5.1	10.3	4.7	11.2
1993	1	4.7	13.0	4.8	8.5
1994	1	5.0	-	5.7	9.1
1995	1	2.4	7.9	2.9	-
1996	1	3.2	13.9	-	14.7
1997	1	2.8	13.2	-	13.4

The data of Table 2 show that in the years 1992-1997 vegetative components of trees (new needles and shoots) still actively accumulated radiocaesium. The maximum value of TF is observed in 1996. Subsequently we observed a tendency to decrease. It is too early to report a stable decrease in the transfer factor values. However, we can only say that the input of ^{137}Cs in components of pine has become constant. The dynamics of transfer factor in wood is the same. Its values are as follows:

    1992 - 0.00243,   1993 - 0.00271,

    1994 - 0.00279,   1995 - 0.00286,

    1996 - 0.00306,   1997 - 0.00275 m^2/kg.

The process of stabilization was started from 1993 (10% of the maximum value). According to our estimates it will continue until 2000. The dynamics of the TF into wood has close correlation with the TF in vegetative organs of the tree.

According [3], the rate of uptake of nutrient elements including radionuclides from the soil to woody plants depends on the plant age. We studied the ^{137}Cs content in wood of pine trees of four classes (5 to 20, 21 to 40, 41 to 60 and 61 to 80) to include in the model age peculiarities of the transfer and accumulation in pine wood . With this object in mind we selected pine stands of different ages occurring under the same site conditions. For each plot the ^{137}Cs content was estimated on the basis of the analysis of six soil samples taken from the upper 20 cm layer. The same wood samples were used to determine both the K contents and the ^{137}Cs specific activities. Transfer factors of ^{137}Cs from soil into wood were calculated and averaged for each age class.

The study of the age dependence of the ^{137}Cs uptake rate shows that the transfer factors of ^{137}Cs from soil into wood (both full section and post-accident annual rings) decrease as the stand ages and are 6 to 7 times lower for older stands as for young-growth ones (Table 3).

TABLE 3. Means transfer factors (TF) of ^{137}Cs into wood of pine trees of each age class and potassium contents in the same wood samples

The age, years	Transfer coefficient (full section), m^2/Kg, 10^{-3}	Transfer coefficient (post accident annual rings), m^2/kg, 10^{-3}	Contents of K, %
0-20	3.46 ± 1.03	3.46 ± 1.05	0.245 ± 0.026
21-40	3.00 ± 0.48	3.53 ± 0.95	0.165 ± 0.005
41-60	1.42 ± 0.35	2.48 ± 0.56	0.136 ± 0.008
61-80	0.46 ± 0.10	0.59 ± 0.10	0.107 ± 0.006

It should be noted that as the tree ages, the consumption of potassium changes in much the same way as that of radiocaesium [3]. Therefore, for comparison Table 3 gives the contents of the potassium in the same wood samples taken from pine trees of different ages. The ^{137}Cs contents are closely correlated with those of K, the coefficient of correlation being 0.96+/-0.03. By this is meant that the input of potassium into wood is similar to that of radiocaesium; the uptake of ^{137}Cs accounts for the physiological peculiarities of nutrition of pine trees. We can infer that the rate of nutrient uptake by trees regularly decreases as they age, resulting in lower contents of ^{137}Cs in wood of mature stands. The ^{137}Cs accumulation is the most intensive in young trees.

The age dependence of the transfer factor of ^{137}Cs from soil into wood of pine trees was approximated by a function of the type:

$$Y = 0.429 * 10^{-3} * a^{1.22} * \exp(-0.073 * a)$$

where Y = the ^{137}Cs transfer factor from soil into pine wood and a is the age of the pine tree.

The age dependence of the input of ^{137}Cs from soil into pine wood shows once again that the study of this process in forest ecosystems entails great difficulties. The age dependence is of great importance in understanding of the processes of radionuclide redistribution among the compartments of phytocoenosis.

In order to make the age dependence more precise, we studied the behaviour of ^{137}Cs in trees of the main deciduous species. The studies were carried out at deciduous stationary plots of long-term observation.

The results obtained show that the transfer factors of ^{137}Cs into trees of these species have age dependence. This dependence demonstrates physiological peculiarities of nutrition and growth of the tree. Naturally, each species has a peculiar type of this dependence. For comparison, we shall normalize the ^{137}Cs transfer factor by dividing the transfer factor of ^{137}Cs in wood of trees of different ages by that in wood of young trees. Such normalized of ^{137}Cs transfer factors are shown for deciduous species in Fig.1. The age dependence of the ^{137}Cs transfer factors can be described by equations. The analytical type of the age dependence for deciduous species is given in Table 4.

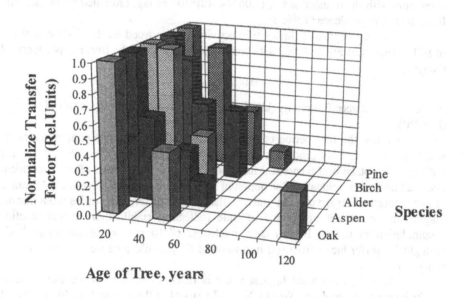

Fig.1 The age dependence of the normalized transfer factors of ^{137}Cs from soil into wood of different species

TABLE 4. The analytical type of the age dependence for deciduous species

Species	Analytical form
Birch	$Y = (-0.5 * \ln(a) + 2.5) * 0.000178$
Aspen	$Y = (-0.7 * \ln(a) + 3.0) * 0.00218$
Alder	$Y = (-0.7 * \ln(a) + 3.2) * 0.00218$
Oak	$Y = (-0.4 * \ln(a) + 2.1) * 0.00373$

In formulas of the Table 4:

Y = the ^{137}Cs transfer factor, m^2/kg, and a = the age of the tree, years.

With the systematic approach to the establishment and study of forest plots we could select stands occurring under the same site conditions. The ^{137}Cs deposition in soil was a variable factor for these stands. Such a method of selection of plots enabled us to define the character of influence of the radionuclide deposition on accumulation by woody plants.

The investigation was concerned with plots established in mossy pine forests (II age class, yield class I, free of undergrowth) occurring on soddy podsolic, sandy soils.

The mean of transfer factor of ^{137}Cs from soil into pine wood appeared to be equal to $(2.63+/-0.32)*10^{-3}$ m^2/kg with a small relative error. This value was in good agreement with the transfer factor, $(3.00+/-0.48)*10^{-3}$ m^2/kg, estimated experimentally from the age dependence (Table 3).

The dependence of the specific activity of pine wood on the ^{137}Cs deposition in soil is shown in Fig.2. These data were approximated by the linear dependence of the type:

$$Y=0.0029*X-0.54,$$

where Y = the specific activity (kBq/kg) and X = the ^{137}Cs deposition in soil (kBq/m^2).

During the first period of time after the fallout, intensive contamination of wood is observed. This process is dependent upon age of the tree. The rate of the radionuclide uptake is at theoretical maximum within 4 to 14 years after the accident and then decreases gradually. The notion of a maxima is a matter of convention and is of just theoretical importance. In practice, the transfer factor value varies around some magnitude over a definite period of time indicating stabilization of radionuclide accumulation in plants. In this case, a change of 10 percent of the value ^{137}Cs biological transfer factor from soil into wood of 60-year-old pine trees occurs over 14 years.

Once the theoretical highest value is reached, the rate of the radionuclide uptake by the tree decreases. Within 24 to 27 years after the accident the high value of the ^{137}Cs biological transfer factor is reduced by half.

The maximum value of the ^{137}Cs transfer factor varies from mature to young-growth pine stands in the ratio 1:15. In other words, the highest radionuclide content should be expected in young-growth stands.

The processes involved in the model influence variously the dynamics of the biological transfer factor. The accumulation of the radionuclides by the tree through out its life, transfer of radionuclides to the root system at the cost of their migration deep into the soil and increase of the amount of plant-available forms with depth favour increasing the transfer factor value; in contrast the redistribution of radionuclides among tree components, radionuclide transfer to the soil through the litter decay and variation in the transfer factor value as the tree ages contribute to a

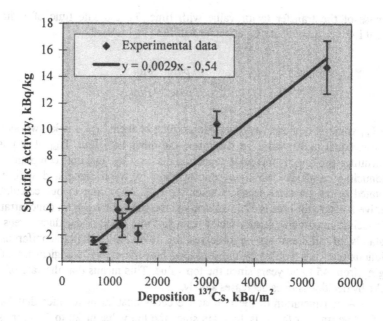

Fig.2. The dependence of the specific activity of pine wood on the ^{137}Cs deposition

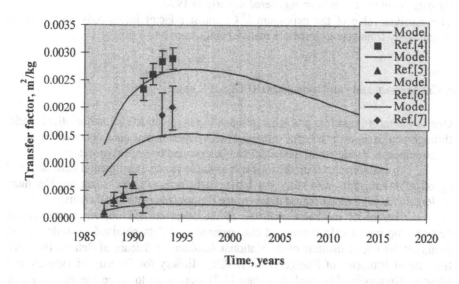

Fig.3. Dynamics of the ^{137}Cs biological transfer factor from soil to wood of pine trees of different age and its experimental measurements taken by different authors

decrease of the transfer factor value with time. To estimate time of radionuclides present in tree components, an effective half life (T_{eff}) is used:

$$T_{eff} = \frac{T_{phys} * T_{biol}}{T_{phys} + T_{Biol}}$$

The T_{eff} value is related to two half-times. One of them, T_{phys}, is known as the ^{137}Cs half-life, equal to 30 years. To determine the other half-time, T_{biol}, it is possible by performing prolonged full-scale experiments or, by making computations. The radionuclide residence half-time concentration in wood associated with processes unrelated to the physical decay is taken as T_{biol}. According to our calculations the effective half-life for pine is 12 to 14 years to the time of maximum concentration. The radionuclide maximum transfer factor from soil into wood of deciduous trees growing in stands of different age is observed in 1996-2001. Unlike coniferous stands, deciduous ones are typified by rather a long half-life of the biological transfer factor ranging from 45 to 50 years since the top value. This means that the rate of the ^{137}Cs uptake by deciduous trees decreases slowly.

A consideration of the radioactive decay enables us to infer that half-life of the effective transfer factor is 18 years since the top value or 28 to 32 years since the accident.

Alder stands occurring on bogs have their distinctive pattern. Here, the top transfer factor value has been registered as early as 1990.
The absolute value of the maximum ^{137}Cs transfer factor into woody plants varies significantly from young-growth to mature stands from 2 to 7 times.

4. Calibration and validation of FORESTLIFE model

One of the most important problems in modeling is to validate a mathematical model through comparison of the data computed with experimental ones obtained by direct measurements. Validation of the model can be executed by several processes.

To validate model predictions it is possible to use experimental data obtained by other investigators who carry out independent research. Such data are the most preferable for validation because these reduce the possible methodical errors.

The model validation with use of the other method implies a comparison between the model calculation and the experimental data obtained by independent groups at the Forest Institute of the National Academy of Science of Belarus (NASB), Institute of Problems of Energetics of NASB, Ministry for Forestry of Belarus and Moscow University. The published data [4-7] were used to make the comparisons. Model calculations were done under the conditions of these experimental observations. Recall that age, tree species and forest type are the initial data for FORESTLIFE model calculations. The results of the comparison are given in Fig.3. These show that the computed data are in good agreement with experimental ones.

The FORESTLIFE model permits prediction of contamination levels of trees to be harvested in different zones of contamination.

Predictions were made for the ^{137}Cs contamination levels of wood of pine trees occurring in the area that received 40 Ci/km2 with regard to age peculiarities of accumulation of ^{137}Cs by woody plants. These predictions and permissible ^{137}Cs contents in industrial and cultural forest products are shown in Fig.4.

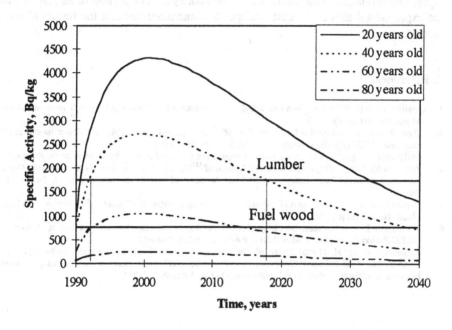

Fig.4. The predictions and permissible ^{137}Cs contents in industrial and cultural forest products.

The calculated curves plotted for different age class match pine age in 1993. The mature pine stands (80 years old) occurring in area where the contamination density is 40 Ci/km2 are suitable for every kind of economic use. However, the predicted values of the contamination levels of wood on 40-year-old pine trees are higher than the current permissible contents in fuel wood, round and construction timber (740 Bq/kg) in the years 1992-2037. Thus these are not suitable for economic use. A higher contamination level as against the permissible content will be observed in waney-edged timber (1850 Bq/kg) in the years 1992-2018.

5. Conclusion

The phenomenologic prediction model FORESTLIFE was created on the basis of the idea that a forest ecosystem is a living organism with its usual laws of the nutrition, growth and circulation of elements and energy. The parameters of the model were determined using the data base obtained by field observations of the radionuclide redistribution on the network of forest plots. The mathematical dependences between necessary parameters were obtained. The model permits the prediction of the radionuclide contamination levels for all components of forest stands of different types. The validation of the model was carried out by the comparison of its results with the experimental data of the other independent measurements and the results of other models.

6. References

1. Ipatyev V.(ed.) (1994) Forest and Chernobyl. Forest Institute of National Academy of Science of Belarus, Minsk, Belarus. (in Russian).
2. Dvornik A.M. and Zhucenko T.A. Behaviour of ^{137}Cs in pine stands of Belarus Polesye: modeling and prediction. ANRI (J.Radioecology) 3/4, 59-66. Moscow (in Russian).
3. Molchanov A.A.(Ed.) (1974) Productivity of organic and biological mass of forest. Moscow (in Russian).
4. Perevolotsky A.N., (1995) Regularities of migration of ^{137}Cs in the basic components of pine stands Gomel Polesye. PhD tesis, Forest Institute of National Academy of Science of Belarus, Gomel, Belarus. (in Russian).
5. Baraboshkin A.V. (1995) Report of radioactive monitoring department of Ministry for Forestry of Belarus, Minsk, Belarus (in Russian).
6. Tikhomirov F. et al. Radionuclides migration in natural and semi-natural ecosystems Report 1991-1992. //C.E.C.Project ECP-5, Doc.ENEA-DISP/ARA-MET(1992)6. Pp.49-108.
7. Grebenkov A. et. al. (1996) Study of Decontamination and Waste Management Technologies for contaminated Rural and Forest Environment. // The radiological consequences of the Chernobyl accident. Proceedings of the first international conference. - Minsk. 1996. Pp.221-224.

MODEL FOR PREDICTING THE LONG-TERM RADIOCESIUM CONTAMINATION OF MUSHROOMS

M. STEINER, S. NALEZINSKI, W. RÜHM, E. WIRTH
Institute for Radiation Hygiene, Federal Office for Radiation Protection, Ingolstädter Landstr. 1, D-85764 Oberschleißheim/Neuherberg, GERMANY

1. Introduction

Since 1987, a coniferous forest in Bavaria has continuously been monitored for radiocesium. About 300 soil samples and about 400 samples of different fungal species were analyzed for ^{134}Cs and ^{137}Cs. Based on this extensive data set a radioecological model for the long-term prediction of radiocesium contamination of mushrooms has been developed. The present paper summarizes some essential results of the work of the last decade. Details are published elsewhere [1-4].

The identification of key processes which govern radiocesium dynamics in forest ecosystems proved to be an important step to keep the model as simple as possible without losing predictive power. In view of the fact that fungal mycelia occupy distinct layers of forest soil, the vertical migration of radiocesium in forest soil turned out to be a such key parameter which determines the temporal evolution of activity levels in mushrooms. Accordingly, an appropriate radioecological model comprises the following components:

1. Migration of radiocesium in forest soil is described using a compartment model. Five compartments represent the five uppermost horizons of forest soil. This model facilitates a prognosis of the time-dependent vertical profile of radiocesium in forest soil.
2. Fungal mycelia were localized using the isotopic ratio $^{137}Cs/^{134}Cs$. The approach is based on the idea that the isotopic ratio in fungal fruit bodies should reflect the isotopic ratio of that soil horizon, from which radiocesium is predominantly taken up.
3. Transfer parameters soil/fungi were derived as concentration ratios, explicitly referring to that soil layer where fungal mycelia are located.

2. Vertical migration of radiocesium in forest soil

The vertical migration of radiocesium in forest soil (cambisol on calcareous moraine) is satisfyingly well described using a compartment model. It consists of five compartments, representing the five uppermost horizons (L, Of, Oh, Ah, and B) of undisturbed forest

I. Linkov and W.R. Schell (eds.), Contaminated Forests, 195–201.
© 1999 *Kluwer Academic Publishers. Printed in the Netherlands.*

soil. To keep the model as simple as possible, litterfall is neglected as source of activity for the L horizon, as well as root uptake by plants as a loss of activity in all layers. The corresponding system of differential equations for the decay-corrected activity levels of radiocesium can be written as follows:

$$\frac{d}{dt} L(t) = -\lambda_1 \cdot L(t) \tag{1}$$

$$\frac{d}{dt} Of(t) = \lambda_1 \cdot L(t) - \lambda_2 \cdot Of(t) \tag{2}$$

$$\frac{d}{dt} Oh(t) = \lambda_2 \cdot Of(t) - \lambda_3 \cdot Oh(t) \tag{3}$$

$$\frac{d}{dt} Ah(t) = \lambda_3 \cdot Of(t) - \lambda_4 \cdot Ah(t) \tag{4}$$

$$\frac{d}{dt} B(t) = \lambda_4 \cdot Ah(t) - \lambda_5 \cdot B(t) \tag{5}$$

L(t) is the decay-corrected activity of [134]Cs in the L horizon, given in (Bq m^{-2}) as function of time, Of(t) is the decay-corrected activity of [134]Cs in the Of horizon as function of time, etc.

The dynamic behavior of cesium from the Chernobyl fallout is well reproduced by the solutions of eqs. 1 - 5. The results of a least-squares fit to the experimental data on [134]Cs activities in different horizons are brought together in Table 1. From top of the forest floor, the resulting ecological half-lives $T_{1/2,i} = \ln 2/\lambda_i$ increase with soil depth, from about 3 years in the L horizon to about 8 years in the Ah horizon. The B horizon acts as an effective sink for radiocesium.

TABLE 1. Ecological half-lives for different horizons of a coniferous forest in Bavaria. The half-lives were calculated from the fitted time constants λ_i via $T_{1/2,i} = \ln 2/\lambda_i$.

Horizon	L	Of	Oh	Ah
λ_i (y^{-1})	0.25 ± 0.05	0.18 ± 0.04	0.16 ± 0.04	0.09 ± 0.06
$T_{1/2}$ (y)	2.8 ± 0.5	3.8 ± 0.8	4.4 ± 1.2	7.7 ± 4.9
$T_{1/2}$ (y cm^{-1})	4.0 ± 0.7	2.5 ± 0.5	2.4 ± 0.7	7.0 ± 4.5

The present vertical distribution of ^{137}Cs originating from nuclear weapons fallout is in fairly good agreement with the calculated predictions of the migration model, suggesting that the model facilitates a reliable prognosis of radiocesium activities in forest soil, at least for 25 years from the date of deposition onwards. For further details see [1].

Fig. 1 shows the future development of the ^{134}Cs activity until 2036 as predicted by the soil compartment model. In order to elucidate the pure migration behavior, radiocesium activities are corrected for physical decay for 1 May 1986.

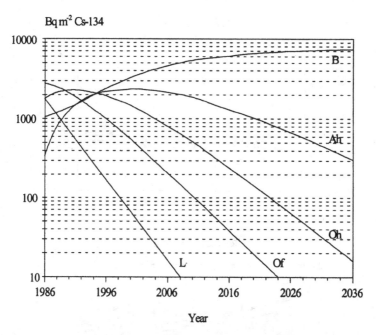

Fig.. 1. Extrapolation of the ^{134}Cs activities, expressed in (Bq m^{-2}) and decay-corrected for 1 May 1986, for the five uppermost soil horizons of a coniferous forest in Bavaria as predicted by the compartment model described in the text.

3. Major mycelium location

Fungal mycelia were localized using the isotopic ratio ^{137}Cs/^{134}Cs, corrected for physical decay. The approach is based on the idea that the isotopic ratio in fungal fruit bodies should reflect the isotopic ratio of that soil horizon, from which radiocesium is predominantly taken up.

The time-dependent isotopic ratio ^{137}Cs/^{134}Cs turned out to be a "fingerprint" of the different layers of forest soil at the investigated site. In the uppermost horizons L and Of the ^{137}Cs/^{134}Cs ratios are practically constant over time, reflecting the isotopic ratio 1.75 of the Chernobyl fallout in south Bavaria. Until 1994 the ratios decreased from 2.0 to 1.85 in the Oh horizon, from 2.4 to 2.1 in the Ah horizon, and from 2.5 to about 2.3 in the B horizon, respectively.

As examples, Fig. 2 presents the isotopic ratios ^{137}Cs/^{134}Cs as functions of time measured in samples of *Lepista nebularis* and *Russula cyanoxantha* together with the 95 percent confidence bands. The ^{137}Cs/^{134}Cs ratios for different soil horizons are also shown for comparison.

^{137}Cs/^{134}Cs ratios measured in fruit bodies of the saprophytic species *Lepista nebularis* are approximately constant over time and scatter around 1.75 for all years. Obviously, this species has a superficial mycelium located in the L and/or Of horizon.

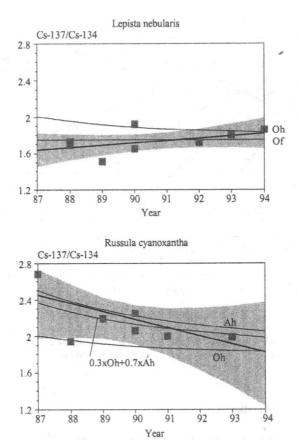

Fig. 2. ^{137}Cs/^{134}Cs ratios as functions of time in two mushroom species, decay-corrected for 1 May 1986. The black boxes denote measurements. The thick full line represents linear regression curves and the shaded areas are the corresponding 95 percent confidence bands. The predictions of the compartment model for the isotopic ratios of different horizons are shown for comparison.

As can be seen from Fig. 2, the ^{137}Cs/^{134}Cs ratios in samples of *Russula cyanoxantha* are significantly higher compared to the cesium ratios in the Oh horizon for all years, but significantly lower than the cesium ratios in the Ah horizon. This finding indicates that the symbiont *Russula cyanoxantha* is supplied with radiocesium from both Oh and Ah horizon.

Analogously, the location of fungal mycelia, i.e. that soil layer from which radiocesium is predominantly taken up, was identified for 14 different fungal species. In summary, fungal species can be subdivided into four different groups according to the location of their mycelia: group 1 L/Of horizon, group 2 L/Of/Oh horizon, group 3 Oh horizon, group 4 Oh/Ah horizon. Fungal mycelia of symbiotic species tend to be located in deeper soil layers compared to saprophytic species. For further details see [2].

4. Transfer parameters

In view of the fact that fungal mycelia occupy distinct layers of forest soil, transfer parameters soil/fungi for radiocesium should explicitly refer to that soil layer, from which radiocesium is predominantly taken up. Concentration ratios CR defined as

$$CR = \frac{^{134}Cs_{fruit\,body}(t)}{^{134}Cs_{soil\,horizon}(t)} \qquad (6)$$

turned out to be suitable transfer parameters. Hereby $^{134}Cs_{fruit\,body}(t)$ and $^{134}Cs_{soil\,horizon}(t)$ denote the activity levels, expressed as Bq kg^{-1} dry weight, in fungal fruit bodies and the soil layer, which the mycelium inhabits, respectively.

Concentration ratios as defined above differ by about two orders of magnitude, even at the same sampling site. At the investigated Bavarian sampling site concentration ratios vary from 0.5 ± 0.1 for fruit bodies of *Lepista nebularis* up to 42 ± 5 for fruit bodies of *Hydnum repandum*. On average concentration ratios for symbiotic species are about one order of magnitude higher than those for saprophytic species. For further details see [3].

5. Predicted future radiocesium contamination in fungal fruit bodies

A detailed analysis revealed that the time-dependent activity levels in fungal fruit bodies reflect the time-dependent activity level of that soil horizon, from which radiocesium is predominantly taken up. From extensive time series of field measurements, which cover a decade, it was deduced that the transfer parameters themselves remain constant over time, at least from three years after deposition onwards. A comparison of concentration ratios soil/fungi for radioactive ^{134}Cs and ^{137}Cs and stable ^{133}Cs further indicates that the concentration ratios will stay constant in future within a factor of two or less [4].

The future temporal evolution of activity levels in fungal fruit bodies is determined by the vertical migration of radiocesium and the location of mycelia in forest soil. As can be seen from Fig. 3, four groups of mushrooms can be distinguished: Twenty-five years after deposition, activity levels of radiocesium will be about 1% (L/Of horizon), 5% (L/Of/Oh horizon), 15% (Oh horizon) and about 140% (Oh/Ah horizon) of the initial value.

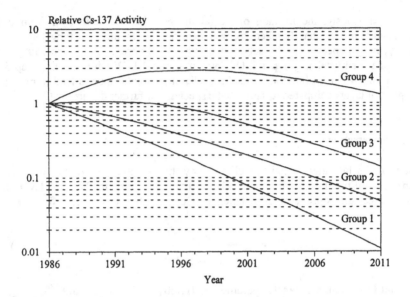

Fig. 3: Estimated activity levels of ^{137}Cs as a function of time for four different groups of fungi, taking migration in soil and physical decay into account (Location of mycelia: group 1 L/Of, group 2 L/Of/Oh, group 3 Oh, group 4 Oh/Ah). Activity levels are normalized to an initial value of 1.

Previously this radioecological model has been applied to different species of berry plants (blueberry, blackberry, raspberry, strawberry). First results indicate that radiocesium is predominantly taken up from the organic horizons. Also for berry plants transfer parameters remain constant with time from several years after deposition onwards.

6. Conclusions

The present paper describes a radioecological model for the long-term dynamics of radiocesium in forest soil and fungal fruit bodies. It has been shown that the identification of key processes which govern radiocesium dynamics in forest ecosystems proved to be an important step to keep the model as simple as possible without losing predictive power. In view of the fact that fungal mycelia occupy distinct layers of forest soil, the vertical migration of radiocesium in forest soil was demonstrated to be a such key parameter which determines the temporal evolution of activity levels in mushrooms. The same basic idea was published by Schell et al. [5], who applied a modification of their model FORESTPATH [6] to two fungal species, with the location of the mycelia taken from literature. However, the radiocesium concentration in fruit bodies was assumed to be equal to the radiocesium concentration in the corresponding soil horizon, i.e. the concentration ratio was assumed to be 1. In addition to a compartment model for the vertical migration of radiocesium in forest soil, the present paper describes an approach to localize fungal mycelia in forest soil and suggests suitable transfer parameters soil/fungi. These three components are considered to be inevitable for quantitative estimates of the dynamic long-term contamination of fungal fruit bodies.

The appealing feature of this comparatively simple radioecological model is the fact that the long-term dynamics of radiocesium is inherently included within the model itself. The model parameters remain constant over time, which facilitates extrapolation into future and allows reliable long-term predictions.

7. References

1. Rühm, W., Kammerer, L., Hiersche, L., and Wirth, E. (1996) Migration of ^{137}Cs and [134]Cs in different forest soil layers, *J. Environ. Radioactivity 33*, 63-75; Erratum *J. Environ. Radioactivity 34*, 103-106.
2. Rühm, W., Kammerer, L., Hiersche, L., and Wirth, E. (1997) The ^{137}Cs/[134]Cs ratio in fungi as an indicator of the major mycelium location in forest soil, *J. Environ. Radioactivity 35*, 129-148.
3. Rühm, W., Steiner, M., Kammerer, L., Hiersche, L., and Wirth, E. (1998) Estimating future radiocaesium contamination of fungi on the basis of behaviour patterns derived from past instances of contamination, *J. Environ. Radioactivity 39*, 129-147.
4. Rühm, W., Yoshida, S., Muramatsu, Y., Steiner, M., and Wirth, E. (1998) Distribution patterns for stable [133]Cs and their implications with respect to the long-term fate of radioactive [134]Cs and ^{137}Cs in a natural ecosystem. Submitted to *J. Environ. Radioactivity*.
5. Schell, W.R., Linkov, I., Belinkaia, E., and Morel, B. (1996) Application of a dynamic model for evaluating radionuclide concentration in fungi, Proceedings of 1996 International Congress on Radiation Protection, April 14-19, 1996, Vienna, Vol. 2, p. 752-754.
6. Schell, W.R., Linkov, I., Myttenaere, C., and Morel, B. (1996) A dynamic model for evaluating radionuclide distribution in forests from nuclear accidents, *Health Physics 70*, 318-335.

A MODEL FOR THE TRANSFER AND RECYCLING OF CS-137 WITHIN A DECIDUOUS FOREST ECOSYSTEM

E.M. SEYMOUR, P.I. MITCHELL and L. LEÓN VINTRÓ
Department of Experimental Physics, University College Dublin, Dublin 4, IRELAND

D.J. LITTLE
Department of Environmental Resource Management, UniversityCollege Dublin, Dublin 4, IRELAND

Abstract

The distribution of fallout ^{137}Cs in the component parts of a semi-natural sessile oak forest sited at Brackloon in Co. Mayo, Ireland is examined in detail. Field data confirm that soil is the major repository for ^{137}Cs in this deciduous ecosystem, with approximately 94% of the total inventory in the upper organic horizons. In contrast, the wood, roots and litter combined, constituting virtually 98% of the forest biomass, contain only 2–3% of the caesium inventory, with the balance in the mineral soil. Differences between the distribution observed at Brackloon and those reported for coniferous ecosystems elsewhere are discussed. In addition, a compartmental model is proposed which simulates the accumulation and recycling of ^{137}Cs within this deciduous environment and enables its distribution to be projected over a timescale of 100 years. Model-derived data indicate that the key processes controlling re-distribution include the rates of fixation of ^{137}Cs within the Of and Oh layers, and its rate of removal from the latter. Overall, the data highlight the need to consider both labile and non-labile forms of ^{137}Cs, and to differentiate the main soil horizons.

1. Introduction

Numerous studies have shown that forest ecosystems efficiently accumulate and recycle a wide range of pollutants including anthropogenic radionuclides. Moreover, it is now recognised that radiocontaminants, e.g. ^{137}Cs ($T_{\frac{1}{2}}$ = 30 y), once deposited, will persist in such environments for many decades, if not longer. In the past, the main thrust of research has been on coniferous forests, with the result that a significant database now exists on ^{137}Cs accumulation and recycling in these systems [1–9]. Considerably fewer studies of deciduous forests have been reported, the most notable being those of Olson [10] on *Liriodendron*, Witherspoon and Taylor [11] on *Quercus rubra*, both at Oak Ridge in Tennessee, Croom and Ragsdale [12] on *Quercus laevis* in Atlanta, Georgia,

203

I. Linkov and W.R. Schell (eds.), Contaminated Forests, 203–215.

Rauret et al. [13] on *Quercus ilex* in Tarragona, Spain, and Antonopoulos-Domis et al. [14] on *Quercus conferta* in Thessaloniki, Greece. These studies indicate that the proportion of the initial deposition intercepted by foliage in coniferous forests is generally higher than in the case of deciduous forests. In fact, following the Chernobyl accident, coniferous forests were shown to have trapped 60–90% of the deposited [137]Cs [2,5,15], while deciduous forests intercepted 10–40% [5].

Upon deposition to foliage, ^{137}Cs is either leached by precipitation to the underlying forest litter or absorbed by the foliage with some being retracted and stored by the tree prior to leaf-fall [16]. The rates at which these processes proceed are quite different. For example, studies of (coniferous) pine and spruce stands have shown that the initial half-time for the removal of ^{137}Cs by leaching is in the order of weeks [2,11,17], whereas for deciduous foliage it is closer to a few days [11]. In the longer term, leaf-fall is considered by many to be the dominant mechanism, with half-times of four years for coniferous needles and one year for deciduous leaves being proposed [18]. On the other hand, Bunzl and Schimmack [2], who made quantitative measurements of this process in a spruce stand in Bavaria, reported that only 7% of the intercepted caesium was actually transferred to the soil by leaf-fall.

Litter is of key importance in radionuclide cycling in forests as it can be contaminated by direct deposition as well by the processes mentioned above. The rate of litter decomposition and the subsequent release of plant-available radionuclides are important factors in the control of the rate of recycling. Deciduous leaf-fall is an annual occurrence and litter decomposition in such an environment has been shown to be rapid, with one study suggesting a half-time for ^{137}Cs of just under a month [19]. By contrast, litter decomposition in coniferous forests appears to be much slower, half-times of between 5 and 35 months being reported [17,20].

The retention of caesium in the organic layers of forest soils has been well documented [3,5,6,7,21,22]. Coughtrey and Thorne [23] reported that caesium fixation increases with moisture content and organic matter, both of which are characteristically high in forest soils. This entrapment or lack of mobility has been attributed to a number of factors including the presence of clay minerals and microbial immobilization.

It is recognised that microflora and mesofauna play an important role in the immobilization of caesium in forest soils. Microflora transport nutrients into the upper horizons and this process is accelerated by mesofauna, particularly microarthropods and enchtraeids [24]. In coniferous forests, Rafferty et al. [25] found evidence to suggest that fungi have a significant function in the incorporation of ^{137}Cs into litter during decomposition. Thus, fresh litter is continually invaded by mycelia from deeper horizons, thereby extending the residence half-time for ^{137}Cs in this compartment.

Data on transfer factors from soil to vegetation, though useful, are insufficient on their own to enable predictions of uptake by trees, as a significant fraction of the ^{137}Cs present may be in a non-labile form. In addition, the wood may act as reservoir for caesium, at first translocating it to new growth and later retracting it into the trunk and roots at the end of the growing season.

The primary objective of the combined field and modelling study described here was to adapt the generic model FORESTPATH [18] in order to simulate the dynamic behaviour of ^{137}Cs in a deciduous forest ecosystem and predict its long-term behaviour. Other objectives included the identification, via a sensitivity analysis, of the key parameter(s) controlling the distribution of ^{137}Cs in this environment and the determination of representative residence half-times for ^{137}Cs in a deciduous setting. To this end, ^{137}Cs levels and inventories were determined in the various compartments of a semi-natural sessile oak (*Quercus petraea*) forest ecosystem located at Brackloon, County Mayo in the west of Ireland. Samples examined included the major components of oak (i.e., wood, bark, leaves, stems and roots), lichens and mosses, indigenous vegetation species (*Vaccinium myrtillus, Blechnum spicant, Luzula sylvatica*), and soil from each of the main soil horizons.

2. Materials and methods

2.1. MODEL DESCRIPTION

The structure of our model (*Figure 1*) differs from that of FORESTPATH in three important respects. Firstly, the 'organic layer' compartment in the generic model has been sub-divided in our model into litter, Of and Oh horizons. This was thought necessary due to the complexity of the organic layer and the variation in timescales for decomposition and translocation within each of these sub-compartments. Secondly, the tree and understory compartments have also been sub-divided, each having sub-compartments representing leaves, wood and roots. This allows the distribution of ^{137}Cs within the above-ground vegetation to be determined. Thirdly, our model recognises that caesium is present in both labile and non-labile forms in soil, and we have, therefore, sub-divided the three soil compartments to take account of this reality. Surface run-off and erosion have been discounted as we have concluded, on the basis of the measured ^{137}Cs inventory from weapons fallout, that the net loss by these processes is virtually nil at the Brackloon site.

The model uses as input the extended (in time) deposition from atmospheric nuclear weapons testing in the 1950s and 1960s, and the acute deposition from the Chernobyl accident (*Figure 2*), and was calibrated using measured ^{137}Cs concentrations in the various forest materials (compartments).

The residence half-times derived with the aid of the model are site-specific and effectively take account of the main ecological processes excluding radioactive decay. Transfer of ^{137}Cs between compartments is described mathematically by a series of coupled first order differential equations, while the computational approach used for all simulations is a fourth-order Runge-Kutta integration.

2.2. SITE DESCRIPTION

The study site is located within Brackloon Wood, County Mayo in the west of Ireland, at an altitude of between 30–75 m above sea-level. The site has been the subject of intensive monitoring and research since 1991 under the EC's programme for the protection of the Community's forests against atmospheric pollution [26]. Mean annual precipitation is 1380 mm, of which from 10 to 19% has been shown to be intercepted by the canopy [26]. Stemflow at the site has been reported to be very small [26]. The soil is classified as a humus-iron podzol developed on weathered schist and gneiss [26].

206

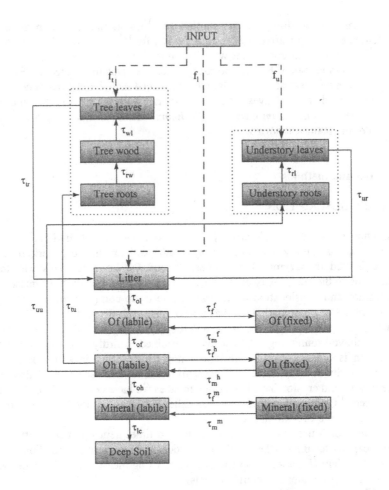

Fig. 1. Compartmental structure of the model

Soil texture, pH, cation exchange capacity and organic matter content at the site are given in TABLE 1. The tree canopy is dominated by oak (*Quercus petraea*), most of which are at least 150 years-old, with a rich ground cover of bilberry (*Vaccinium myrtillus*), hard fern (*Blechnum spicant*) and woodrush (*Luzula sylvatica*). The boughs of the oak trees are laden with mosses and lichens, especially lungwort (*Lobaria pulmonaria*).

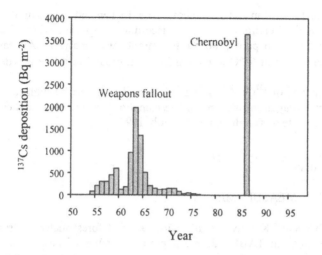

Fig. 2. ^{137}Cs deposition history at Brackloon Wood (site co-ordinates: 53°45.51'N 09°33.44'W)

TABLE 1. Typical soil characteristics at the Brackloon site (*source*: ref. 26)

Horizon	C.E.C. mmol ± 100 g^{-1}	pH	Organic matter (%)	Coarse sand (%)	Fine sand (%)	Silt (%)	Clay (%)	Bulk density (g cm^{-3})
Of	44.2	4.5	92.2	-	-	-	-	-
Oh	23.1	4.0	92.1	-	-	-	-	-
E	2.1	4.6	1.5	3	36	51	9	0.74
Bhs	8.3	3.9	4.5	-	-	-	-	-
Bs	6.9	4.3	3.9	8	34	46	12	1.38
C	1.9	4.8	1.9	12	20	39	28	1.04

2.3. SAMPLING AND ANALYSIS

Soil samples were extracted and divided into different layers according to their genesis, as indicated in TABLE 1. The samples were oven-dried at 105°C for 24 hours to obtain the dry soil bulk density. Any large stones were removed and weighed separately. The samples were then ground to a fine powder and thoroughly mixed to ensure homogeneity.

The understory vegetation was removed from the soil and sorted according to species. Each plant was separated into leaves, stems and roots, and a composite sample of each category prepared. Oak trees were sampled by collecting buds, leaves, stems, branches, bark and roots from representative trees on the site. In addition, an oak tree typical of the stand in the study area (ca. 200 years old) was felled. Cross-sections were sampled at heights of 1.3 m, 7 m and 13 m above base. Lichen and moss samples were collected from this and other trees and sorted according to species, where feasible. Unidentified specimens were pooled and a sub-sample was analysed. All samples (excluding soils) were dried at 65°C to constant weight, ground to a fine powder and homogenized.

Massic activities of ^{137}Cs, ^{134}Cs and ^{40}K were determined by high-resolution gamma spectrometry using an intrinsic n-type germanium detector. All activities were decay-corrected to the date of sampling, i.e., 8–9 July 1997.

3. Results and Discussion

3.1. EXPERIMENTAL MEASUREMENTS

Measured ^{137}Cs, ^{134}Cs and ^{40}K activities in a wide range of forest materials from the Brackloon site are given in TABLE 2. An interesting feature of these data is the relatively high activities found in soils, lichens and mosses compared to the other materials sampled.

3.1.1. Soil

The total forest inventory of ^{137}Cs (per unit area) was found to be 6730 Bq m^{-2}, with 97.5% in the combined forest floor (litter) and soil layers. This percentage is almost identical to the figure given by Antonopoulos-Domis et al. [14], who found 97% of the total ^{137}Cs inventory in the same compartments under oak (*Quercus conferta* Kit.). It is, however, somewhat higher than the percentages reported for coniferous [4,6,9] or coniferous-dominated [5] forests elsewhere.

It proved possible to detect ^{134}Cs ($T_{\frac{1}{2}}$ = 2.06 y) in the surface organic soil horizons due to the comparatively high levels of Chernobyl-sourced radiocaesium present. Thus, the relative contributions from weapons testing in the atmosphere and from the Chernobyl accident to the total ^{137}Cs inventory within the Of and Oh horizons could be determined using the ^{134}Cs/^{137}Cs ratio of 0.53 measured in fresh fallout over Ireland following the Chernobyl accident [27]. This showed that approximately 57% of the total ^{137}Cs in these horizons is attributable to weapons fallout with the balance arising from Chernobyl. The contribution from weapons fallout to the total inventory, at approximately 3800 Bq m^{-2}, is consistent with that predicted for these latitudes on the basis of the measured annual rainfall at Brackloon. It is also in good agreement with published data on the geographical distribution of weapons fallout ^{137}Cs throughout Ireland [28].

Our data show that 94% of the total ^{137}Cs inventory at the site currently resides in the top 10 cm of the soil column (*Figure 3*) even though the bulk of the deposition took place more than three decades earlier. The top 10 cm is highly organic and comprises the Of and Oh soil layers, each with an organic matter content of about 92% (TABLE 1). Predictably, downward migration is very slow, with little radiocaesium escaping to the underlying mineral layers. The small fraction that is leached is strongly sorbed in the first few centimetres of mineral soil, beneath which radiocaesium levels appear to diminish exponentially with depth.

TABLE 2. Mean ^{137}Cs, ^{134}Cs and ^{40}K massic activities (±1σ, dry wt.) in forest materials sampled at Brackloon, Ireland in July 1997

Compartment	Description	Activity (Bq kg^{-1})			^{137}Cs/^{134}Cs
		^{137}Cs	^{134}Cs	^{40}K	
Trees	*Quercus petraea*:				
	buds	5 ± 1		93 ± 16	
	leaves	57 ± 1		447 ± 12	
	stems	20 ± 1		170 ± 10	
	branches	14 ± 1		105 ± 12	
	trunk	6.6 ± 0.5		10 ± 7	
	bark	25 ± 1		89 ± 4	
	roots	58 ± 1		113 ± 11	
Lichens	*Lobaria pulmonaria*	74 ± 1		137 ± 9	
	Usnea spp.	9.0 ± 0.5		113 ± 8	
	Parmelia perlata	107 ± 1	1.2 ± 0.2	43 ± 10	90 ± 15
	Parmelia laevigata	367 ± 4	3.4 ± 0.3	72 ± 11	110 ± 10
	Cladonia squamosa	256 ± 3	3.1 ± 0.3	38 ± 10	80 ± 8
	Mixed species	12 ± 1		82 ± 21	
Mosses	*Isothecium myosuroides*	322 ± 3	3.6 ± 0.2	96 ± 9	90 ± 5
	Mixed species	144 ± 2	1.4 ± 0.2	73 ± 11	100 ± 15
Understory	*Vaccinium myrtillus*:				
	berries[†]	3.0 ± 0.2		24 ± 4	
	stems	22.0 ± 0.5		85 ± 6	
	leaves	23 ± 1		224 ± 19	
	roots	15.0 ± 0.5		33.0 ± 0.5	
	Luzula sylvatica:				
	leaves	27 ± 1		578 ± 23	
	roots	118 ± 1	1.2 ± 0.2	578 ± 23	100 ± 15
	Blechnum spicant:				
	leaves	45 ± 1		373 ± 17	
	roots	114 ± 1	0.6 ± 0.1	184 ± 8	190 ± 45
Soil	litter	20 ± 1		92 ± 5	
	Of (0–9 cm)	363 ± 2	3.6 ± 0.1	161 ± 6	100 ± 3
	Oh (9–10 cm)	187 ± 1	1.0 ± 0.1	504 ± 5	190 ± 20
	E (10–18 cm)	14 ± 1		655 ± 20	
	Bhs (18–30 cm)	5± 1		580 ± 9	
	Bs (30–60 cm)	2.0 ± 0.2		407 ± 9	
	C (60–80 cm)	1.0 ± 0.1		472 ± 7	

[†] Fresh wt.

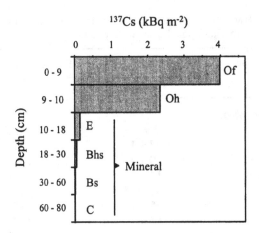

Figure 3. Vertical distribution of ^{137}Cs in the soil column at Brackloon (July 1997)

3.1.2. *Trees*

Of the total ^{137}Cs inventory, only 2.4% was found in the oak trees. This compares well with the percentage (2.2%) reported by Antonopoulos-Domis et al. [14] for *Quercus conferta* Kit. In coniferous forests, the percentage is higher, ranging from 6.7% to 14% [5,9]. For mixed forests, Mamikhin [29] estimated the proportion to be 0.3% in 1987, a year after Chernobyl, increasing to 1.7% by 1990.

It is interesting to compare the proportions of ^{40}K and ^{137}Cs in oak-tree components, including mosses and lichens. Our data show that wood constitutes the main ^{40}K reservoir (63.4%) with smaller amounts in the leaves (27.6%) and in the roots (7.1%). Virtually the same percentage of ^{137}Cs is held in the wood (66.4%), though the distribution pattern is different for the leaves (10.7%) and the roots (14.9%). This contrasts with Sombre's observation that ^{40}K and ^{137}Cs have similar distribution patterns in trees and that leaves are the main reservoir of ^{137}Cs [17]. The percentages of ^{40}K and ^{137}Cs in mosses and lichens combined were 1.9% and 8.0%, respectively.

3.1.3. *Understory*

The understory or ground vegetation contained only 0.02% of the total ^{137}Cs inventory. Others have reported values of 0.005% for oak forests [14] and 0.4–1% for spruce forests [5,7]. These low values are not surprising given that the biomass in this compartment is very small in most forests. Our measurements indicate that *Luzula sylvatica* and *Blechnum spicant* exhibit similar activities, with roots having a higher activity than leaves in both cases. In contrast, all components of *Vaccinium myrtillus* display comparable activities. Assuming all three species have the same biomass distribution, the total ^{137}Cs inventory in *Vaccinium myrtillus* appears to be about a factor of three lower than in the other two. This is consistent with the findings of other workers [4,7] and suggests lower uptake and/or higher transfer of ^{137}Cs from roots to stems and leaves in *Vaccinium myrtillus* compared with other species.

3.2. MODEL RESULTS

3.2.1. *Residence half-times for* [137]*Cs*

The model has been used to determine residence half-times for ^{137}Cs in various forest compartments and these are given in TABLE 3. For the purposes of comparison a set of generic values compiled from the published literature by Linkov [18] has also been included.

As data on fixation and mobilization rates in the different soil horizons at Brackloon were not available, we used as a starting point parameter values derived by Croom and Ragsdale [12] for what they defined as the lower soil compartment (i.e.,>5 cm depth). These values were adjusted in order to reproduce the percentages of labile and non-labile ^{137}Cs in the Of, Oh and mineral soil compartments as determined in extraction experiments carried out by other workers in similar soils [17,21,22,30,31]. Thus, we have assumed for modelling purposes that 90% of the ^{137}Cs in the Of horizon, 97% in the Oh horizon and 88% in the mineral soil is in a fixed, non-labile form at the present time. Similarly, we used a generic value [18] for the rate of leaching to deeper mineral horizons and adjusted accordingly. Root uptake by oak trees was assumed to occur from the Oh horizon, as the greater proportion of nutrient uptake is by fine roots distributed through this layer. In general, the site-specific parameter values derived for Brackloon are within the range of previously reported values [18].

3.2.2. *Predicted inventories in forest compartments*

Model-predicted ^{137}Cs inventories in different forest compartments for the period 1950–2050 are shown in *Figure 4*. The plots indicate that the Of and Oh soil compartments have between them retained the bulk of the deposition over the years, holding 94% of the total inventory at the present time.

Following the Chernobyl pulse which, in the case of Ireland was virtually a 24-hour event, the inventories in the tree and understory compartments increased sharply and then fell quickly in a matter of months due to leaching by precipitation and leaf-fall. Thereafter, levels in the understory continued to fall, while a small increase in the tree compartment took place as a result of root uptake approximately 8 years after the initial pulse. As litter decomposed and became part of the Of horizon, the inventory in the Of layer increased, peaking about 3–4 years after deposition, while levels in new litter fell rapidly. The Oh layer exhibited similar behaviour to the Of layer, with the inventory peaking approximately 8–9 years after deposition. The inventory in the mineral compartment has also peaked, 10–11 years post Chernobyl. Model predictions indicate that by the year 2050 the proportions of the total inventory in the Of and Oh layers will be similar, with about 6% distributed through the mineral layers.

It is worth noting that in the four decades which have elapsed since contamination by weapons fallout first began, almost 44% of the deposited ^{137}Cs at Brackloon has been eliminated by radioactive decay.

TABLE 3. Interception fractions and residence half-times for ^{137}Cs (note that the interception fraction assigned to trees includes interception by mosses and lichens growing on the trees)

Parameter	Notation	This study	Generic value Source: Linkov [18]
Interception fraction:			
tree	f_t	0.6	0.6
understory	f_u	0.2	0.2
litter	f_l	0.2	0.2
Soil fixation half-time (y):			0.64 (0.15–5)
Of	τ_f^f	4.3	
Oh	τ_f^h	0.79	
Mineral	τ_f^m	2.1	
Soil mobilization half-time (y):			1.1 (0.5–69.3)
Of	τ_m^f	100	
Oh	τ_m^h	69	
Mineral	τ_m^m	100	
Organic layer removal half-time (y):			3 (1–50)
litter	τ_{ol}	0.5	
Of	τ_{of}	5.7	
Oh	τ_{oh}	8.9	
Leaching to deep soil half-time (y):	τ_{lc}	450	400 (100–6000)
Tree half-time (y):			
root uptake	τ_{tu}	5	1 (0.5–50)
root-to-wood	τ_{rw}	1.7	
wood-to-leaf	τ_{wl}	6	
removal (leaf-fall)	τ_{tr}	0.7	0.72 (0.5–5)
Understory half-time (y):			
root uptake	τ_{uu}	85	10 (1–100)
root-to-leaf	τ_{rl}	1	
removal (leaf-fall)	τ_{ur}	0.3	0.23 (0.10–0.36)

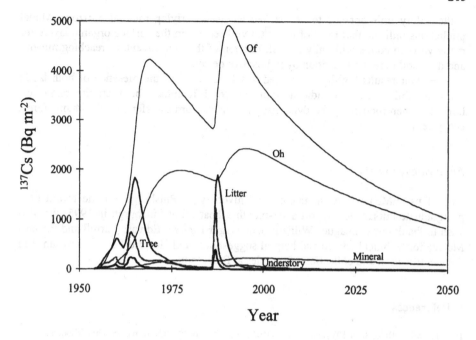

Fig. 4. Model-predicted ^{137}Cs inventories in forest compartments at Brackloon to year 2050

3.2.3. *Sensitivity Analysis*

A simple test was carried out to evaluate the sensitivity of the model to variations in chosen parameter values. Each parameter was varied in turn by an order of magnitude above and below its model optimized (or selected) value while maintaining the input values of the other parameters. The maximum and minimum concentrations obtained in this way were used to calculate the variability index as defined by Schell et al. [32]. The results of the analysis showed clearly that the most sensitive parameters controlling the distribution of ^{137}Cs in this deciduous forest ecosystem are the rates of ^{137}Cs fixation within the surface organic layers and its rate of removal from the Of horizon.

4. Conclusions

Field measurements confirm that soil is the major repository of ^{137}Cs in a deciduous forest ecosystem, with 94% of the total inventory retained within the Of and Oh horizons. In contrast, the wood, roots and litter, which together constitute almost 98% of the forest biomass, contain only 2–3% of the total inventory.

The bulk (>90%) of the ^{137}Cs in the surface organic horizons is present in a non-labile form, most likely as a result of complexation with decomposing organic matter. Very little appears to be leached to the underlying mineral horizons. Model predictions indicate that most of the ^{137}Cs will remain in the surface organic layers for many years to come, with only a small fraction of the total inventory reaching mineral and deep soil before elimination by radioactive decay.

Our results highlight the need to take account of the presence of both labile and non-labile forms of radiocaesium and to differentiate between the main soil horizons when modelling the dynamics of radiocaesium transfer in a deciduous forest ecosystem.

Acknowledgements

One of us (EMS) wishes to thank the University of Pittsburgh for the award of a postgraduate bursary to support a six-month sabbatical at Pittsburgh in 1995. We also wish to thank our colleagues William Schell, Igor Linkov, Edward Farrell and Edward McGee for technical advice and helpful suggestions, and Howard Fox for assistance in the identification and collection of samples at Brackloon.

5. References

1. Tobler, L., Bajo, S., and Wyttenbach, A. (1988) Deposition of ^{134,137}Cs from Chernobyl fallout on Norway spruce and forest soil and its incorporation into spruce twigs, *J. Environ. Radioactivity* 6, 225-245.

2. Bunzl, K. and Schimmack, W. (1989) Interception and retention of Chernobyl-derived ^{134}Cs, ^{137}Cs and ^{106}Ru in a spruce stand, *Sci. Tot. Environ.* 78, 77-87.

3. Thiry, Y. and Myttenaere, C. (1993) Behaviour of radiocaesium in forest multilayered soils, *J. Environ. Radioactivity* 18, 247-257.

4. Fawaris, B.H. and Johanson, K.J. (1994) Radiocesium in soil and plants in a forest in central Sweden, *Sci. Tot. Environ.* 157, 133-138.

5. Melin, J., Wallberg, L., and Suomela, J. (1994) Distribution and retention of cesium and strontium in Swedish boreal forest ecosystems, *Sci. Tot. Environ.* 157, 93-106.

6. Raitio, H. and Rantavaara, A. (1994) Airborne radiocesium in Scots pine and Norway spruce needles, *Sci. Tot. Environ.* 157, 171-180.

7. Strandberg, M. (1994) Radiocesium in a Danish pine forest ecosystem, *Sci. Tot. Environ.* 157, 125-132.

8. Rühm, W., Kammerer, L., Hiersche, L., and Wirth, E. (1996) Migration of ^{137}Cs and ^{134}Cs in different forest soil layers, *J. Environ. Radioactivity* 33(1), 63-75.

9. McGee, E.J., Johanson, K.J., Fawaris, B.H., Synnott, H.J., Nielsen, S.P., Horrill, A.D., Kennedy, V.H., Barbayiannis, N., Veresoglou, D.S., Colgan, P.A., and McGarry, A. Chernobyl fallout in a Swedish spruce forest ecosystem, *J. Environ. Radioactivity* (submitted).

10. Olson, J.S. (1965) Equations for cesium transfer in a *Liriodendron* forest, *Health Phys.* 11, 1385-1392.

11. Witherspoon, J.P. and Taylor Jr., F.G. (1969) Retention of a fallout simulant containing ^{134}Cs by pine and oak trees, *Health Phys.* 17, 825-829.

12. Croom, J.M. and Ragsdale, H.L. (1980) A model of radiocesium cycling in a sand hills-turkey oak (*Quercus laevis*) ecosystem, *Ecological Modelling* 11, 55-65.

13. Rauret, G., Llauradó, M., Tent, J., Rigol, A., Alegre, L.H., and Utrillas, M.J. (1994) Deposition on holm oak leaf surfaces of accidentally released radionuclides, *Sci. Tot. Environ.* 157, 7-16.

14. Antonopoulos-Domis, M., Clouvas, A., Xanthos, S., and Alifrangis, D.A. (1997) Radiocesium contamination in a submediterranean semi-natural ecosystem following the Chernobyl accident: measurements and models, *Health Phys.* 72(2), 243-255.

15. Sokolov, V.E., Ryabov, I.N., Ryabtsev, I.A., Tikhomirov, F.A., Shevchenko, V.A., and Taskaev, A.I. (1990) Ecological and genetic consequences of the Chernobyl atomic power plant accident, SCOPE-RADPATH Meeting, Lancaster, UK.

16. Myttenaere, C., Schell, W.R., Thiry, Y., Sombré, L., Ronneau, C., and van der Stegen de Schrieck, J. (1993) Modelling of Cs-137 cycling in forests: recent developments and research needed, *Sci. Tot. Environ.* 136, 77-91.

17. Sombré, L., Vanhouche, M., de Brouwer, S., Ronneau, C., Lambotte, J.M., and Myttenaere, C. (1994) Long-term radiocesium behaviour in spruce and oak forests, *Sci. Tot. Environ.* 157, 59-71.

18. Linkov, I. (1995) Radionuclide Transport in Forest Ecosystems: Modeling Approaches and Safety Evaluation, PhD Thesis, University of Pittsburgh, 171 pp.

19. Sauras, T., Roca, M.C., Tent, J., Llauradó, M., Vidal, M., Rauret, G., and Vallejo, V.R. (1994) Migration study of radionuclides in a Mediterranean forest soil using synthetic aerosols, *Sci. Tot. Environ.* 157, 231-238.

20. Bunzl, K., Kracke, W., and Schimmack, W. (1992) Vertical migration of plutonium-239, + -240, americium-241 and caesium-137 fallout in a forest soil under spruce, *Analyst* 117 (March), 469-474.

21. Andolina, J. and Guillitte, O. (1990) Radiocesium availability and retention sites in forest humus, in G. Desmet, P. Nassimbeni, and M. Belli (eds.), *Transfer of Radionuclides in Natural and Semi-Natural Environments*, Applied Elsevier Science, pp. 135-142.

22. Thiry, Y., Vanhouche, M., Van Der Vaeren, P., de Brouwer, S., and Myttenaere, C. (1994) Determination of the physico-chemical parameters which influence the Cs availability in forest soils, *Sci. Tot. Environ.* 157, 261-265.

23. Coughtrey, P.J. and Thorne, M.C. (1983) *Radionuclide Distribution and Transport in Terrestrial and Aquatic Ecosystems*, Vol I. A.A. Balkema, Rotterdam.

24. Brückmann, A. and Wolters, V. (1994) Microbial immobilization and recycling of ^{137}Cs in the organic layers of forest ecosystems: relationship to environmental conditions, humification and invertebrate activity, *Sci. Tot. Environ.* 157, 249-256.

25. Rafferty, B., Dawson, D., and Kliashtorin, A. (1997) Decomposition in two pine forests: the mobilisation of ^{137}Cs and K from forest litter, *Soil Biol. Biochem.* 29(11/12), 1673-1681.

26. Boyle, G.M., Farrell, E.P., and Cummins, T. (1997) Intensive Monitoring Network - Ireland, Forest Ecosystem Research Group Report No. 18, University College Dublin.

27. Mitchell, P.I., Sanchez-Cabeza, J.A., Ryan, T.P., McGarry, A.T., and Vidal-Quadras, A. (1990) Preliminary estimates of cumulative caesium and plutonium deposition in the Irish terrestrial environment, *J. Radioanal. Nucl. Chem. (Articles)* 138(2), 241-256.

28. Ryan, T.P. (1992) Nuclear Fallout in the Irish Terrestrial Environment, PhD Thesis, National University of Ireland, 206 pp.

29. Mamikhin, S.V., Tikhomirov, F.A., and Shcheglov, A.I. (1997) Dynamics of ^{137}Cs in the forests of the 30-km zone around the Chernobyl nuclear power plant, *Sci. Tot. Environ.* 193, 169-177.

30. Shand, C.A., Cheshire, M.V., Smith, S., Vidal, M., and Rauret, G. (1994) Distribution of radiocaesium in organic soils, *J. Environ. Radioactivity* 23, 285-302.

31. Poiarkov, V.A., Nazarov, A.N., and Kaletnik, N.N. (1995) Post-Chernobyl radiomonitoring of Ukrainian forest ecosystems, *J. Environ. 335Radioactivity* 26, 259-271.

32. Schell, W.R., Linkov, I., Myttenaere, C., and Morel, B. (1996) A dynamic model for evaluating radionuclide distribution in forests from nuclear accidents, *Health Phys.* 70(3), 318-.

MODELLING RADIOCAESIUM BIOAVAILABILITY IN FOREST SOILS

A.V. KONOPLEV, A.A. BULGAKOV and V.E. POPOV
SPA «Typhoon», Obninsk 249020, RUSSIA

R. AVILA
SSI, Stockholm 17116, SWEDEN

J. DRISSNER, E. KLEMT, R. MILLER and G. ZIBOLD
University of Applied Sciences, Weingarten 88241, GERMANY

K.-J. JOHANSON
Swedish University of Agricultural Sciences, Uppsala 75007, SWEDEN

I.V. KONOPLEVA
Institute of Agricultural Radiology & Agroecology, Obninsk 249020, RUSSIA

I. NIKOLOVA
N. Pouskarov Institute of Soil Sciences and Agroecology, 1080 Sofia, BULGARIA

1. Introduction

The experience of radiation accidents, the Chernobyl accident in the first place, has shown that the methodology used to develop and estimate the effectiveness of countermeasures to reduce negative consequences of accidental contamination of the environment by radionuclides does not meet the requirements of fast response in case of an emergency. In order to overcome this shortcoming, Geographic Information Systems (GIS) and computer systems for decision-making support in case of emergency are being actively developed [1]. An essential part of such systems are mathematical models of migration and transformation of radionuclides in the environment and their impact on man and components of terrestrial and aquatic ecosystems. The accuracy of prediction by these models in determining the effectiveness of decisions making is largely dependent on the accuracy of estimating input parameters [2]. In this context, it is important to create an expert system for estimation of input model parameters which would form a separate module in the decision-making system. Special emphasis should be put on those parameters to which the predictions are more sensitive. An important component in calculation of radiation

217

I. Linkov and W.R. Schell (eds.), Contaminated Forests, 217–229.
© *1999 Kluwer Academic Publishers. Printed in the Netherlands.*

dose is the estimation of the soil-plant transfer of radionuclides. For this purpose, the concentration ratio CR is normally used which is defined as the ratio of radionuclide concentrations in the plant and in the upper soil layer. The values of CR determined experimentally in different conditions for a particular plant can differ by a factor hundred or even thousand [3]. Accordingly, using a plant-average CR leads to a considerable uncertainty in prediction. For this reason, numerous attempts were made to develop methods for estimating site-specific values of CR. A great effort went into deriving empirical relations between the accumulation of radionuclides in plants and soil properties [4]. An essential drawback of purely empirical characteristics is that they are not generally valid. In order to be applied to soils different from those that they were obtained, characteristics should be justified based on information about mechanisms of sorption-desorption of radionuclides in soil and their soil-plant transfer.

Today it is generally accepted that one of the key factors governing soil-plant transfer of radionuclides is the exchangeability of radionuclides in soil [5,6]. Although this hypothesis is rather well founded, the analysis of experimental data has shown that it is not sufficient [6]. This is probably explained by the fact that along with the fraction of exchangeable form of the radionuclide in soil, the value of CR is influenced by other factors. The most important of them are the concentration of the radionuclide and major cations in the soil solution [7,8,9]. For caesium isotopes, that are among the most significant radionuclides, two hypotheses were proposed accounting for the effect of the composition of soil solution on radionuclides transfer to plant. According to the first hypothesis, the radiocaesium concentration in plant is proportional to the ratio of its concentration in soil solution and the potassium concentration. The comparison with experimental data, however, has shown that there is no meaningful correlation between the concentration ratio and ^{137}Cs/K ratio in soil solution [10]. Another hypothesis, which seems more sound, is that the concentration factor is proportional to the fraction of radionuclide in the root exchange complex dependent on the composition of soil solution [8]. Using this hypothesis a model of radiocaesium bioavailability in soils was proposed [9,11]. The objective of this work is to parameterise CR through soil characteristics on the basis of a process level model and to test this model against the data sets on radiocaesium soil-plant transfer in forest ecosystems in Germany, Russia, Sweden and Switzerland.

2. Modelling radiocaesium soil-plant transfer in forest ecosystems

In the recent years, increasing attention has been paid to the role of fungi mycelium in radiocaesium migration in forest ecosystems. This is associated with, first of all, high ability of fungi to accumulate radiocaesium. In some papers it is stated that up to 50% and more of the radiocaesium inventory in soil can be bound in the mycelium [12]. It is argued that the role of mycelium in radiocaesium migration can be twofold. On the one hand radiocaesium uptake by mycelium can lead to its biological fixation and consequently effective reduction in its concentration in the soil solution. On the other hand, because of the symbiosis of mycelium and roots of forest plants (mycorrhizae) the mycelium can serve as a direct source of radiocaesium to plants. That is why the dynamics of soil to plant radiocaesium transfer in forest ecosystems can, in principle, be described by a model which is schematically shown in Fig. 1.

Yet, the problem with the use of such a model is the lack of reliable data on radiocaesium transfer from mycelium to plants. This process is poorly understood and can not be parameterized yet. Besides the dynamic model does not allow deriving an integral parameter accounting for bioavailability of radiocaesium in the soil. This can be done within an equilibrium model having in mind the restrictions in its applicability.

2.1. EQUILIBRIUM MODEL OF RADIOCAESIUM SOIL-PLANT TRANSFER

The conceptual scheme of the equilibrium model of radiocaesium soil-plant transfer in forests is presented in Fig. 2. At present it is generally accepted that radiocaesium is sorbed by soils very selectively and the retention behaviour of radiocaesium is controlled by micaceous clay minerals. The highly selective sites are located at the expanded edges of the clay particle interlayers and are called „frayed edge sites" - FES. The ability of soil to sorb radiocaesium selectively is characterised by the capacity of frayed edge sites [FES] or radiocaesium interception potential, RIP, which is a product of [FES] and selectivity coefficient of Cs in relation to correspondent competitive ion [13,14,15].

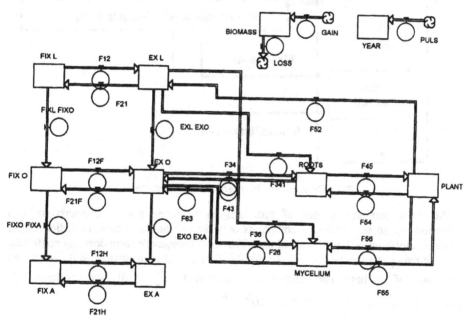

Fig.1. Conceptual representation of radiocaesium soil plant transfer in forest ecosystems with a consideration of fungi involvement in the process

220

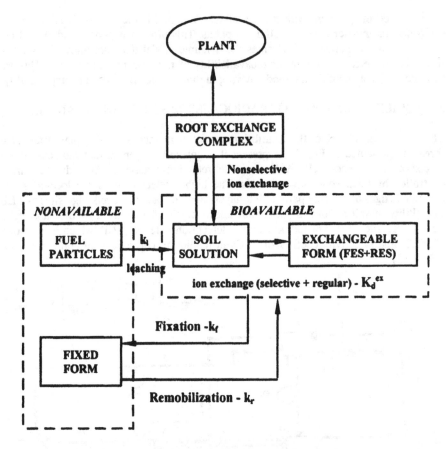

Fig. 2. Conceptual model of radiocaesium soil-plant transfer

The main competitive ions of radiocaesium in forest soils are potassium and ammonium. In environmental conditions only the exchangeable form of radiocaesium is in equilibrium with its dissolved part. The nonexchangeable form does not contribute to the radiocaesium content in soil solution. Therefore it is reasonable to introduce the notions of exchangeable distribution coefficient K_d^{ex} [16] and exchangeable radiocaesium interception potential $RIP^{ex}(K)$ [17]:

$$RIP^{ex}(K) = K_d^{ex}(Cs) \times [K]_w = K_c^{FES}(Cs / K) \times [FES]$$ (1)

where the exchangeable distribution coefficient is the ratio of exchangeable caesium concentration in the solid phase to the caesium concentration in soil solution at equilibrium; K_c^{FES} (Cs/K) is the selectivity coefficient of caesium in relation to potassium on FES.

The proposed model of radiocaesium soil-plant transfer is based on the following assumptions:

1) Radiocaesium is taken up by the plant from soil solution and its concentration in plant is a linear function of the radionuclide loading in the root exchange complex, which is determined by cation composition of the soil solution. Root exchange complex is not selective, i.e. the selectivity coefficient of Cs in relation to K and NH$_4$ equal unity [8];

2) Only the exchangeable portion of ^{137}Cs inventory in soil is involved in the immediate exchange with the soil solution [6,16];

3) The exchangeable radiocaesium interception potential (RIP^{ex}) and the cation composition of soil solution govern the ^{137}Cs concentration in soil solution according to ion exchange equilibrium [17].

The model includes the ion exchange sorption-desorption process on the soil, the ion exchange at the root exchange complex and the uptake from the root surface to the plant itself. The fixed part of the radionuclides is not involved in the immediate transfer to the plant. Here the fixation and remobilization includes not only the processes on clay particles but also the processes involving mycelium and other organic components in the soil. The corresponding rate constants account for integral values of all processes under consideration.

Prior to plant uptake, the radionuclide must pass through the cell wall free space, characterised by a specific cation exchange capacity [18]. This assumption differs from the approaches which are usually applied to the radionuclide soil-plant transfer, and which are based on a linearity of the concentration in the plant with the radionuclide concentration in the soil solution. The distinguishing feature of the proposed model is that Cs loading in the root exchange complex is influenced by the cationic composition of the soil solution. The root exchange complex is associated with carboxylic groups, and thus it is not selective. This approach has recently been used for radiocaesium transfer in the soil solution-plant system [8]. As the root exchange complex is nonselective, the selectivity coefficients of the K/NH$_4$ and the Cs/K pair both equal unity. The same assumption can be used for the Ca/Mg pair. Consequently, K$^+$ and NH$_4^+$ can be treated as one ion M$^+$, and Ca^{2+} and Mg^{2+} as one ion, represented by M^{2+}. On the basis of these assumptions one obtains [8]:

$$[^{137}Cs]_{plant} = k \cdot \frac{Z_{M^+}}{[M^+]_w} \cdot [^{137}Cs]_w \qquad (2)$$

where Z_{M^+} is the fraction of M$^+$ in the root ion exchange complex (potassium window); $[M^+]_w$ is the concentration of M$^+$ in the soil solution; $[^{137}Cs]_w$ is the ^{137}Cs concentration in the soil solution; $[^{137}Cs]_{plant}$ is the ^{137}Cs concentration in the plant; k is a proportionality constant between the uptake and the radiocaesium level in the exchange complex. It reflects the efficiency of the transport process across the membrane. At the same time, the concentration of ^{137}Cs in the soil solution is determined by the sorption-desorption equilibrium of the radionuclide in the "soil-soil solution" system. This means that radiocaesium occurs in the soil solution in a simultaneous equilibrium with two cation exchangers: the selective exchanger of the soil and the nonselective one of the root surface. The ^{137}Cs concentration in the soil solution at ion exchange equilibrium of radiocaesium on the FES can be presented as follows:

$$[^{137}Cs]_{ss} = \frac{\alpha_{ex} \cdot [^{137}Cs]_{tot}}{RIP^{ex}(K)} \cdot ([K]_w + K_c^{FES}(NH_4/K) \cdot [NH_4]_w) \qquad (3)$$

where α_{ex} is the portion of exchangeable ^{137}Cs in the soil; $[^{137}Cs]_{tot}$ is the total concentration of ^{137}Cs in the soil; $RIP^{ex}(K)$ is the radiocaesium interception potential of the soil in relation to potassium; $[K]_w$ and $[NH_4]_w$ are concentrations of correspondent cations in the soil solution; $K_c^{FES}(NH_4/K)$ is the selectivity coefficient of ammonium in relation to potassium at FES.

The "potassium window" Z_{M^+} in the nonselective root exchange complex can be expressed in terms of cation composition of the soil solution. For most soils, $Z_{M^{2+}} \gg Z_{M^+}$, and Z_{M^+} is proportional to the following parameter $\frac{[K]_w + [NH_4]_w}{\sqrt{([Ca]+[Mg])}}$ of the soil solution. Substituting this and (3) in (2) and taking into account that $CR=[^{137}Cs]_{plant}/[^{137}Cs]_{tot}$, we obtain that at the equilibrium conditions the CR is a linear function of the availability factor A [9]:

$$CR=B \times A \qquad (4)$$

where B is the parameter dependent on the plant characteristics (especially the capacity of the root exchange complex). The availability factor A is proportional to the fraction of radiocaesium in the root exchange complex and is parameterised through the composition of the soil solution and sorption properties of soils as follows [9]:

$$A = \frac{\alpha_{ex}PNAR}{RIP^{ex}} \qquad (5)$$

where PNAR is the adsorption potassium-ammonium ratio [19]:

$$PNAR = \frac{[K]_w + K_c^{FES}(NH_4/K)[NH_4]_w}{\sqrt{[Ca]_w + [Mg]_w}} \qquad (6)$$

where $[K]_w$, $[NH_4]_w$, $[Ca]_w$, $[Mg]_w$ are the concentrations of the corresponding cations in the soil solution, $K_c(NH_4/K)$ is the selectivity coefficient of potassium exchange for ammonium at selective adsorption sites (FES).

In fact, equation (5) can serve as a basis for prediction of CR using characteristics of the soil and the soil solution. Actually the availability factor A is the combination of three basic soil parameters: a) portion of exchangeable form of ^{137}Cs (α_{ex}) characterising the fixation ability of soil; b) exchangeable radiocaesium interception potential (RIP^{ex}) characterising the ability of the soil to sorb radiocaesium selectively and reversibly; and c) PNAR, a parameter of the cation composition of the soil solution.

2.2. SIMPLIFIED METHOD OF CHARACTERISATION OF RADIOCAESIUM BIOAVAILABILITY IN SOILS

The proposed method can be used for obtaining fairly accurate soil-specific estimates of the CR. For doing this, however, one should know such soil characteristics as the content of mobile forms of ammonium cation and exchangeable radiocaesium interception potential (RIP^{ex}). The values of these parameters are known for a limited number of soils. Methods for their theoretical estimation have not been developed yet and the methodology used for experimental determination of RIP^{ex} is rather complicated. Using this method for mapping the CR over extensive territories would involve significant and probably unacceptable costs. Therefore, a simpler method is proposed in this article for parameterisation of the availability factor of radiocaesium to plants using only those soil characteristics which can either be found in the literature or estimated using known correlation ratios.

This simplified parameterisation of the availability factor can be derived with the equation calculating radiocaesium concentration in the soil solution proposed in [16,20]:

$$[^{137}Cs]_w = \frac{[^{137}Cs]_{ex}[K]_w}{K_c^{eff}[K]_{ex}} \quad (7)$$

where $[^{137}Cs]_{ex}$, $[K]_{ex}$ is the concentration of the exchangeable form of ^{137}Cs in soil (Bq/kg) and potassium (meq/kg), respectively; K_c^{eff} is the effective selectivity coefficient of the potassium cation exchange for caesium cation in the soil absorbing complex [20].

Using (7) we get

$$A^* = \frac{a_{ex}PAR}{K_c^{eff}(Cs/K)[K]_{ex}} \quad (8)$$

where PAR is the potassium adsorption ratio (mM$^{1/2}$),

$$PAR = \frac{[K]_w}{\sqrt{[Ca]_w + [Mg]_w}} \quad (9)$$

The equation (8) is similar to that derived in [11]. The only difference is that equation (8) includes the effective selectivity coefficient. This may allow a more accurate estimation of the availability factor using the classification of soils by values of $K_c^{eff}(Cs/K)$ proposed in [21]. A better accuracy can be achieved if the value of $K_c^{eff}(Cs/K)$ is measured experimentally.

The value of the PAR can be expressed through the EPR (exchangeable potassium ratio), which equals to the ratio of the exchangeable potassium and the sum of exchangeable calcium and magnesium in soil [22]:

$$PAR = 9.5(EPR - 0.036) = 9.5\{\frac{[K]_{ex}}{[Ca]_{ex} + [Mg]_{ex}} - 0.036\} \tag{10}$$

By combining equation (8) and (10) we finally get:

$$A^* = \frac{9.5\, a_{ex}}{K_c^{eff}(Cs/K)} \{\frac{1}{[Ca]_{ex} + [Mg]_{ex}} - \frac{0.036}{[K]_{ex}}\} \tag{11}$$

As follows from equation (1), LgCR is the linear function of LgA*, the coefficient of LgA* is one:

$$LgCR = LgB + LgA^* \tag{12}$$

The presented model is based on the assumption, that only the exchangeable form of the radionuclide is potentially involved in the transfer to plant. At the same time, the portion of exchangeable radionuclide, determined by sequential extractions, characterises a state of dynamic equilibrium between the processes of fixation and remobilization of radiocaesium. In the long term, the total inventory of the radionuclide in soil can be potentially bioavailable by virtue of existence of remobilization. Therefore, the stationary state of exchangeable radiocaesium not always corresponds to a biologically available portion of the radionuclide. Apparently, a technique that is more adequate would be extraction from soil during a time interval close to the time scale of the soil-plant transfer.

3. Experimental testing of the model

3.1. MATERIALS AND METHODS

Experimental studies with the purpose of testing the model were carried out in several forest sites. Description of the experimental sites under study and soil types are presented in Table 1.

Fern (*Pteridium aquilinum* (L.) Kuhn and *Dryopteris filix-mas* (L.) Schot) was chosen as a reference plant for this study. For some sites data on bilberry (*Vaccinium myrtillus*) and blackberry (*Rubus fruticosus*) were also obtained. The roots of these plants are located in the humus layer of the soil (O_h or A_h). At the same time an analysis

Table 1. Listing of the soils studied, the symbols used and soil type*

Sampling site	Notation	Type of forest	Soil type*	Texture
GERMANY, Baden – Wuerttemberg				
Muensingen	Mu	spruce	Luvisol	sandy loam
Eggwald 1	Eg1	spruce	podzolic Luvisol	silty loam
Eggwald 2	Eg2	spruce	podzolic Luvisol	silty loam
Eggwald 3	Eg3	spruce	podzolic Luvisol	silty loam
Konstanz	Kon	spruce	Cambisol	sandy loam
Wolfach	Wol	spruce	Cambisol	sandy loam
Riedlingen	Ried	spruce	stagnosolic Cambisol	mottled loam
Isny	Is	spruce	Luvisol	sandy loam
RUSSIA, Bryansk region				
Tokovische	Tok	Mixed	humic Gleysol	loamy sand
St.Bobovichy	StB	Deciduous	Podsol	sand
Peat bog	PB	Deciduous	Histosol	
SWEDEN, Uppsala region				
Uppsala 3	Up3	Coniferous	ferric Podsol	light loam
Uppsala 11	Up11	Coniferous	ferric Podsol	sandy loam
Uppsala 13	Up13	Coniferous	ferric Podsol	sandy loam
Uppsala 21	Up21	Mixed	Podsol	sand
Uppsala 23	Up23	Coniferous	Podsol	fine gravel
SWITZERLAND, Ticino #				
Torricella 1	Tor1	Deciduous		
Torricella 2	Tor2	Deciduous		
Porza 1	Por1	Deciduous		
Porza 2	Por2	Deciduous		

* According to FAO-Unesco soil map as given in State Forestry soil map
# soil type and texture not given

of radiocaesium vertical distribution shows that most part of its inventory is located in these horizons. The characteristics of the soil, involved in equation for availability parameter A, were also measured for the root zone layer. For this purpose water and 1N ammonium acetate extractions were carried out. Major cations were measured in the extracts by AAS, and ^{137}Cs was measured in the ammonium extract for obtaining α_{ex}. Echangeable ammonium was measured colorimetrically in the 2M KCl extraction using indophenyl-blue method [23]. PNAR was calculated on the basis measured cation concentrations in the water extract. The modified technique of [15] has been used to measure $RIP^{ex}(K)$. Detailed description of the procedure is presented in [9,17]. The selectivity coefficient $K_c^{FES}(NH_4/K)$ was calculated as a ratio of $RIP^{ex}(K)$ and $RIP^{ex}(NH_4)$.

3.2. RESULTS AND DISCUSSION

The values of the soil parameters (for the root horizon) involved in equation (5) and the specific concentration ratios CR of two species of fern measured in the investigated forest sites are presented in Table 2.

Table 2. Characteristics of soils under study and concentration ratios for fern

Sampling site	α_{ex}, %	$RIP^{ex}(K)$, meq/kg	PNAR, $(mM)^{1/2}$	CR_r (fern), kg/kg
Mu	5,8	78	2,88	0,98*
Eg1	10	25	5,9	2,89*
Eg2	6,4	21	9,79	2,83*
Eg3	2,5	55	2,43	0,21*
Kon	10,5	26	1,92	0,59*
Wol	11,5	87	7,66	1,77*
Rie	12	26	2,74	1,58*
Is	11,1	22	4,53	3,71*
Tok	23,5	51	4,32	2,84*
StB	45,2	87	4,16	9,40**
PB	61	28	3,60	
Up 3	19	43	3,06	5,78**
Up 11	27	27	1,14	3,31**
Up 13	37	31	0,90	
Up 14	14	40	1,65	3,85**
Up 21	2,7	324	0,80	0,04**
Up 23	1,4	694	0,89	0,00**
Tor1	2,8	267	0,70	0,03**
Tor2	4,0	412	1,35	0,03**
Por1	1,2	264	1,15	0,05**
Por2	1,9	188	1,53	0,05**

* Dryopteris filix-mas, ** Pteridium aquilinum

In all cases the slope of the line calculated by the least square method is close to unity. The satisfactory agreement of the theoretical and experimental dependencies indicates that the proposed method can be used for estimation of site-specific soil-plant concentration. Part of the parameters in equation (11) for the availability factor can be found by expert judgement or measured using a rather simple procedure (α_{ex}, K_c^{eff}), while other parameters ($[Ca]_{ex}$, $[Mg]_{ex}$, $[K]_{ex}$) are important agrochemical indicators and for many regions they can be found in the literature. This allows using the proposed method for mapping the radiocaesium availability to plants. Incorporation of such maps into Geographic Information Systems (GIS) will lead to essential reduction in uncertainty of dose and risk assessments.

As demonstrated by the experimental data obtained recently [7, 24, 25] the radiocaesium transfer to plants is dependent not only on the value of the PAR, but also on the absolute potassium concentration in the soil solution. The radiocaesium concentration in plants is decreasing in almost inverse proportion to the potassium concentration in the nutrient solution with the increase of the last to 1 mM, following which it remains practically unchanged [7, 24]. This effect may be rather noticeable in arable soils deficient in potassium, but in forest soils the potassium concentration

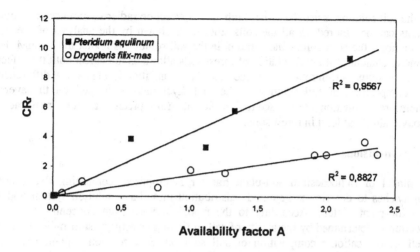

Fig. 3. Model testing: experimental dependence of CR_r for two species of fern on availability factor A in various forest sites of Germany, Russia, Sweden and Switzerland

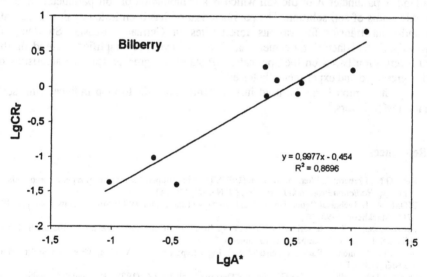

Fig. 4. Testing of the simplified parameterisation: experimental dependencies of $LgCR_r$ for bilberry on Lg A* calculated according equation (11).

in the soil solution is usually rather high and its effect on radiocaesium accumulation in plants can be ignored. In all the soils studied, as shown by the analysis of the water extractions, the potassium concentration in the soil solution is higher than 1 mM. High mineral content of the soil solution of forest soils allows us to neglect both the effect of the potassium concentration on radiocaesium accumulation in plants, the difference in the composition of soil solution at the root layer and in the soil on the average. Therefore, equation (12) seems sufficient for predicting the radiocaesium bioavailability at least in forest soils.

4. Conclusions

A model of radiocaesium soil-plant transfer is proposed, combining two separate approaches to describe radionuclide distribution in soil-soil solution system and soil solution-plant system. According to the model, radiocaesium concentration in soil solution is determined by the ability of soil to fix and selectively sorb radiocaesium as well as the cationic composition of soil solution. Soil solution - plant transfer is determined by Cs concentration in the soil solution and also by cationic status of the soil solution.

It is derived that the soil-plant transfer factor should be a linear function of availability parameter A of the soil which is a combination of soil parameters and the cationic status of soil solution. The parameterisation has been tested against field data on soil-plant transfer for various forest sites in Germany, Russia, Sweden and Switzerland. Satisfactory agreement has been achieved. A simplified method of the parameterisation based on the application of standard agrochemical characteristics of soil is presented and experimentally tested.

The approach can be used in radioecological GIS to map radiocaesium soil-plant transfer factors.

5. References

1. Kelly G.N., Ehrhardt J., Shershakov V.M. (1996). Decision support for off-site emergency preparedness in Europe. *Radiation Protection Dosimetry*, 64, No. 1/2, 129-142.
2. BIOMOVS II Technical Report No. 17. An Overview of the BIOMOVS II Study and its Findings. SSI, 17116 Stockholm, 1996, 36 p.
3. IAEA Technical Reports series No. 364 (1994). Handbook of parameter Values for the prediction of radionuclide transfer in temperate environments. 74p.
4. International Union of Radioecologists (1992). Eight Report of IUR Working Group on Soil-to-Plant Transfer. Balen, Belgium.
5. Oughton D.H., Salbu B., Riise G., Lien H., Oestby G., Noeren A. (1992). Radionuclide mobility and bioavailability in Norwegian and Soviet soils. Analyst, 117, 481-486.
6. Konoplev A.V., Viktorova N.V., Virchenko E.P., Popov V.E., Bulgakov A.A., Desmet G.M. (1993). Influence of agricultural countermeasures on the ratio of different chemical forms of radionuclides in soil and soil solution.- *Science of the Total Environment*, 137, 147-162.
7. Shaw G., Hewamanna R., Lillywhite J., Bell J.N.B. (1992). Radiocaesium uptake andranslocation in wheat with reference to the transfer factor concept and ion competition effects. *J. Environ. Radioactivity*, 16, 167-180.
8. Smolders E., Sweeck L., Merckx R., Cremers A. (1997). Cationic interactions in radiocaesium uptake from solution by spinach. *J. Env. Radioactivity*, 34, № 2, 161-170.

9. Konoplev A.V., Drissner J., Klemt E., Konopleva I.V., Miller R., Zibold G. (1997). Characterisation of soil in terms of radiocaesium availability to plants. Proceedings of XXVII Annual Meeting of ESNA . Ghent (Belgium), 29 August - 2 September 1997. Working Group 3: Soil-Plant-Relationships, 163-169.

10. ECP-2 (1996). The transfer of radionuclides through the terrestrial environment to agricultural products, including the evaluation of agrochemical practices. (Ed.: G. Rauret & S. Firsakova). Final Report. European Commission EUR 16528 en. 182 p.

11. Konoplev A.V., Drissner J., Klemt E., Konopleva I.V., Zibold G. (1996). Parameterisation of radiocaesium soil-plant transfer using soil characteristics. Proceedings of XXVIth Annual Meeting of ESNA, Working Group 3: Soil-Plant Relationships. Busteni (Romania), 12-16 September 1996, pp. 147-153.

12. Desmet G., Nassimbeni P., Belli M. (Ed.) (1990). Transfer of radionuclides in natural and semi-natural environments. Elsevier Applied Sciece, 693 p.

13. Sawhney B.L. (1972). Selective sorption and fixation of cations by clay minerals: a review. *Clays Clay Miner.*, **20**, 93-100.

14. Cremers A., Elsen A., De Preter P., Maes A. (1988). Quantitative analysis of radiocaesium retention in soils. *Nature*, **335**, No. 6187, 247-249.

15. Sweeck L., Wauters J., Valcke E., Cremers A. (1990). The specific interception potential of soils for radiocaesium. In.: "Transfer of radionuclides in Natural and Semi-natural environments".(Ed. G. Desmet, P.Nassimbeni, M.Belli), Elsevier Applied Science, 249-258.

16. Konoplev A.V., Bulgakov A.A., Popov V.E., Bobovnikova Ts.I. (1992c) Behaviour of long-lived radionuclides in a soil-water system. *Analyst*, **117**, 1041-1047.

17. Konoplev A.V., Konopleva I.V. (1998). Determination of radiocaesium characteristics of reversible selective sorption by soils and bottom sediments. *Russian Geochemistry* (in press).

18. Haynes R.J. (1980). Ion exchange properties of roots and ionic interactions within the root apoplasm: their role in ion accumulation by plants. *The Botanical Review*, **46**, No. 1, 75-99.

19. Wauters J., Elsen E., Cremers A. (1996). Prediction of solid/liquid distribution coefficients of radiocaesium in soils and sediments. Part III: A quantitative test of K_D predictive equation. *Applied Geochemistry*, **11**, 601-606.

20. Konoplev A.V., Bulgakov A.A. (1995). Modelling of the transformation of speciation processes of Chernobyl origin ^{137}Cs and ^{90}Sr in the soil and in bottom sediments. In: Environmental Impact of Radioactive Releases. Proceedings of an International Symposium on Environmental Impact of Radioactive Releases, IAEA, Vienna, 8-12 May 1995, 311-321.

21. Konoplev A.V., Bulgakov A.A. (1998). Prediction of ^{90}Sr and ^{137}Cs distribution in natural soil-water systems. In: Chernobyl Lessons Learned (Environment and Countermeasures). Sandia National Laboratory, Albuquerque, NM, USA (in press)

22. Richards L.(Ed) (1954) Diagnosis and improvement of saline and alkali soils. US DA Handbook 60, NY.

23. Krom M.D. (1980). Spectrophotometric determination of ammonia: a study of a modified Berthelot reaction using salicilate and dichloroisocyanurate. *Analyst*, **105**, 305-316.

24. Smolders E., Kiebooms L., Buysse J., Merckx R. (1996a). ^{137}Cs uptake in spring wheat (*Triticum aestivum* L. cv Tonic) at varying K supply. I. The effect in solution culture. *Plant and Soil*, **181**, 205-209.

25. Smolders E., Kiebooms L., Buysse J., Merckx R. (1996b). ^{137}Cs uptake in spring wheat (*Triticum aestivum* L. cv Tonic) at varying K supply. II. A potted soil experiment. *Plant and Soil*, **181**, 211-219.

MODELING INTERMITTENT PROCESSES IN RADIONUCLIDE MIGRATION IN SOIL SYSTEMS

K. A. HIGLEY,
Department of Nuclear Engineering, Oregon State University
100 Radiation Center, Oregon State University, Corvallis, OR 97331-
5902, USA

Abstract

A mathematical model was developed to provide insight into possible mechanisms of plutonium migration into soil. Repeated studies on an old plutonium contaminated grasslands soil have suggested that this radionuclide rapidly migrated several centimeters into the soil and then remained relatively place bound for two decades. Soil core samples collected within 4 y of the original atmospheric deposition revealed an exponentially decreasing profile of plutonium with increasing depth in soil. Approximately 50% of the inventory was found in the top 3 cm of soil, and greater than 90% in the top 12 cm. However, plutonium was also detected as deep as 20 cm in the soil profile. Twenty-five years later, the same, statistically indistinguishable, exponentially decreasing concentration versus depth gradient (slope) was observed. These results raised the question of how this plutonium initially moved to depth and then remained immobile throughout the intervening years. Investigations into the plutonium chemistry at this locale relative to other plutonium contaminated sites in the United States provided limited insight into movement mechanisms. Leach tests and soil-plutonium particle size associations, also yielded little information. A conceptual model was developed of the soil system. A simple one-dimensional diffusion equation was tested based on the simple soil model. The diffusion model was tested to predict near surface migration. It could not adequately address the long-term distribution observed in the field samples. Additional models were constructed to assess the effect of other processes on contaminant distribution. Chronic soil loss mechanisms were evaluated. Current evidence suggests that intermittent activities, such as animal intrusion may also be responsible for contaminant mixing into soil. The mathematical models were developed using the software Stella created by High Performance Systems. The final model considers erosion, intermittent flow, cracking, and biological migration processes in the physical migration of strongly sorbed contaminants such as plutonium. The model incorporates intermittent processes as a function of time of year, as well as preferential flow of water through near surface cracks on an episodic basis.

I. Linkov and W.R. Schell (eds.), Contaminated Forests, 231–238.

1. Introduction

Predicting the long-term behavior of radionuclides in the environment is necessary in order to assess the long-term risk. Several tools have been applied to this problem. Early methods relied on use of steady-state assumptions to predict risk (ie., empirically derived observations of nuclide concentration in one environmental compartment relative to another - the concentration ratio approach). As the science of radioecology advanced, the use of dynamic models based on first-order kinetics was utilized to describe radionuclide cycling in environmental systems[1]. The dynamic models provided greater insight into movement of selected nuclides. However, the model predictions did not always match observed reality. Modifications to these simplistic box and arrow systems are needed to improve long-term predictive abilities. In this example, the predictions of four simple mathematical models are compared to an observed environmental distribution of plutonium in a grassland ecosystem.

2. Background

The Rocky Flats Environmental Technology Site (RFETS) in Colorado, USA was previously used for the production of plutonium triggers for nuclear weapons. Radiological remediation activities in the late 1960's resulted in contamination of grassland from windblown plutonium-laden soil. The area has been largely untouched since the original event occurred 30 years ago. Repeated studies at the grassland site have suggested that this radionuclide rapidly migrated several centimeters into the soil and then remained relatively place bound for two decades. Soil core samples collected within 4 y of the original atmospheric deposition revealed an exponentially decreasing profile of plutonium with increasing depth in soil. Approximately 50% of the inventory was found in the top 3 cm of soil, and greater than 90% in the top 12 cm. However, plutonium was also detected as deep as 20 cm in the soil profile. Repeated investigations at the RFETS show that the peak activity concentration has remained in the 0-3 cm profile over the last 30 y [2]. Very little is found in biological material. The 1989 data exhibit the same slope (activity concentration versus soil depth) as the 1973 data (*Figure 1*). It is also apparent that the plutonium has not moved deeper into the soil, but the overall inventory has (apparently) decreased. The question is then where is the plutonium? Has it moved from the site, or is the apparent decreased inventory an artifact of the field sampling process?

These results raised the question of how this plutonium initially moved to depth and then remained immobile throughout the intervening years. Investigations into the plutonium chemistry at this locale relative to other plutonium contaminated sites in the Unites States provided limited insight into movement mechanisms [3]. Leach tests and soil-plutonium particle size associates, also yielded little information.

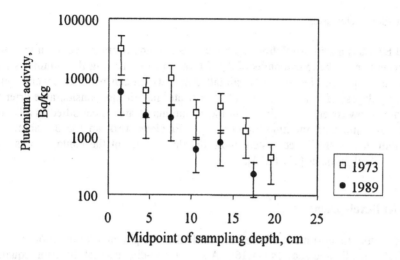

Figure 1. Example of plutonium soil profile data from RFETS.

3. Contaminant Characteristics

Plutonium is a member of the actinium series. Because of its electronic shell structure it has several oxidation states, which, under certain conditions can coexist. The particular species are determined by the presence of oxidizing and reducing agents in the environment as well as the abundance of complexing ligands. Regardless of the form of plutonium initially deposited to soils, it is largely converted to Pu(IV). This oxidation state is generally insoluble, and sorption of plutonium to soils and sediments results in its relative immobility in soils [4]. Observations on the environmental behavior of plutonium show that concentrations in soils and sediments are typically greater than in water or other environmental media by orders of magnitude. More than 99% of the Pu inventory in most ecosystems is found in the soil, particularly in the soil surface.

Because it exists in a strongly sorbed state on surface soils, the primary route of transport of plutonium is through processes governing the distribution and movement of soil. For surface soils, the principle physical transport mechanism have largely been assumed to be wind and water erosion [4]. Lesser consideration has been given to migration through soil cracks or pores. However, pore spaces can be a route for the rapid migration of contaminants into soil [5,6,7,8,9] . Soils with pronounced structure and large continuous macropores can be subject to a phenomenon known as bypass flow [7]. Soils tend to crack repeatedly along the same weak planes, rehealing themselves following wetting, only to recrack again along the same lines [5,8,10].

4. Site Characteristics

The RFETS has a semi-arid climate; annual precipitation averages 38.5 cm yr^{-1}, with 50% occurring in the four months of April through July. During the summer season the site is subject to high-intensity rainfall events which can result in several cm of rain being deposited in a localized area. The winter months are considerable drier but with some snow cover. The soil type of the contaminated area is classified as a Denver Kutch clay loam with montmorillonitic clay. Previous work on the distribution of plutonium in the RFETS ecosystem indicated that > 99% of the contaminants were attached to soil particles [11, 12,13].

5. Model Development

Kinetic (compartment) models have been applied to the plutonium migration problem at RFETS with limited success [14,16]. A simple one-dimensional diffusion equation was tested for its ability to predict near surface migration and could not adequately address the long-term distribution observed in the field samples [14]. Recent events have suggested that intermittent activities, such as soil cracking, or 100-year rainfall events, may be responsible for episodic contaminant migration [15]. In an attempt to improve our predictive abilities of plutonium migration in this specific system, a series of mathematical models were developed. The models were written using the software Stella created by High Performance Systems. The models considered intermittent flow, cracking and biological migration processes in the physical migration of strongly sorbed contaminants such as plutonium. They incorporated soil cracking as a function of time of year, as well as preferential flow of water through near surface cracks. Biological mixing mechanisms, such as might occur with earthworm movement, were also considered in this model structure.

      The conceptual model of soil used as the basis for developing the mathematical models to predict contaminant migration is a simple three-compartment system (see *Figure 2a*). It consists of surface soil, labile soil, and deep soil compartments. The compartments are loosely based on physical characteristics and the degree of biological activity within each. The surface soil is prone to resuspension of litter-and soil-attached plutonium. This layer is potentially also subject to loss via erosion. The labile soil contains the bulk of the plant root mass. The deep soil is considered a probable sink for contaminants. This simple model formed the basis for all kinetic models which were constructed. A brief review of each follows.

5.1. DIFFUSION MODEL

In previous work a simple one-dimensional diffusion model was tested for its ability to predict near surface migration [16]. The structure of the model is shown in *Figure 2a*. As modeled, all activity was initially on the soil surface at time t=0. The loss rate constant was arbitrarily set at 0.05 mo^{-1}. The results are shown in *Figure2b*. The

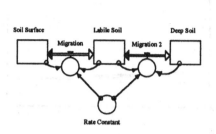

Figure 2a. Conceptual model of the diffusion equation. The double headed arrows indicate bi-directional flow.

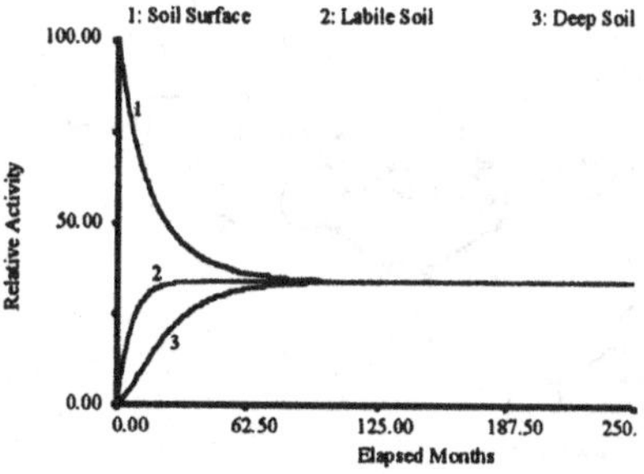

Figure 2b. Diffusion model output showing activity vs time. All compartments attain equal activity over time.

activity on the surface decreases as the plutonium diffuses to depth. Ultimately the activity at the surface is projected to equal that in the underlying soils due to simple diffusion processes. The limitation of this model is that it does not adequately simulate the long-term distribution observed in the field samples. It may be because the model does not account for structural inhomogeneities in the soil or intermittent processes. It is an extreme oversimplification of a natural soil system.

5.2. EROSION MODEL

A soil erosion component was added to the simple diffusion model *(Figure 3a,b)*. Erosion was presumed to be a seasonally varying event. Colorado is known for high intensity rainstorms during the summer monsoon season. The erosion rate was arbitrarily varied from a summer maximum to a winter minimum according to the following function:

$$Erosion = 0.005 - 0.005\cos((2\pi \cdot time)/12)$$

Where *'time'* is the elapsed time in the simulation. All other parameters in the model remained constant from the earlier model. The result is very similar to the diffusion model. All compartments approach equivalent activity over time. However, in this model the overall activity decreases due to erosive losses. As before, it does not adequately mimic the soil profile found in *Figure 1*.

5.3. ANIMAL INTRUSION MODEL

Animal impacts on movement of the soil between layers were added to the model *(Figure 4a,b)*. All other parameters were kept at previous values in the simulation. The animal activity was assumed to be seasonal, with a maximum in the summer months

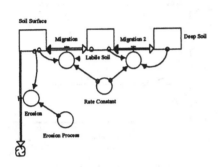

Figure 3a. Conceptual model for erosive processes.

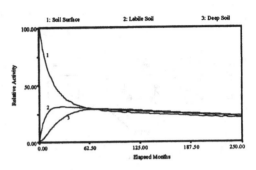

Figure 3b. Model output showing activity vs time. Activity in all compartments decrease from erosion.

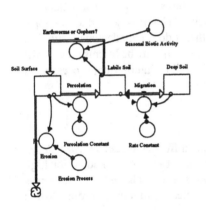

Figure 4a. Conceptual model for animal intrusive processes.

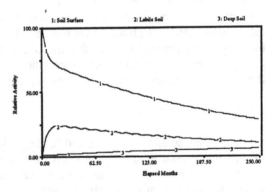

Figure 4b. Model output showing activity vs time. Note that the surface compartment does not equilibrate with the subsurface system. All compartments also decrease in inventory due to erosive processes..

(this actually is incorrect for many earthworms which are most active in the spring and fall when the soil is moist). The animal intrusion rate was described by the equation:

$$Intrusion = 0.15 - 0.01\cos((2\pi \cdot time)/12)$$

This varying intrusion rate moves material from the labile soil compartment to the surface on a seasonally varying level. In conjunction with erosion, the end result of this simulation is a decreasing inventory with time. However, note that the surface and labile soil compartment activities do not rapidly equilibrate as was predicted by both the diffusion and erosion models

5.4. INTERMITTENT/EPISODIC EVENTS

The model was further modified to include the addition of intermittent events such as might occur during extreme rainfall events (*Figure 5a,b*). Examples include contaminants migrating through macropore structures. The model was set to simulate a massive pulse on a multi-year cycle, using the following equation:

Storm = pulse(0.3,7,84)

which describes a pulse event of magnitude 0.3, occurring in the seventh month (for one time step) and repeating every 84 months. While the pulse results in a rapid change in inventory in both the surface and labile compartments, the system rapidly returns to its previous condition. This model suggests pulse events may not be as significant as originally envisioned by the author.

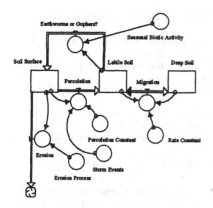

Figure 5a. Conceptual model for intermittent processes.

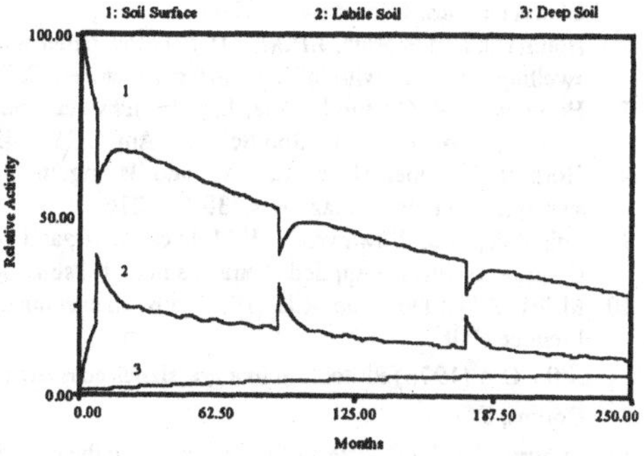

Figure 5b. Model output showing activity vs time. Note that the surface compartment does not equilibrate with the subsurface system. All compartments decrease due to loss of material by erosion.

6. Summary and Conclusions

A series of simple models were evaluated for their ability to produce a response similar to that observed from field studies. The models were designed to evaluate simple diffusion, chronic erosion, animal intrusion, and intermittent storm events. The models which included animal intrusion and intermittent events best reproduced the observed data (and suggest that erosion may be a significant factor in removal of plutonium). However, before either of these models can be accepted as a reasonable interpretation of physical events, additional evidence will be needed. Studies are currently underway involving soil thin sectioning and radiography which will be used to further evaluate these models.

7. References

1   Whicker, F.W.; Schultz, V. (1982) Radioecology: nuclear energy and the environment. Boca Raton , FL, CRC press, Inc; (2 vol).

2   Webb, S.B. (1992) A study of plutonium in soil and vegetation at the Rocky Flats Plant. Master's thesis, Colorado State University, Fort Collins, CO.

3   Higley, K.A., (1994) Vertical movement of actinide-contaminated soil particles. Ph.D. Dissertation, Colorado State University, Ft. Collins, CO.

4   Watters, R.L.; Edginton, D. N.; Hakonson, T.E.; Hanson, W.C.; Smith, M.H.; Whicker, F.W.; Wildung, R.E. (1980) Synthesis of the research literature. In: Hanson, W.C., (ed). Transuranic elements in the environment. U.S. Department of Energy / National Technical Information Service; DOE/TIC-22800; pp.1-44.

5   Bouma, J and Dekker, L.W. (1978) A case study of infiltration into dry clay soil. I. Morphological Characteristics. Geoderma 20:27-40.

6   Bouma, J. and Wösten, J.H.M. (1979) Flow patterns during extended saturated flow in tow, undisturbed swelling clay soils with different macrostructures. Soil Sci. Soc Am.J. 43:16-22.

7   Booltink, H.W.G.; and Bouma, J., (1991)Physical and morphological characterization of bypass flow in a well-structured clay soil. Soil Sci. Soc. Am. J. 55:1249-1254.

8   Horn, R.; Taubner, H,; Wuttke, M.; and Baumgartl, T.(1994); Soil physical properties related to soil structure. Soil and Tillage Res. 30:187-216.

9   Shipatalo, M.J.; Edwawreds, W.M; Dick, W.A.; and Owens, L.B. (1990)Initial storm effects on macropore transport of surface-applied chemicals in no-till soil. Soil Sci. Soc. Am. J. 54:1530-1536.

10 Miller, R.W.; Donahue, R.L. (199) Soils, an introduction to soils and plant growth. Englewood Cliffs, NJ, Prentice Hall.

11 Little, C.A.(1976) Plutonium in a grasslands ecosystem. PhD dissertation. Colorado State University, Fort Collins, CO.

12 Tamura, T. (1977) Effect of pretreatment on the size distribution of plutonium in surface soil from Rocky Flats, In: Transuranics in Desert Ecosystems, Nevada Applied Ecology Group. NVO-1871 UC-11: 173-186.

13 Langer, G. (1986) Dust transport- wind blown mechanical resuspension, July 1983 to December 1984. Rockwell International, RFP-3914.

14 Higley, K.A. (1992). Assessment of a model to determine the transport and fate of Am-241 and Pu-239,240. Master's thesis. Colorado State University, Fort Collins, CO.

15 Personal communication, Iggy Litaor, RFETS, July, 1995.

16 Higley, K.A., (1993) How good is good enough, when do you go with what you've got? in, R.L. Kathryn, D.H. Denham, K. Salmon (eds), Environmental Health Physics Research Enterprises, WA, pp. 267-280.

MODELLING OF ^{137}Cs BEHAVIOUR IN FOREST GAME FOOD CHAINS

S. FESENKO and S. SPIRIDONOV
Russian Institute of Agricultural Radiology and Agroecology, 249020, Obninsk, RUSSIA

R. AVILA
*Swedish Radiation Protection Institute ,
17116 Stockholm, SWEDEN*

1. Introduction

After the Chernobyl accident it became clear that radioactive contaminated forests can give doses to the population comparable to the doses from radioactive contaminated agroecosystems. Mathematical models of the migration of radionuclides in forest ecosystems are used to evaluate the consequences of the forest contamination and the effectiveness of potential countermeasures.

Over the last 30 years various groups of researchers developed several models which describe the behavior of radionuclides in forests. A recent review of these models [1] has pointed out that most published models describe the transfer processes with rate constants, which are often highly variable and include more than one process, making their interpretation and estimation difficult. The models described in the literature, except FORESTPATH [2], are site specific and the characteristics of the forests used for calibrations are usually not provided, which limits the applicability to other sites. It should also be noted that until now there is no model that describes the dynamics of radionuclide activity concentrations in forest game. This is probably due to lack of experimental data and knowledge of the dietary habits and habitat of wild animals.

In the present publication a model for prediction of ^{137}Cs concentrations in forest products consumed by man with a focus on meat from forest game is presented. Some examples of its application for simulation of the radionuclide transfer through forest food chains are given.

2. Description of the FORESTGAME model

2.1 CONCEPTUAL MODEL

FORESTGAME is a dynamic compartment model connected to a model of biomass dynamics. The model is focused on the prediction of ^{137}Cs accumulation by forest game.

I. Linkov and W.R. Schell (eds.), Contaminated Forests, 239–247.

The following assumptions regarding transfer processes of ^{137}Cs migration in the forest ecosystem were made:

(a) The deposits of ^{137}Cs are distributed between the forest canopy (leaves and bark), understorey and forest litter. The deposits retained by leaves and bark are removed from their surfaces due to weathering processes (by wind and rain) and the shedding of leaves and needles;

(b) The radionuclides deposited on the surface of forest litter may penetrate to zones where the mycelia of mushrooms and plant roots are located. The radionuclides located in the litter and root zone are redistributed between fractions of different properties. The radionuclide availability for root uptake decreases with time as a result of its fixation and leaching from horizons where mycelia or roots of plants are located;

(c) The radionuclides existing in available form in the root zone are transferred to the living biomass of wood and to new leaves (needles). Every year ^{137}Cs accumulated in leaves is transferred to the litter creating a fresh litter layer. Two wood compartments, available and unavailable, are distinguished. The available wood corresponds to the living wood, i.e. the active annual ring containing xylem and phloem channels. A continuous exchange between the living tree and the rest of the tree leads to penetration of ^{137}Cs into the inner part of the wood;

(d) The radionuclides existing in available form in the zone where mycelia of mushrooms and roots of plants are located are transferred to mushrooms and other parts of the understorey, which may be consumed by the population and by forest game. The localization of roots and mycelia in the soil strongly influences the uptake of radionuclides from soil by mushrooms and forest plants [3];

(e) Grass, leaves, understorey and mushrooms are consumed by game. The fraction of different types of forest feeds consumed by game depends on the season of the year. The radionuclide concentration in game products depends on the feed consumption rate, on the concentration of radionuclides in the feeds and on the features of the radionuclide metabolism in the organism of the animal.

Each forest management unit is represented in the model by 20 state variables. State variables represent the quantity of radionuclide (Bq m^{-2}) in each specific compartment (Fig 1.). The following compartments are considered: bark, leaf surface, leaf inner parts, wood (available and unavailable pools), litter horizons L-O$_f$ and O$_h$ (available and unavailable pools), soil horizons A and B (available and unavailable pools), bushes, mushrooms, berries, grass (dwarf shrubs, grasses, ferns etc.), muscle tissue of moose and roe deer.

In the model, four categories of forests corresponding to two types of trees and two types of soil are adopted (Table 1). This is based on experimental evidence [4,5] indicating a different behavior of ^{137}Cs in forests with automorphic and semi-hydromorphic soils.

TABLE 1 Classification of forests adopted in the model

	Automorphic soil	Semi-hydromorphic soil
Coniferous forest	CA	CH
Deciduous forest	DA	DH

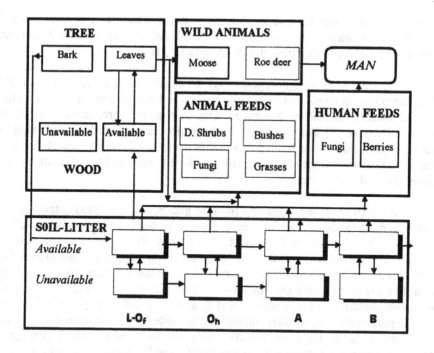

Fig. 1 Conceptual scheme of the FORESTGAME model

An important feature of the model is that it describes the long-term cycles of ^{137}Cs migration in the forest as a composition of many annual cycles. During the annual cycles the radionuclides are partially removed from the soil in the period of plants growth and subsequently a portion is returned to the soil surface during senescence and by weathering processes.

2.2 MATHEMATICAL MODEL

The mathematical formulation of the model corresponds with the so-called linear compartment models. A set of 20 coupled linear differential equations, equation 1, describes the net accumulation of the radionuclide in the compartments over time. The initial distribution of the radionuclides in the compartments is user-defined.

$$\frac{dA_i(t)}{dt} = \sum_{j, i \neq j} k_{ji} A_j - \sum_{n, n \neq i} k_{in} A_i - \lambda A_i, \tag{1}$$

where:

$A_i(t)$ is the total ^{137}Cs activity accumulated in the compartment i at time t (Bq/m^2); $A_j(t)$ is the total ^{137}Cs activity accumulated in the compartment j at time t (Bq/m^2); k_{ji} is the rate constant for the transfer from the j-th compartment to the i-th compartment (day^{-1}); k_{in} is the rate constant for the transfer from the i-th compartment to the n-th compartment (day^{-1}); λ is the ^{137}Cs radioactive decay constant (day^{-1}).

2.2.1 Subdivision of litter-soil layers. One feature of the forest soils is a clear sub-division into an upper, mainly organic horizon and a lower mineral horizon, differing in characteristics such as density, pH, clay content, moisture, nutrient status, biological activity, etc. [3]. For this reason, in the model the soil-litter layer is divided into four sub-layers, each corresponding to a soil horizon: L+Of, Oh, A, and B. Each soil layer is further divided into two compartments for the unavailable and available fractions of the total ^{137}Cs content.

2.2.2 Root uptake and transfer from the understorey vegetation to the soil. Root uptake by mushrooms and understorey vegetation is described as a function of root distribution in soil, available fraction of ^{137}Cs in soil, the soil-to-plant concentration ratio (CR) and the biomass growth rates. It is assumed that during biomass growth there is equilibrium between the amount of available ^{137}Cs in the soil-litter layer and in vegetation. In order to reduce the uncertainties of the estimations, different species of mushrooms were divided into two groups, A and B, according to the location of the mycelia as it was proposed in [4]. A differentiation is also made between mushrooms consumed by man and by game. The fruits of berries (eaten by man) and the whole plant (game feed) are considered as two different compartments. The ^{137}Cs flux of the radionuclide to the j-th type of understorey vegetation (R^j) in units of Bq/m^{2}d is calculated with the following equation:

$$R^j = \sum_i R_i^j = \frac{dA_j}{dt} = CR^j \cdot \frac{dB^j}{dt} \cdot \sum_i \delta_i^j \cdot \frac{A_i}{\rho_i \cdot x_i}, \tag{2}$$

where:

A_j is the quantity of the radionuclide in the j-th type of understorey vegetation at time t (Bq/m^2); A_i is the quantity of the radionuclide in available form in the i-th layer of the soil-litter system at time t (Bq/m^2); R_i^j is the flux of the radionuclide from the i-th soil-litter layer to the j-th type of understorey vegetation (Bq/m^{2}d); CR^j is the soil-to-plant concentration ratio of the radionuclide for the j-th type of understorey vegetation (Bq/kg per Bq/kg in soil expressed in units of fresh weight); ρ_i is the bulk density of the i-th soil-litter layer (kg/m^3); x_i is the depth of the i-th soil-litter layer (m); δ_i^j is the fraction of active roots of the j-th type of understorey vegetation that is located in the i-th soil-litter horizon (in relative units) and B^j is the above ground biomass of the j type of understorey vegetation at time t (kg/m^2).

The CR^j is defined as the ratio of the concentration of the radionuclide in the plant and the concentration of available radionuclide in soil and have, therefore, the same value for all soil-litter layers.

2.2.3 Biomass growth and senescence. The vegetative season of the understorey vegetation is divided into three periods. The first period begins when the growth of the vegetation starts (t_o) and ends when the vegetation biomass has reached a maximum (t_m) and the process of senescence and dying of the plants starts. The plant growth rate (dB/dt) is calculated on a daily basis assuming a logistic growth. The second period goes from the moment of maximum biomass to the end of the vegetative period. An exponential decrease of the biomass is assumed during this period. After each vegetative season there is a phase of physiological rest of the plants, which lasts until the beginning of the next vegetative period. In the model this was formulated as follows:

Period of biomass growth $(t_0 < t = t_m)$

$$\frac{dB^j}{dt} = k_g \cdot \left(\frac{B^j_{max} - B^j)}{B^j_{max}} \right) \qquad B^j(0) = B^j_{min} \tag{3}$$

Period of senescence and physiological rest $(t > t_m)$

$$\frac{dB^j}{dt} = -k_s \cdot B^j , \; if \; B^j > B_{min} ,$$

$$\frac{dB^j}{dt} = 0 , \; if \; B^j \leq B_{min} \tag{4}$$

where,
B_j is the biomass of the j-th type of understorey vegetation (kg/m²); B^j_{max} is the maximum biomass that can be reached by the j-th type of understorey vegetation (kg/m²); B^j_{min} is the minimum biomass of the j-th type of understorey vegetation (kg/m²); k_s is the rate constant of biomass decrease for the j-th type of understorey vegetation (d⁻¹) and k_g is the growth rate constant for the j-th type of understorey vegetation (d⁻¹).

The dynamics of mushrooms biomass is calculated in the same way, but different vegetative periods are considered for summer and autumn mushrooms.

2.2.4 Wild game food chain. The dynamic of radionuclide quantity in muscle tissue of moose and roe deer is simulated with equation 5. Five main components of the game diet were considered: parts of trees, shrubs (bushes), fungi, dwarf-shrubs and grass. The dynamics of radionuclides in these products as well as quantity of feed consumed are critical for the estimation of radionuclide concentration in the meat of game. The diet of game is considered in the model on a monthly basis..

$$\frac{dA^j}{dt} = F^j \cdot N^j \cdot Q^j \cdot \sum_i q_i^j C_i - \left(\lambda_b^j + \lambda \right) \cdot A^j , \tag{5}$$

where:

A^j is the activity level in the animal of j-th species (Bq/m²), N^j is the number of animals of the j-th species per square meter (m⁻²); F^j is the fraction of incorporated ¹³⁷Cs going to the meat for the j-th species of game; Q^j is the total daily food ingestion by the j-th species of game (kg/d), q_i^j is the fraction of the i-th type of animal feed in the diet of the j-th species of game (relative units); C_i is the ¹³⁷Cs activity concentration in the i-th type of animal feed (Bq/kg); $\lambda_b{}^j$ is the rate constant of biological elimination from meat of the j-th species of game (d^{-1}) and λ is the ^{137}Cs rate constant of radioactive decay (d^{-1}).

3. Estimation of the model parameters and validation of the model predictions

A set of values consisting of a best estimate for each category of forest and a general interval of variation for all types of forests was estimated for the model parameters. Some parameter values were found directly in the literature [1-5,7], but for the majority of parameters the values where estimated by model calibration using experimental data and new knowledge gathered during a monitoring program carried out in the Bryansk region of the Russian Federation.

The main approach being used for testing the model is the comparison of the model predictions with independent experimental data. Up to now the model has been tested against the experimental data obtained in Obersschwaben area (Germany) and the Harbo area (Sweden) [7,8]. The results of the comparison indicate that the model simulates correctly the seasonal pattern of ^{137}Cs activity concentrations in roe deer meat. A good agreement between measured and calculated absolute values for all seasons of the year was observed [8].

4. Sensitivity analysis

The model includes many parameters with different values for the types of forests considered. However, a limited number of these parameters influence significantly on the radionuclide levels in meat of game within the hunting period. The sensitivity of the model outputs to variations of input parameters for a period 2 to 30 years after the initial deposition was studied. The average activity concentrations in roe deer and moose meat during the hunting season in central Sweden were the studied model outputs. A sensitivity index (SI) was calculated for each parameter. One example of the obtained results is given in Fig. 2.

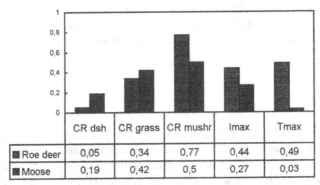

	CR dsh	CR grass	CR mushr	Imax	Tmax
■ Roe deer	0,05	0,34	0,77	0,44	0,49
■ Moose	0,19	0,42	0,5	0,27	0,03

Fig. 2. SI for parameters of ^{137}Cs intake by roe deer and moose (10 years after deposition)

For roe deer the CR of mushrooms and the parameters defining the horizontal (T_{max}) and vertical position (I_{max}) of the peak of mushroom ingestion have the highest SI (0,78). The parameter maximum ingestion rate of mushrooms (I_{max}) has also an important effect on the activity levels in moose meat.

5. Selected results

The model can be used for making estimations that are helpful for evaluating the consequences of a radioactive contamination of the forest and for selecting reasonable countermeasures. Some examples of such estimations are presented below. The dynamics of ^{137}Cs levels in different horizons of the soil-litter layer for a deposition density of 1 kBq m^{-2} is shown in Fig. 3. The corresponding values of ^{137}Cs concentrations in different forest products are presented in Fig. 4. It is clear that the irregular dynamics of "available" amounts of ^{137}Cs in different litter-soil horizons determines the disproportional change of the radionuclide content in different parts of the understorey: berries, mushrooms, bushes, grass, and in game. Fig. 5 illustrates the possibilities of the model to predict seasonal variations of ^{137}Cs content in meat of moose and roe deer. The results presented in Fig. 5 confirm the conclusion that the composition of game diet has a great influence on the dynamics of ^{137}Cs concentrations in meat of game. The presence of fungi in the diet of roe deer results in two peaks of high level of concentration ^{137}Cs in meat of roe deer in summer and in autumn. The contribution of fungi to the diet of moose is less [7] and therefore the level of contamination of the moose meat is lower. The maximal concentrations of ^{137}Cs in meat of moose and roe deer are observed in different months. The decrease of activity levels is more pronounced in roe deer meat than in moose meat.

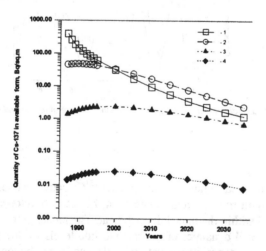

Fig. 3. Dynamics of available forms of ^{137}Cs in different horizons of litter-soil system: 1- forest litter horizons L and O_f; 2- forest litter horizons H; 3- organic horizon of soil (horizon A); 4 - mineral horizon of soil (horizon B)

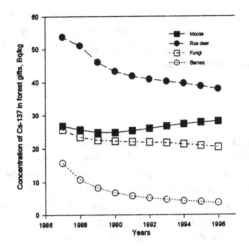

Fig. 4. Dynamics of ^{137}Cs concentrations in forest products (deposition density is 1 kBq m^{-2})

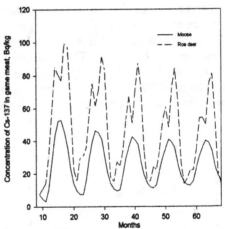

Fig. 5. Seasonal dynamic of ^{137}Cs concentration in muscle of moose and roe deer

6. Conclusion

A model that can be used to interpret and predict ^{137}Cs activity concentrations in forest products, especially in roe deer and moose meat, has been developed. Several new approaches were used within the model development. The model permits evaluation of the long-term and seasonal dynamics of the radionuclide levels in different forest components. Comparisons of experimental data with the model predictions have shown that the model can explain the main regularities of the dynamics of ^{137}Cs levels in forest products.

7. Acknowledgements

This work has been partially supported by SSI contract RYS 6.15.

8. References

1. Avila, R., Moberg, L., Hubbard, L. Modelling of radionuclide migration in forest ecosystems. (1998) SSI report 98:07 ISSN 0282-4434.
2. Schell W. R., Linkov I., Myttenaere C., Morel B. (1996) A dynamic model for evaluating radionuclide distribution in forests from nuclear accidents. Health Physics, 70, 3, 318-335
3. Cheglov A.I. Biogeochemistry of technogenic radionuclides in forest ecosystems in central region of the east-European plain. (1997) Doctoral Thesis. MSU, Moscow. 403 p. Pathway analysis and dose distributions. (1996) Joint Study Project No 5. Final report. EUR 16547 EN, 130 p.
4. Behaviour of radionuclides in natural and semi-natural environments. (1996) Experimental collaboration project No 5. Final report. EUR-16531 EN. 147 p.
5. Ipatyev V.A. et al. Forest and Chernobyl. Forest ecosystems after the Chernobyl NPP accident, 1986-1994. (1994) Minsk, 252 p. (In Russian)
6. Avila, R., Fesenko, S., Spiridonov, S., Johansson, K.J. (1999) FORESTGAME: A dynamic model to predict seasonal and long term changes of ^{137}Cs activity concentrations in forest food chains. (in press)
7. Johanson K.J., Bergstrom R. Radiocaesium transfer to man from moose and roe deer in Sweden. (1994) The Science of Total Environment, 157, pp. 309-316.
8. Lindner G., Drissner J., Herrmann T., Hund M., Zech W., Zibold G., Zimmerer R. Seasonal and regional variations in the transfer of cesium radionuclides from soil to roe deer and plants in a prealpine forest. (1994) The Science of the Total Environment, 157, pp. 189-196.

MAPPING OF RADIOACTIVELY CONTAMINATED TERRITORIES WITH GEOSTATISTICS AND ARTIFICIAL NEURAL NETWORKS

M. KANEVSKI, R. ARUTYUNYAN, L. BOLSHOV, V.DEMYANOV,
S.CHERNOV, E. SAVELIEVA, V. TIMONIN
*Institute of Nuclear Safety (IBRAE), B. Tulskaya 52, 113191 Moscow,
RUSSIA*

M. MAIGNAN, M.F. MAIGNAN
University of Lausanne, 1015 Lausanne, SWITZERLAND

This work presents a brief review of spatial data analysis methods and their application to radioactive contamination of territories. Two methods are described in the paper and applied to real data on soil contamination with strontium 90 (Sr90) and cesium 137 (Cs137) – cokriging model and general regression neural networks. Cokriging is a geostatistical predicator based on multivariate linear regression, which allows inclusion of data on correlated variables in the joint estimation procedure, in order to improve the prediction quality and to reduce estimation errors. General regression neural network is a nonparametric estimator, which is fast and produces high quality results on extremely variable Chernobyl data. Such research with adaptation of these methods to the characterization of radionuclides data has been under progress for 5 years by the senior authors group.

1. Introduction

The present review deals with the description and application of methodology for analysing and modelling of spatially distributed radioecological data. The methodology is based on adapting recent developments in geostatistics and artificial neural networks. Previous studies, concerning radioactively contaminated territories following the Chernobyl accident used a variety of tools and of methods, as published in [1-6]. It was shown that highly variable and spotty contamination at different geographical scales (ranging from some meters to hundreds of kilometres) complicates the analysis, interpretation and presentation of results. Among the important aspects of the study presented here is its multidisciplinary and multivariate nature. Usually, data bases, includes information on contaminated populated sites, contain hundreds of records on different aspects of the problem, both qualitative and quantitative. From different points of view the Chernobyl case study is unique with many lessons to be learned. Selection of the appropriate model depends on the quality and the quantity of the data and of the final objectives of the study.

I. Linkov and W.R. Schell (eds.), Contaminated Forests, 249–256.
© 1999 *Kluwer Academic Publishers. Printed in the Netherlands.*

The basic problems of decision-oriented mapping can be described in short as follows. There are measurements of several variables at different, usually nonregular, points in space. In general, the quality and the quantity of information related to the problems are as follows: different variables differs – there are both "cheap" and "expensive" data. The problem take into account available information and knowledge as follows: 1) estimate/interpolate/predict information/data at points where measurements are not available; 2) draw a map of contamination in the region under study; 3) take into account measurements errors; 4) perform spatial estimations of several correlated variables when the number and the quality of the data differ; 5) do probabilistic mapping; estimate local probability density function and probability of exceeding some definite levels of contamination; 6) describe spatial uncertainty and variability.

There are two methodological approaches for the spatially distributed data treatment: deterministic (e.g., inverse distance squared, triangulations, splines, multiquadric equations) and statistical (e.g., geostatistics, stochastic simulations, fractal interpolations and modeling). Generally speaking one kind of model explicitly takes into account spatial continuity described by different measures – semivariograms, generalized covariance functions, etc. [7]; the other kinds of models do not explicitly take into account spatial continuity.

2. Geostatistics

Geostatistics deals with statistical treatment of data. It is assumed that measured data $z(u)$ are realizations of a random field $Z(u)$. It should be noted that selection of random fields in order to model a regionalized variable is a matter of analytical convenience. This does not imply that the phenomenon under study is indeed fully random. In order to use geostatistics we have to determine structure (covariance function) of this field by using the available data.

The most famous geostatistical model is kriging. Kriging belongs to the class of BLUE (best linear unbiased estimator) or BLUP (best linear unbiased predictor). It means that 1) mean predicted must be equal the real mean (known – simple kriging; unknown – ordinary kriging), and 2) squared error of estimation is minimal. In case of several correlated variables (Z_i) measured at locations x_a the geostatistical prediction at the point (x_0) (co-kriging) is the following:

$$Z^{*}_{i0}(x_0) = \sum_{i=1}^{N}\sum_{a=1}^{n_i}\omega^i_a Z_i(x_a)$$

where the regression weights w_a^i are defined from the following system of equations:

$$\sum\sum\omega^j_\beta\gamma_{ij}(x_\alpha - x_\beta) + \mu_i = \gamma_{ii_0}(x_\alpha - x_0) \quad i = 1,\ldots N, \alpha = 1,\ldots n_i$$

$$\sum_{\beta}^{n_i} \omega_{\beta}^i = \delta_{ii_0} \quad i = 1,...,N$$

where μ_i are Lagrange multipliers; cross-variograms γ_{ij} are defined as the mean value (by i and j) of the product ($\mathbf{h}$ – a separation vector between points):

$$2\gamma_{ij}(\mathbf{h}) = E[(Z_i(\mathbf{x}+\mathbf{h}) - Z_i(\mathbf{x}))(Z_j(\mathbf{x}+\mathbf{h}) - Z_j(\mathbf{x}))]$$

where $E[.]$ is the mathematical expectation operator.

Structural functions – variograms ($i=j$) and cross-variograms – are studied by using original data and then are modeled with theoretical models. Analysis is performed by taking into account anisotropy of structural functions.

For checking the validity of different semivariogram models cross-validation technique is widely used. This approach can be used for inter comparisons between different methods as well. The "leave-one-out" strategy is the basis of cross-validation: we estimate a value of known datum with the help of all other data (known datum is extracted from original data base during estimation). The same procedure is done for all data. The difference between the real and the estimated values gives us a "map of errors". Cross-validation approach does not depend on deterministic or statistical treatment and that it is useful in all cases. By using traditional interpolators with the help of cross-validation we can select the "best" model-dependent parameters (e.g. power and search radius in inverse distance squared model).

Geostatistics provides statistical tools for:

1. Calculating the most accurate predictions, according to well-defined criteria, based on measurements and other related information. It uses the spatial-correlation structures of spatial functions.

2. Quantifying the accuracy of these predictions. Kriging yields a measure for accuracy of its maps in the form of the estimation variance , which can be mapped in conjunction with the estimated values.

3. Selecting the parameters to be measured, and where and when to measure them, if there is an opportunity to collect more data (monitoring networks optimization and design). By studying the variance map, the user can identify the least-accurate regions that should be targeted for further observations.

Thus, as a result of geostatistical analysis, at least two maps can be prepared: 1) map of estimates 2) map of estimated variance, which is an indication of the quality of 1). Combination of these two maps leads to the so-called "thick" isolines – isolines with finite thickness. This kind of isoline describes the uncertainty of decision oriented mapping.

There are many geostatistical models forming a kriging family: simple kriging, ordinary kriging, lognormal kriging, universal kriging, cokriging, moving window kriging, disjunctive kriging, indicator kriging, etc [7]. The indicator kriging is a nonparametric method for the estimations of local probability density functions (pdf). The local pdf can be used in probabilistic mapping (e.g., mapping of probability of exceeding some predefined/intervention levels) and in a consequent cost-benefit analysis based on loss functions.

Recent achievements in geostatistics deal with conditional stochastic simulations and risk mapping [7]. There are two major differences between estimations and simulations: the main objectives of the interpolators are to provide "best" local estimates z(u) of each unsampled value without specific regard to the resulting spatial statistics of the estimates. In case of simulations the resulting global features and statistics (the same first two experimentally found moments – mean and covariance or variogram, as well as the histogram) of the simulated values take precedence over local accuracy. Stochastic simulation is the process of preparing alternative, equally probable, high resolution models of the spatial distribution of z(u). The variable can be categorical, indicating presence or absence of a particular characteristics, or it can be continuous. Kriging, for example, provides a single numerical model, which is the "best" in some local sense. Simulations provide many alternative numerical models $z^l(u)$, each of which is a "good" representation of the reality in some global sense. The difference among these alternative models or realizations provides a measure of joint spatial uncertainty. The application of conditional stochastic simulations to the analysis and modeling of Chernobyl data is presented in [4].

Most of the geostatistical models heavily rely on deep expert analysis (e.g., exploratory variography and modeling of spatial correlation structures) and are based on some theoretical assumptions which rarely can be found in a real world (e.g. second-order spatial stationarity).

3. Artificial neural networks

Artificial neural networks (ANN) are analytical systems that address problems whose solutions have not been explicitly formulated. In this way they contrast to classical computers and computer programs, which are designed to solve problems whose solutions, although they may be extremely complex, have been made explicit. Artificial neural networks consist of numerous, simple processing units (neurons) that we can globally program for computation. We can program or train neural networks to store, recognize, and associatively retrieve patterns; to filter noise from measurement data; to control ill-defined problems. In summary, ANN are designed to estimate sampled functions when we do not know the form of the functions. Unlike statistical estimators, they estimate a function without a mathematical model of how outputs depend on inputs. Neural networks are model-free estimators. They "learn from experience" with numerical and, sometimes, linguistic sample data. It is very important to note that the power of the ANN can be efficiently realized in combination with the other geostatistical methods.

In general, Artificial Neural Networks are a collection of simple computational units (cells) interlinked by a system of connections (synoptic connections). The number of units and connections form a network topology. The number of units can be very large and the connections very complex.

ANN can be superior to other methods under the following conditions: the robustness of ANN is important when the data on which conclusions are to be based are fuzzy (human opinions, ill-defined categories) or are subjected to large errors. The important decision patterns are subtle or deeply hidden. One of the principal advantages of a neural network is its ability to discover patterns in data which are so obscure as to be imperceptible to human researches and standard statistical methods. The data exhibit significant unpredictible nonlinearity.

In the framework of the present work – mapping of radioactively contaminated territories – the basic idea of ANN use is to develop and to train neural networks by using original data/measurements and then to generalize the prediction at the unsampled points. In statistical terminology this is a traditional regression problem. In order to solve this kind of problems the feedforward neural networks (multilayer perceptrons – MLP) and general regression neural networks are widely used [3,5,8]. It was shown that MLP can be efficiently used to model large scale structures and nonlinear trends [5,6]. Hybrid models Neural Networks Residual Kriging/Cokriging – NNRK/NNRCK (ANN+gostatistics) have been proposed as a solution to deal with highly variable data at many scales [5]. The basic idea is to: 1) use ANN to model nonlinear trends, 2) analyze of the residuals, 3) do geostatistical modeling of residuals, 4) obtain final results as a sum of 1) and 3).

Another possibility for spatial data mapping is to use General Regression Neural Networks (GRNN) which are the ANN treatment of the well-known nonparametric Nadaraya-Watson estimators [9]. The basic formula, describing estimate at the point (X,Y), by using measurements Z_i is following:

$$Z_m(X,Y) = \frac{\sum_{i=1}^{n} Z_i \exp(-D_i^2 / 2h^2)}{\sum_{i=1}^{n} \exp(-D_i^2 / 2h^2)} = \sum_{i=1}^{n} \omega_i(X,Y) Z_i$$

where distance between points

$$D_i^2 = (x - x^i)^2 + (y - y^i)^2$$

ANN interpretation of this formula is presented in the Specht's work [9].

The problem during network learning/training by using training data set is to find an unknown smoothing parameter h. In the present study cross-validation was used and the quality of training has been studied by analysing residuals with the help of univariate and spatial statistics. After learning, the network should be tested and validated and then can be used for generalisations, such as making predictions at unsampled points. The smoothing parameter h influences the type of solution.

General regression is based on a well-elaborated mathematical background – multivariate kernel regression methods, which have a long successful statistical

254

history. Finally, it is important to note, that higher moments and variance of the predictions can be also estimated with more elaborated GRNN models.

4. Case Studies

In the present work two case studies based on Chernobyl fallout data are presented.

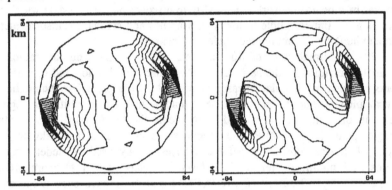

Fig.1. Experimental variogram rose for Cs137 (left) and Sr90 (right).

Geostatistical coestimations (Sr90+Cs137) of soil pollution in the most contaminated Western part of Briansk region. The problem is to map of Sr90 (286 samples) by using additional information on correlated Cs137 (680 samples).

Variogram roses are graphical representation of the variograms for several directions and several lags **h** – separation distances between points. Experimental variogram roses are calculated by using original data. 2D experimental variogram surfaces (isolines of variogram values) calculated by using original data are presented in Fig. 1. Usually, principal directions are used for modeling – fitting of experimental variograms to the known analytical functions. From Fig.1 it is evident that variogram roses of Cs137 and Sr90 have different anisotropic structures.

More dense monitoring network for Cs137 allows a reduced estimation error for coestimation and also enlarge the interpolation are by using more samples of Cs137. The final estimation contours with "thick" isolines and presented in Fig. 2. The boundaries of the "thick" isolines correspond to the values: [estimates ± √(estimation variance)] and reflect uncertainty of the spatial predictions.

GRNN spatial predictions of soil contamination by Chernobyl Cs137 radionuclide in Austria [10]. Part of this data have been extracted for the present scientific study. The optimal bandwidth value h=6 km was chosen by using cross-validation. This value was used for the spatial prediction mapping results of which are presented in Fig. 3. An important stage of the study consist of the analysis of residuals. The statistical analysis show that the residuals can be treated as a white noise. They are not spatially correlated, this corresponds to the so-called nugget effect of variogram model. Thus, all information, described by first and second order moment has been extracted by GRNN.

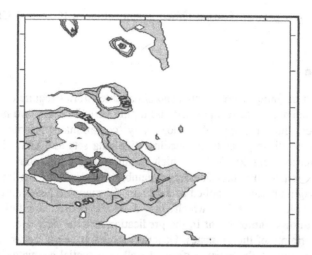

Fig. 2 Coestimations of Sr90. Isolines at 0.5 and 1 (Ci/sq.km) along with their regions of uncertainty.

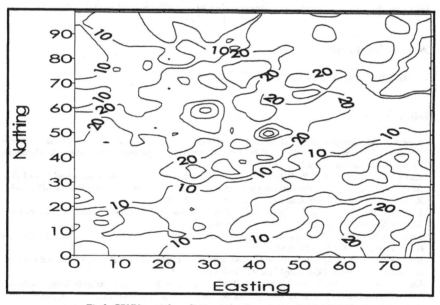

Fig.3. GRNN spatial predictions of soil contamination by Cs137.

All results have been obtained and prepared by using Geostat Office software tools (http://www.ibrae.ac.ru/~mkanev). The Geostat office includes all necessary software needed for the spatial analysis and modeling of 2 dimensional environmental data: from the batch statistical analysis, description of monitoring networks, structural analysis and modeling (variography), geostatistical interpolations and simulations to artificial neural networks predictions and final mapping on Geographical Information

Systems (GIS). At present, Geostat Office is compatible with several GIS including MapInfo GIS.

5. Conclusion

The problem of mapping radioactively contaminated territories require complex multi-method approach due to the complicated and highly variable data. Two main methods were used for the mapping of radioactively contaminated territories: 1) two dimensional geostatistics, adapted to medium size data sets and under hypothesis of spatial stationarity, and 2) GRNN which is an iteration on non-linear distance weighting. Geostatistical approach offers a number of powerful tools for spatial data analysis. Cokriging predictor applied in the paper for mapping Sr90 allows the use of additional samples of Cs137, which has a more dense monitoring network. Coestimation allows improvement in the predication and the decrease the estimation error, i.e. uncertainty of the estimates. General regression neural networks applied to predication mapping in the paper performed well as a spatial estimator, to learn fast and to model the complex spatial structure of the pattern. The case studies are based on two real data sets on surface contamination by the Chernobyl radionuclides.

6. Acknowledgments

The work was supported in part by INTAS grant 96-1957.

7. References

1. Kanevsky, M., Arutyunyan, R., Bolshov, L. et al. (1997). «Geostatistical Portrayal of the Chernobyl Accident». In: E.Y. Baafi, N.A. Shofield (Eds.) Geostatistics Wollongong'96, volume 2. Dordrecht: Kluwer Academic Publishers 1043-1054.
2. Kanevsky, M., Arutyunyan, R., Bolshov, L., et al. «Chernobyl Fallout: Review of Advanced Spatial Data Analysis». In «geoENV I – Geostatistics for Environmental Applications» A. Soares, J. Gomez-Hernandez, R. Froidevaux (Eds.) Kluwer Academic Publishers, 1997, 389-400.
3. Kanevsky, M. (1995). «Use of artificial neural networks for the analysis of radioecological data». News of Russian Academy of Sciences, N.3: 26-33 (in Russian).
4. Kanevsky, M. (1995). «Stochastic simulations of Chernobyl fallouts variability. News of Russian Academy of Sciences», N.3: 47-55 (in Russian)..
5. Kanevsky, M., Arutyunyan R., Bolshov L., Demyanov V., Maignan M. (1996) «Artificial neural networks and spatial estimations of Chernobyl fallout». Geoinformatics. 7: 5-11.
6. Kanevski M., Demyanov V., and Maignan M. (1997) Mapping of Soil Contamination by Using Artificial Neural Networks and Multivariate Geostatistics. In: Artificial Neural Networks ICANN '97. Proceedings W. Gerstner, A. Germond, M. Hasler, J.-D. Nicould (eds.). Lecture Notes in Computer Science, Springer, p. 1125.
7. Goovaerts, P. (1997) «Geostatistics for Natural Resources Evaluation», Oxford University Press.
8. Haykin S. (1994) «Neural Networks. A Comprehensive Foundation». New York, Macmillan College Publishing Co., 696 p.
9. Specht, D. (1991) «A General Regression Neural Network». IEEE Trans. on Neural Networks, 2: 568-576.
10. «Cesium Belastung der Boden Oesterreichs» (1996) Monographien Band 60 Wien, Bundesministerium fuer Umwelt.

MODELLING OF RADIONUCLIDES AND TRACE ELEMENTS IN FOREST ECOSYSTEMS

Report of the Working Group on Modelling[1]

WILLIAM R. SCHELL
Department of Environmental and Occupational Health,
University of Pittsburgh, Pittsburgh, PA 15261, USA
22822 Mariano Drive, Laguna Niguel, CA 92677

YVES THIRY
Belgian Nuclear Research Centre, SCK-CEN,
Boeretang 200, Mol 2400, BELGIUM

EILEEN M. SEYMOUR
Department of Experimental Physics, University College Dublin,
Belfield, Dublin 4, IRELAND

1. Introduction

The main objective of this Working Group was to review and recommend optimal models that could be applied to forest ecosystems to assess the consequences of a potential future accident or routine release of radionuclides. Such models could be used to assist environmental managers and policy regulators in estimating risk to humans and to the environment, to assess the ecological and socioeconomic consequences in guiding policy decisions and to formulate remediation practices. To this end, the Working Group focused on a number of issues including:

- Review of current modelling methods and their underlying databases
- New approaches to modelling
- Identification of areas not addressed by existing models, i.e., compartments, processes, pollutants, etc.
- Recommendations for improving modelling capability.

2. Discussion

2.1. REVIEW OF MODELS

Presentations made by members of the Working Group provided the basis for reviewing modelling methods and their applications. Existing models have been

---

[1] Members of the Working Group: W.R. Schell (Chairperson), Y. Thiry (Rapporteur), E. Seymour (Rapporteur), R. Avila, D. Burmistrov, A. Dvornik, S. Fesenko, V. Golizkov, K. Higley, N. Kaletnik, M. Kanevsky, S. Nalezinski, F. Neves, A. Orlov, A. Sidorov, E. Steinnes, A. Venter, T. Zhuchenko

I. Linkov and W.R. Schell (eds.), Contaminated Forests, 257–261.

expanded to include compartments such as mushrooms, grasses, herbs and berries. These plants form an important part of the food chain as they tend to concentrate radionuclides and are essential components of the diet of wild animals such as roe deer and moose which, in turn, are consumed by man.

Different predictive models dealing mainly with the biogeochemical processes controlling the fluxes of radionuclides between soil and forest ecosystem were presented. Each model is based on independent datasets generally relating to site specific studies. RIFE consists of one equilibrium screening model (RIFEQ) with a fully stochastic capability, a simplified (5 compartment) dynamic model RIFE1 which can be used in stochastic or in deterministic modes and an advanced model RIFE2 (10 compartments) with only deterministic capabilities. Each of these three models were calibrated with field data derived from forest sites in Western Europe. FORESTLAND is a dynamic model also consisting of a set of coupled models which describe the radiocaesium mobilisation during an acute or steady-state contamination phase. Independent models were presented that considered biomass growth, age and type of forest for internal and external dose evaluations. The main transfer processes have been identified using the "interaction matrices" where most of the parameters were evaluated using a large dataset collected in the Bryansk region in Russia. FORESTLIFE and FORESTDOSE are two other complementary predictive models. These models were developed from data collected in the vast contaminated forest areas in Southeast Belarus. FORESTLIFE is a radioecological model focusing on the contamination of wood. FORESTDOSE is a radiation dose model for calculating external exposure in forests and contribution of forest products to the internal dose. The generic model FORESPATH calculates a time series of radionuclide inventories distributed between specified main forest compartments using coupled ordinary differential equations. The model is flexible in that the number of forest compartments and relevant parameters can be specified by the user. The model also can be adjusted for dose calculations. FORESTPATH is currently being operated in both deterministic and stochastic modes. It can be used to evaluate general patterns of radionuclide distribution in the ecosystems with highly uncertain data (a generic application) or in a time-series of measurements for the specific compartments in a more studied ecosystem (a site-specific application). After calibration with data characterising a deciduous forest in Ireland, the results obtained from using a modified version of the FORESPATH model were presented and illustrated with prediction of radiocaesium distribution between various components of oak and individual vegetative species.

Other specific models were presented showing that compartment models can be suitably expanded to include specific key storage reservoirs such as mushrooms, grasses, herbs and berries. These plants form an important part of the food chain as they tend to concentrate radionuclides and are essential components of the diet of man or wild animals such as roe deer and moose. The influence of fungi on radiocaesium concentration in game meat was studied and modelled using the FORESTGAME model. The concentration of ^{137}Cs in the meat of roe deer exhibited two peaks each year corresponding to the occurrence of fungi in both summer and autumn while only one peak was observed in moose tissue whose diet does not depend as strongly on fungi.

This shows not only the importance of considering fungi but also of biomass dynamics and dietary habits of animals. Another site-specific model considered the location in the soil profile of both fungal mycelia and of fine plant roots in the uptake of radiocaesium. The temporal evolution of activity levels in mushrooms reflects the migration of radiocaesium in soil as described in a coupled compartment model. At the soil level, the need to take into account the presence of both labile and non-labile forms of radiocaesium was recognised. An equilibrium model of radiocaesium transfer from forest soil to plant showed that a radiocaesium transfer factor is a linear function of the lability factor which depends on exchangeability, radiocaesium fixation potential and cation composition of the soil solution.

2.2. NEW APPROACHES IN MODELING

Accuracy in predicting contaminant consequences using current models is largely dependent on the abundance of data values and the suitability of estimating input parameters. New mathematical methods for obtaining information from limited experimental data were discussed. Baysian updating methods were presented and applied to specific data sets from contaminated forests in Chernobyl and from the Techa River in the Urals. To try to understand the biogeochemical and hydrological transport of radionuclides, similar modelling methods have been employed. Both discrete and stochastic methods have been proposed for sensitivity analyses related to model structure and parameter uncertainty. Specifically, incorporation of the spatial migration of radionuclides was discussed. Such methods include cellular automata and dispersion where GIS can be employed effectively.

Mapping of radionuclide transport and concentration in the landscape can be approached using methods of GIS, geostatistics and artificial neural networks. This is a stochastic method of probability mapping using Monte Carlo methods in determining uncertainties, variability, risk mapping and artificial intelligence for estimating the best data to be used in predictions. Such methodology can provide better ways both to estimate parameters and to guide more efficient sampling methods in the field.

The biogeochemical processes controlling the transfer of radionuclides were discussed and an approach to model optimization was proposed. Integrated conceptual models have been expanded to include basin-wide studies where interaction among terrestrial, interface and aquatic reservoirs can be evaluated. Such concepts can provide an integrated assessment when connecting compartments can be included.

2.3. LIMITATIONS AND GAPS IN MODEL USE

While a significant database now exists on radiocaesium (and strontium, to a lesser extent) accumulation and modelling in forest ecosystems, much less work has been carried out on determining the processes by which long-lived radionuclides such as plutonium, americium, uranium and iodine interact in this environment. As a result, the understanding of their behaviour in forest ecosystems is limited when compared to what is known about the cycling of these radionuclides in agricultural and marine ecosystems. In addition, the presence of other pollutants such as heavy metals should be taken into account. These may determine the extent of radionuclide migration as they will be competing with each other and with nutrients for participation in processes such as root uptake and organic complexing processes.

Some of the limitations and gaps in our use of models for forest ecosystems include:

- Only limited information in the contaminated zone is available on radionuclide uptake in animals such as rodents, voles, birds, insects, squirrels, deer, wild boar and elk where the transfer of radionuclide can be concentrated.
- Additional work is needed on the bioavailability of radiocaesium (and other pollutants) in soils
- Biomass dynamics and dietary habits of wildlife needs to be taken into account
- Atmospheric input source component is not included in most models
- Models generally are site specific
- Transfer parameters / residence half-times encompass a large number of processes
- Forests are modelled as isolated ecosystems
- Interaction between precipitation, runoff and infiltration either is not correctly modelled or is excluded from most models
- No account is taken of other effects such as fires, chemical releases, soil erosion
- Most important of all is validation of models by independent datasets.

3. Recommendations

The members of the Working Group made a series of specific recommendations, as follows:

- In order to allow other modellers to validate their models, inclusion of site details (i.e., soil physical and chemical characteristics, forest type, etc.) and biomass values plus raw data are needed.
- Further work is needed on bioavailability of radiocaesium, other radionuclides and trace metals in soils.
- The atmospheric source term component is needed in models.
- Make heavy metal analysis as well as radionuclide analysis, if possible, and examine what is known about the behaviour of heavy metals in forest ecosystems as they will also be competing in ecosystem interactions.
- Heavy metals often are more important in the long term because there is no "decay" factor.
- Include in overall models the aquatic and terrestrial ecosystems as well as the interface between the two, i.e., integrated conceptual model.
- Examine the transport and decomposition rate of litter from forest to interface and into aquatic ecosystems.
- Use methods of cellular automata, artificial neural network, GIS, Baysian updating to realise a holistic approach.
- Utilize model-directed sampling in the field to reduce parameter uncertainty.
- Initiate model-model and model-data intercomparison in determining comparable robustness of different models.

Of particular importance is the continuation of model development work in predicting future radionuclide levels and in validating the several models using data from field experiments and measurements in the contaminated forests. The pathways and accumulation processes of several radionuclides and heavy metals should be investigated in addition to ^{137}Cs. Collaborations in field measurements and interpretations should be strongly encouraged amongst the investigators attending this workshop. New measuring instruments should be made available to the local investigators by European and American countries. It is in the interest of the world to continue such studies as the possibility of another nuclear accident is real. The USA terminated most of its radioecological work in 1982 when global fallout from nuclear weapons testing reached a low level and, as we have now learned, this was a short sighted decision.

Finally, the Working Group recommended that NATO and non-NATO countries provide additional financial aid and backing for CIS partner countries to continue their work and to collaborate on further studies of this unique global environmental radioecology laboratory. Arrangements should be made to share crucial data in a major data base. This must, of course, include steps to safeguard intellectual property rights. The first step could be for a major institution to set up a central WEB site from which all relevant papers can be accessed.

Part 3

Remedial Policies and Risk Assessment

EUROPEAN COMMUNITY RESEARCH ACTIVITIES ON FOREST ECOSYSTEMS WITHIN THE 4TH FRAMEWORK OF THE NUCLEAR FISSION SAFETY PROGRAMME

G. VOIGT
GSF-Institute of Radiation Protection, Neuherberg, GERMANY

G. DESMET
European Commission, DG XII, Brussels, BELGIUM

1. Introduction

After the Chernobyl accident, it has been obvious that agriculturally produced foods are not the only important food types which contribute to the radiation dose received by man. Especially the transfer of radionuclides to 'wild food' products (e.g. edible fungi, freshwater fish, game animals) is often much greater than that to agriculturally produced foodstuffs. In addition it has been recognised that these natural food products can contribute significantly to the internal dose. In addition, contamination of forest ecosystems may be sources for considerable external radiation doses to occupational or forestry workers, and to the public using forests for commercial, leisure or food collecting reasons.

Unfortunately, factors and parameters describing and influencing the cycling and fluxes of radionuclides in forests and semi-natural ecosystems have hardly been addressed in pre-accident years. Since the Chernobyl accident, however, different studies have been initiated to increase scientific knowledge and to provide a better understanding of the mechanisms, particularly in the CIS countries (ECP 2, 3 and 5).

Even today, more than 10 years after the accident, increases of radionuclides in certain natural food products have been observed. In general, the ecological half-lives in semi-natural environments are reported to correspond to the physical decay constant of radiocaesium. Therefore, forest ecosystems act as a long-term source for radioactive exposures which have to be considered, especially in regards to the implementation of countermeasures . For these reasons within the 4th framework of the Nuclear Fission Safety Programme of the European Commission, Directorate General DG XII, several projects have been launched to continue or complete already established research activities in this field. In the following, the most important projects launched and funded by EC-DG XII covering forest or semi-natural environments are outlined. This compilation does not include any TACIS, ISTC or any other national funded projects.

I. Linkov and W.R. Schell (eds.), Contaminated Forests, 265–270.

2. EC-DG XII actions on forest ecosystems

An overview of the different EC projects is given in Figure 1 with their acronyms.

2.1 SEMINAT

This project deals with fluxes and long-term dynamics of radionuclides in semi-natural environments. It is co-ordinated by M. Belli from ANPA, Rome, Italy. A detailed introduction into the project is given by her presentation, and is described later in the text and in the contribution of Maria Belli.

2.2 LANDSCAPE

In the project LANDSCAPE, an integrated approach to describing radionuclide flow in semi-natural environments and their exposures to man is taken. The objective of this study is to derive basic criteria for radiation exposures to man from plant and animal food products for different time scales in representative semi-natural ecosystems of Europe. The project is co-ordinated by Leif Moberg SSI (Swedish Radiation Protection Institute) in Stockholm, Sweden.

2.3 FORECO

This INCO-Copernicus project deals with the classification of countermeasures to be applied in forest ecosystems. Therefore, fluxes of radiocaesium are considered using GIS technology. A list of 20 relevant countermeasures have been drafted after the Chernobyl accident in the Chernobyl affected CIS countries of Belarus, Ukraine and Russia. A matrix to integrate basic knowledge such as type of forest soil types etc. with effective practical countermeasures is under construction. Special focus is given to an evaluation of 'ecological quality' which is included in the matrix. In addition external exposures and possible countermeasures are assessed. FORECO is co-ordinated by M. Belli from ANPA (Agencia Nationale Per l'Ambiente) Rome, Italy

2.4 SAVE

The objectives of SAVE (Spatial Analysis of Vulnerable Ecosystems in Europe) are the spatial and dynamic prediction of radiocaesium fluxes into European foods. The major tasks therefore are: to quantify the variation in plant uptake of radiocaesium from major European soil types, to integrate dynamic models of transfer of deposited radiocaesium to food products with spatially varying input information in a GIS format, to produce critical load maps showing deposition levels at which intervention levels will be exceeded for selected food products, to document variation in general dietary habits, with special reference to the intake of radiocaesium, between and within European countries, relevant to radiation protection purposes, and to compare the usefulness of countermeasures with regard to spatially varying parameters. The end product will be designed as a user-friendly system for decision makers which will provide geographically based models in GIS to identify areas in Europe which are vulnerable to radiocaesium contamination, and finally in identifying operative parameters in support of environmental restoration strategies. The project is co-ordinated by Brenda Howard ITE (Institute of Terrestrial Ecology) Grange-Over-Sands, UK.

Fig. 1   Overview of EC DG XII funded projects

 EUROPEAN COMMISSION
DG XII · Science, Research and Development
Directorate F · Energy

SEMINAT
Cycling and long-term dynamics
-litter, organic matter, minerals
Bio-availability and ecological half-lives

LANDSCAPE
Dynamic ecosystems and landscape models
Biological processes and links
Temporal and spatial variation if food intake
(GIS)

FORECO
Cycling and fluxes in forest ecosystems
Countermeasures
Classification, recommendations
External doses

SAVE
Vulnerable ecosystems:
environmental parameters (soil-plant-animal-consumer)
Spatial analysis
Transfer functions and soil mapping (GIS)

EPORA
Catchment and run-off
Chemical pollutants and ecosystem dynamics;
Distribution dynamics

RECOVER
Phytoremediation: Crops for energy production;
Total fluxes of radionuclides
Energy input versus output: economic cost-effectiveness

CESER
Ecosystem dynamics and countermeasures
Duration of side-effects
Strategic impact (GIS)
Direct and indirect costs and benefits to
economic output, environmental quality
and human health

RESTORE/RECLAIM
Dose distribution, land use, living habits
Possibilities, effectiveness and costs of remedial actions;
Correlation between environmental parameters
and transfer functions (GIS); Socio-economic
impact of remedial actions on rural areas;
Practicality and usefulness of remedial actions
in forests and lakes

TEMAS
Scenarios and Elements of intervention:
physical, chemical, environmental, socio-economicparameters
Effectiveness
Side effects

RODOS
Analysis and Prognosis
Countermeasures
Decision support for nuclear emergencies

2.5 EPORA

The project EPORA (Effect of Industrial Pollution On the Distribution Dynamics of Radionuclides in Boreal Understorey Ecosystems) deals with measurement of radionuclides and heavy metals (Cu, Ni, Cd and Cr) in identical samples as a means to describe and analyse the consequences of pollution. The knowledge of the behaviour of ^{137}Cs, ^{90}Sr and $^{239,240}Pu$ in those parts of arctic and boreal understorey ecosystems, where effects of chemical pollution is small or insignificant, provides an important basis for interpretation of observed changes in the more polluted environments. The main objective, therefore, is to study the potential effect of industrial pollution on the migration of radionuclides in soil, on the association of the radionuclides in the different soil constituents, and on transfer of radionuclides from soil to plants. Two sites for these investigations have been selected: Finnish Lapland and the Kola peninsula. The project is co-ordinated by M. Suomela from STUK (Finnish centre for Radiation and Nuclear Safety) Helsinki, Finland

2.6 RECOVER

The Relevancy of Short Rotation Coppice Vegetation for the Remediation of Contaminated Areas (RECOVER) is tested in Europe (Belgium and Sweden) and in Belarus. Short rotation forestry is evaluated for its radioecological and economic alternative for the remediation of contaminated land. Therefore fluxes and cycling of radiocaesium in willow plants, its fate during combustion, energy balances and cost-benefit analysis are assessed. Co-ordinator of this project is Hildegarde Vandenhove SCK-CEN, Mol, Belgium. She is also co-ordinator of the newly launched project PHYTOR which will test the same approach in Ukraine preferentially in flooded sites to prevent remobilization of radiocaesium in these areas.

2.7 CESER

Within the project CESER (Countermeasures: Environmental and Socio-Economic Responses), co-ordinated by C. Salt of the University of Stirling, Scotland, the secondary long-term effects of applied countermeasures (animal based and soil based) are investigated. Two study sites have been selected to investigated different possible economic (e.g. land use change) and environmental (e.g. erosion, nutrient increase) effects: one in Scotland and one in Finland. The ecosystems considered are semi-natural mainly used for grazing sheep and contain, to a certain extent, forests as well. GIS is used to model and demonstrate the processes in space and time.

2.8 RESTORE/RECLAIM

In RESTORE (Restoration Strategies for Radioactive contaminated Ecosystems) an environmental decision support system is being developed which takes into account the fluxes of radiocaesium and radiostrontium in the Chernobyl affected areas (all ecosystems), and other additional radionuclides in the Semipalatinsk test site in Kazakhstan. Consumption behaviours and the contribution of 'natural foodstuffs' to the diet is considered and appropriate countermeasures proposed. Special focus here is self-help of the affected population by providing easy to use information and advice. Cost-benefit analysis of practical and accepted countermeasures are performed. Rehabilitation of contaminated land and its return to economic and ecological use are of special interest, and are addressed in this project. This project is co-ordinated by G. Voigt, GSF Neuherberg Germany; linked to RESTORE is the inco-Copernicus project RECLAIM, co-ordinated by Per Strand, NRPA, Oslo, Norway which has the same objectives as RESTORE and provides information flow from the CIS countries. In addition to the Chernobyl affected countries and Semipalatinsk, this projects deals also with the radioecological consequences of Mayak releases to the river Techa. Within these two projects a 'mushroom leaflet' has been produced which is tested within the Chernobyl affected population, further details are given later in these proceedings under Beresfored et al..

2.9 TEMAS

The project TEMAS (Techniques and Management Strategies for Environmental Restoration and their Ecological Consequences) will provide a conceptual approach appropriate for decision making on countermeasure strategies. Considered are: urban environments, agricultural and grazing systems, secondary effects of interventions (changes of yield and product quality), forest ecosystems (industrial use only), and waste derived from intervention. Different contamination scenarios have been selected to test the developed models and to recommend countermeasures. The project is co-ordinated by Carmen Vazquez from CIEMAT, Madrid, Spain.

2.10 RODOS

A mechanism to achieve a coherent, harmonised and sensitive response to nuclear emergencies is expected to be developed within RODOS, a Real-time On-line Decision Support System. This support system is thought to be generally applicable across the whole of Europe (i.e. all distances, all times) and for the most important radionuclides which might be released during a nuclear accident (more than 60 radionuclides). It is constructed using different modules, the central one being the radioecological model ECOSYS which describes the transfer from deposition to food chains and exposure to man. The system is made in a way that real actual measurements can be on-line and integrated to improve model predictions and dose calculation, and to implement the most effective countermeasures. Co-ordinator of this project which has more than 20 contractors is J. Erhard, FZK (Forschungszentrum Karlsruhe), Karlsruhe, Germany.

3. Summary

In this contribution ten projects dealing with forest ecosystems launched and financially supported by the EC are described with their major objectives and tasks. The co-ordinators are given to provide further information. Meanwhile also contract deliverables and midterm reports are available and/or results are published which can be obtained via the co-ordinators. Most of the projects will be finalised at the mid to end of 1999 and final reports will be then be open for the scientific community.

COUNTERMEASURES FOR RADIOACTIVELY CONTAMINATED FORESTS IN THE RUSSIAN FEDERATION

A. PANFILOV
Federal Forest Service of Russian Federation
199899 Moscow, RUSSIA

A special system of countermeasures to ensure safe radiological management of contaminated forests has been introduced recently in contaminated territories of the Russian Federation. This contamination consists of long lived radionuclides. These territories include 133 forest management units with 350 forestries (forest management sub-units). The area contaminated due to the Chernobyl and East Ural accidents is estimated to be at about 1 and 0.5 million hectares, respectively. About two million hectares have been contaminated in the Altay region and in the Mount Altay Republic due to weapons testing at the Semipalatinsk Testing area. In addition, the Federated forest Service is responsible for forest management of vast territories contaminated by other, smaller scale nuclear accidents.

The radiological situation in contaminated forests changes very slowly. On the one hand, these forests serve as a type of barrier preventing further migration in the environment (Fig. 1). On the other hand, these forests are the source of additional radioactive hazard for the local population and forest personnel. This has resulted in significant socio-economical losses and demands for extra expenditure for restoration of the contaminated areas.

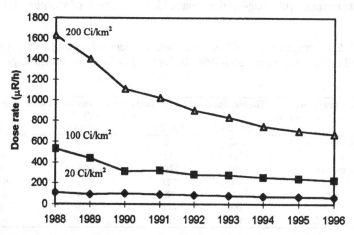

Fig.1. The dynamics of the gamma dose-rate at three stationary key plots (Krasnogorskoe forestry and Bryansk State Body for Forest Management, 150 km from the Chernobyl accident), μR/h

271

I. Linkov and W.R. Schell (eds.), Contaminated Forests, 271–279.
© 1999 Kluwer Academic Publishers. Printed in the Netherlands.

The range of consequences of radioactive contamination of forests is very different in terms of duration, area and intensity of unfavorable impacts on man and ecosystems. Thus the Chernobyl accident has resulted in the ^{137}Cs contamination of 958 thousand hectares of Federal Forests, with 103 forest management units and 319 forestries (Table 1).

Table 1. The area of the Federal forests contaminated by ^{137}Cs due to the Chernobyl accident (thousand hectares, by January 01, 1997)

Administrative unit	Total contaminated area	The distribution of the contaminated areas according to ^{137}Cs deposition in the soil (Ci/km^2)			
		1-5	5-15	15-40	> 40
Bryansk	171.0	103.1	39.7	26.0	2.2
Kaluga	177.8	132.6	43.8	1.4	-
Tula	75.9	64.9	10.9	0.1	-
Orel	74.8	72.7	2.1	-	-
Leningrad region	85.7	85.7	-	-	-
Ruyazan'	70.3	70.2	0.1	-	-
Smolensk	5.0	5.0	-	-	-
Belgorod	15.4	15.4	-	-	-
Voronezh	25.3	25.3	-	-	-
Kursk	21.3	21.3	-	-	-
Lipetsk	15.4	15.4	-	-	-
Tambov	1.7	1.7	-	-	-
Penza	148.4	148.4	-	-	-
Ul'yanovsk	69.4	69.4	-	-	-
Mordovia Republic	1.3	1.3	-	-	-
TOTAL	958.7	832.3	96.7	27.5	2.2

The East Ural Radioactive Trace has contaminated vast areas of Federal Forests with ^{137}Cs and especially [90]Sr (Table 2). At present, almost the entire area of Chelyabinsk and Sverdlovsk (total area 300 thousand hectares) is being studied in terms of radioactive contamination with a wide range of radionuclides. The forests in the Altay region are still contaminated with long-lived radionuclides as a result of weapons testing in the sixties.

TABLE 2. Average content of ^{90}Sr in the compartments of a pine forest in the area of East-Ural radioactive trace (1994-1996. Bq/kg the data by Chelyabinsk Regional Forest Service)

Forest compartments and structural components of pine	The distribution of the contaminated areas according to ^{90}Sr deposition in the soil (Ci/km^2)			
	0.15-3.0	3.0-10.0	10-25	> 25
Mineral soil layer 0-20 cm	363	4023	5459	7105
Forest litter	114	3321	3995	5877
Herbaceous species	296	1241	1395	1646
Wood	87	221	706	no data
Bark	191	773	3157	no data
Branches	168	414	2569	no data
Needles	161	267	1468	no data

The situation is complicated further by the fact that some parts of the forests have been additionally contaminated by a complex mixture of radionuclides due to the long term activities from a range of nuclear installations.

Taking into account the complex dynamics of radionuclide distribution in forest ecosystems, the zones of radioactive contamination in the Federal Forests needs to be revised constantly. To clarify the actual areas of forest contamination, periodic radiological mapping is undertaken by the State Forest Service at 23 Units of the Federation, using standard methods at a scale of 1:100000. For example, it was predicted that 100 years after the Chernobyl accident. the level of contamination in about 10 % of the Plavsk Forest Management Unit will still be sufficient (1-5 Ci/ km²) to warrant Federal social and economic intervention (Table 3).

TABLE 3. Forecast of radioactive deposition in Plavsk forest management unit. Forestries (sub-units) for Plavsk are listed (after Forestry Committee of Tula Region).

Date Forestry	Total area (ha)	Total contaminated area		Deposition 1- 5 Ci/km²		Deposition 5-15 Ci/km²	
		Ha	Sectors	Ha	Sectors	Ha	Sectors
1994							
Plavsk	2259	2259	32	801	15	1458	17
Teplo-Ogarevsk	3282	2024	32	1344	22	680	10
Lipitsk	3927	1729	29	1729	29	-	-
Total:	**9468**	**6012**	**93**	**3874**	**66**	**2138**	**27**
2009							
Plavsk	2259	2209	30	1195	18	1014	12
Teplo-Ogarevsk	3282	1849	26	1386	20	43	6
Lipitsk	3927	1022	17	1022	17	-	-
Total:	**9468**	**5080**	**73**	**3603**	**55**	**1477**	**18**
2024							
Plavsk	2259	1947	27	1465	22	482	5
Teplo-Ogarevsk	3282	1381	21	1318	20	63	1
Lipitsk	3927	328	7	328	7	-	-
Total:	**9468**	**3656**	**55**	**3111**	**49**	**545**	**6**
2039							
Plavsk	2259	1831	25	1831	25	-	-
Teplo-Ogarevsk	3282	1197	17	1197	17	-	-
Lipitsk	3927	-	-	-	-	-	-
Total:	**9468**	**3028**	**42**	**3028**	**42**	-	-
2054							
Plavsk	2259	1491	18	1491	18	-	-
Teplo-Ogarevsk	3282	841	12	841	12	-	-
Lipitsk	3927	-	-	-	-	-	-
Total:	**9468**	**2332**	**30**	**2332**	**30**	-	-
2086							
Plavsk	2259	781	9	781	9	-	-
Teplo-Ogarevsk	3282	116	20	116	20	-	-
Lipitsk	3927	-	-	-	-	-	-
Total:	**9468**	**897**	**29**	**897**	**29**	-	-

It is obvious that the duration of radioactive contamination of some forest area depends primarily on the radionuclide composition of the fallout. Thus, the routine radionuclide releases and accidents at the Mayak installation in the Ural region have resulted in the long-term contamination of natural ecosystems for more than half a century. The forecasted duration of the contamination due to the Chernobyl accident is comparable with the above-said duration.

In contrast, the release of long-lived radionuclides in Tomsk region due to the accident at Siberia Chemical Installation (1993) was practically insignificant. Detailed field studies undertaken in 1993 revealed that the main contributors to the radionuclide composition in this region were ^{95}Zr, ^{95}Nb, and ^{106}Ru.

Immediately after the accident, about 50 thousand hectares of Tugansk Forestry were severely contaminated. Nevertheless, as early as in 1994, the dose rate in this area decreased to within the allowable limits and the total deposition was lower that 1 Ci/km^2 (Table 4).

TABLE 4. Radionuclide content in the soil and average dose-rate in some territories of the Federal forests in Tomsk region at Siberia Chemical Installation (1.5 years after the accident)

Administrative unit, Key plot	Contamination nCi/kg (Bq/kg)				Deposition (Ci/km^2)				Dose rate. µR/h
	Cs-137	Ru-106	Sr-90	Pu-239.240	Cs-137	Ru-106	Sr-90	Pu-239.240	
Tugansk	1.98 (73)	1.08 (40)	0.05 (2)	0.03 (1)	0.4	0.2	<0.1	0.1	11
Egor'evsk-1, Tugansk	1.23 (46)	1.08 (40)	0.03 (1)	0.03 (1)	0.2	0.2	<0.1	0.1	9
Egor'evsk-2, Tmiryas'evsk. Moryakovskoe-3	1.78 (66)	1.08 (40)	0.05 (2)	0.03 (1)	0.3	0.2	<0.1	0.1	11
Timiryas'evsk Bogorodskoe-4	1.12 (42)	1.08 (40)	0.03 (1)	0.03 (1)	0.2	0.2	<0.1	0.1	9
Pervomaiskii Pervomaskoe-5	1.67 (62)	1.08 (40)	0.27 (10)	0.03 (1)	0.3	0.2	<0.1	0.1	9

The problem is that these nuclear accidents have resulted in radioactive contamination of densely populated areas that are extremely dependent on forests for the local economy. The contamination also affects the local sociology and ecology. Therefore it was impossible to terminate forest use in the contaminated areas. At the same time, continuation of routine forest management and use in these areas without special countermeasures would mean a significant increase in the dose rate to the population.

To provide safe radiological forest management in the contaminated areas, the Federal Forest Service has developed and validated a special system of countermeasures. Use of this system makes it possible to significantly diminish the dose to personnel, exclude the use of these forest products where contamination exceeds allowable limits, and provide maximum protection of the forests as a biogeochemical barrier to radionuclide migration from the contaminated areas. The system consists of a network of interdependent protective measures.

The terms "protective measures in forest management" and "forest countermeasures" implies a range of administrative-structural, technological, sanitary, and other measurements and activities intended to diminish or exclude additional irradiation of forestry personnel and the population in areas exposed to radioactive contamination due to nuclear accidents.

The number and type of the countermeasures depend on the physico-chemical properties of the fallout, deposition, type of the forest resource, dose rate of gamma radiation, specifics of forest stand and forest use, regional climate, soil conditions and season.

The protective measures should be adequate for the radiological situation at different stages of the accident. Three main stages are usually distinguished - early, intermediate and final (remedial) with different duration for each stage. We omit the consideration of the early and intermediate stages, since these are rather short compared to the last (remedial) one.

In terms of radioecology, the remedial stage can be subdivided into two periods depending on the specifics of radionuclide migration and distribution in the soil and vegetative cover of the forest ecosystems. The first stage (accumulative) is characterized by intensive radionuclide root uptake from the soil and accumulation in the organs of perennial plants. The second stage (the remedial stage itself) is characterized by the gradual decrease of the radionuclide content in the forest vegetation.

The longevity of the first stage is expected as one half-life of the principal dose forming isotope and the second stage as two half-lives of the same principal dose forming isotope.

There are six major types of countermeasures: administrative, technological, limiting, informational, social, economical and preventive. Administrative countermeasures include:
1. Development of the radiological control service;
2. Mapping of the deposition distribution and development of the radiological classification for the contaminated territory;
3. Radiological monitoring in the contaminated territory;
4. Introduction of the health limits in terms of radionuclide content in the forest resources;
5. Radiological control in the felling areas and other territories used for forest industry purposes.

These measures are mandatory during long periods of time and require essential financial expenditures compared to forests in the non-contaminated areas. The efficiency of these measures is high.

Technological countermeasures are applied to all the zones of radioactive contamination. They include:
1. Regulation of the forest use in different zones of radioactive contamination;
2. Development and use of technologies demanding minimum personnel;
3. Radiological control and introduction of special measures to provide safe working conditions;

4. Radiation control and special measurements for forest-fire prevention under conditions of radioactive contamination;
5. Development and introduction of special standards for regulation of working time and conditions.

The efficiency of the technological countermeasures is determined by the reduction of collective and individual dose burdens to the forest service personnel and the local population. Also, it is important to maintain biological sustainability of the forests and the overall ecological situation. This group of countermeasures is helpful for restoration of the agricultural and industrial activities in the region. It saves jobs and provides conditions for normal forest industrial activity.

Limiting countermeasures are:

1. Temporal prohibition of the forest use at the early and intermediate stages of the accident;
2. Permanent prohibition of the forest use in the severely contaminated zones (^{137}Cs deposition more that 15 Ci/km^2);
3. Prohibition of the use of individual forest resources with radionuclide content exceeding the allowable limits;
4. Regulation of the forest use and processing depending on season and weather conditions.

The limiting countermeasures are used during different stages of the nuclear accident and could be used on a short or long term basis. They are used most often and do not require large additional expense; they are rather effective in terms of reduction of the dose burden to the local population. However, limiting measures cause significant economical damage to the forest industry. For instance, we estimated annual direct losses due the ban of the industrial activities in the five forest management units as 5 -7 million USD.

Informational countermeasures include:

1. Scientific research;
2. Professional education and personnel training;
3. Informational service to the forest managing bodies and local population in terms of radiological situation in the forests.

These countermeasures should be used during all the stages of the accidental situation and complement forest management. Unfortunately, it is difficult to estimate the overall efficiency of informational countermeasures.

Social and economic countermeasures include improvement of working and living conditions and medical service for the forest personnel.

Preventive countermeasures are used in the vicinity of the nuclear installations during their routine exploitation to protect the environment in case of the hypothetical nuclear accident. They include:

1. Forest-planting in the 5-km zone of nuclear installations using radiation-resistant and fire-resistant leafed species;
2. Development of the road network to provide emergency access to all forest territories within the 30-km zone of nuclear installations;
3. Forest classification and licensing of the forest resources in terms of radioecology.

The system of countermeasures is based on the information obtained by the radiation protection division of the Federal Forest Service of the Russian Federation (Fig 2). This information includes data from the periodical radiological mapping and annual monitoring. Radiation monitoring includes annual observation of the radionuclide content of the dominant tree and herbaceous species in permanent control areas. In addition, the radioactive isotope content in mushrooms, berries, and soils is monitored.

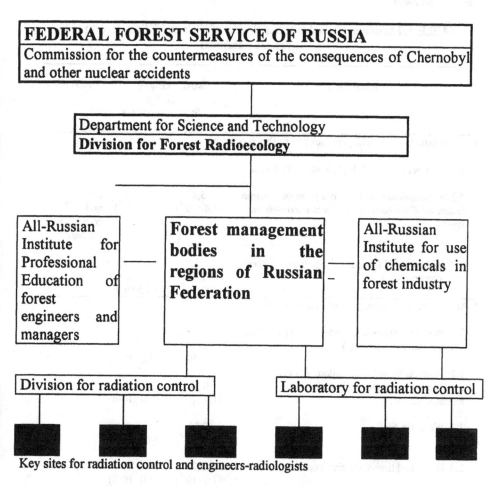

Fig. 2. Structure of the Radiation Protection Service at the Federal Forest Management Bodies of the Russian Federation

In order to protect consumers' rights in the territory of the Russian Federation, the system of mandatory certification (licensing) of live stand and forest products has been introduced (Fig 3). Hygienic standards were established in 1997 (Table 5).

The system of countermeasures presented above was tested on the earlier stages of the nuclear accident of the Siberia Chemical Plant in the Tomsk region in

1993. The application of these countermeasures allowed a reduction in the population dose burdens and a return to the normal forest management in the area of the accident. During the final remedial stage of the nuclear accident this system of countermeasures is applied in the areas of radioactive contamination from Chernobyl and the southern Ural nuclear accidents.

TABLE 5. Maximum allowable levels (MAL) of ^{137}Cs and ^{90}Sr in forest products (hygienic standards)

#	PRODUCT	MAL, kBq/kg (Ci/kg)		NOTES
		^{90}Sr	^{137}Cs	
	1. Alive stand for industrial purposes and further processing			
	1.1. Non-barked wood (logs) at the felling area			
	* *The transportation of the logs from felling area is banned if the bark contamination exceeds the limits*	5,2 $(1,4 \cdot 10^{-7})$	11,1 $(3 \cdot 10^{-7})$	for Cs^{137} and Sr^{90} content in the bark*
	1.2. Barked wood for industrial purposes (except civil construction)	2,3 $(6,2 \cdot 10^{-8})$	3,1 $(8,5 \cdot 10\text{-}8)$	
	2. Alive stand and timber products for civil industry			
	2.1. Products intended for outdoor use (industrial and agricultural tools, etc.)	2,3 $(6,2 \cdot 10^{-8})$	3,1 $(8,5 \cdot 10\text{-}8)$	
	2.2. Products intended for indoor and personal use (furniture, parquet timber, musical instruments, etc.)	0,52 $(1,4 \cdot 10^{-8})$	2,2 $(6,0 \cdot 10\text{-}8)$	
	2.3. Wood for fuel	0,37 $(1,0 \cdot 10^{-8})$	1,4 $(4,0 \cdot 10^{-8})$	
	2.4. Wood and timber for civil construction	5,2 $(1,4 \cdot 10^{-7})$	0,37 $(1,0 \cdot 10\text{-}8)$	
	3. Tree and shrub seeds for forest management utilization	2,5 $(7,0 \cdot 10^{-8})$	7,4 $(2,0 \cdot 10\text{-}7)$	
	4. Fresh phytomass for making green fodders	0,1 $(3,8. \ 10^{-9})$	0,6 $(1,6 \cdot 10^{-8})$	

The system for certification of wood and timber and secondary forest resources to limit ^{137}Cs and ^{90}Sr in the forest products

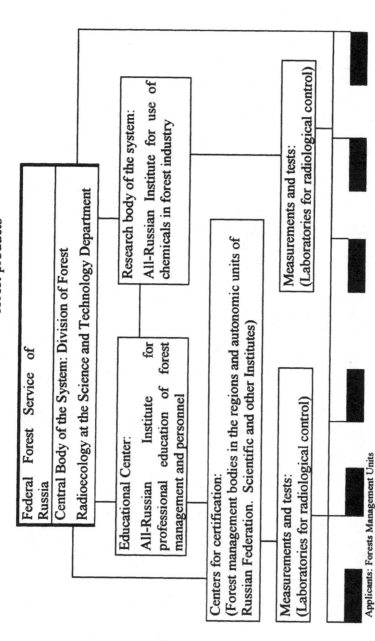

Fig. 3. The system for certification of wood, timber, and secondary forest resources in order to limit ^{137}Cs and ^{90}Sr in forest products

FORESTRY MANAGEMENT IN CONDITIONS OF RADIOACTIVE CONTAMINATION: EXPERIENCE AND PROBLEMS

M. M. KALETNYK
State Forestry Committee of Ukraine
5, Kreshchatik str., Kyiv, 252601, GSP; UKRAINE

This is the second decade that Ukrainian forestry has born the influence of the large-scale environmental catastrophe caused by the Chernobyl nuclear power plant accident. During this period, foresters have accumulated valuable knowledge of forest management in conditions of radioactive contamination, special research groups have been created to study the problem. Despite finding solutions to most issues, there are still many serious social and environmental consequences requiring attention. The solution to this growing range of problems will be more complex as time goes on.

Many sectors of the Ukrainian economy suffered from the accident. Forests, one of the most radioactive sensitive natural environments, continues to be under negative influence of the accident legacy, while the forestry suffers economical losses. The contamination is distributed over the territory of Ukrainian Polessie, where an essential part of wood growing stock of Ukraine is allocated. There is about 3.5 millions hectares of pine stands, having from high to average productivity, each with mushrooms, berries and medicinal herbs. Unfortunately, until now their use has been severely limited due to the contamination.

Besides the above mentioned losses, we need to remember the number of forestry enterprises that were closed. Every year more than 200 thousands hectares of forested lands, where the contamination level exceeds 15 curie per square kilometer, are excluded from forestry use. Every year our loss of timber is more than 300 thousand cubic meters. We should also not forget the loss of other forest uses.

The first organizational challenge was concerned with the radiological monitoring of forests. In the first two weeks after the Chernobyl accident this function was undertaken by a number of dosimeter stations (25), distributed over several (7) administrative regions of the Ukraine. It enabled operative evaluation of the contamination scale and zoning realization, as well as planning of forestry enterprises and evacuation. However, the dosimeter stations' data can not supply forestry management with accurate spatial resolution data on radioactive contamination. Thus, specific survey and mapping methods were needed. They were developed by utilizing experience gained in the former Soviet Union and adopted to the Ukrainian conditions. Such mapping of contaminated regions were used to explain radiation dose. It is clear that the method does not allow control monitoring for forest products, but does provide radiation safety information for forest workers.

I. Linkov and W.R. Schell (eds.), Contaminated Forests, 281–287.
© 1999 *Kluwer Academic Publishers. Printed in the Netherlands.*

TABLE 1. The area of Ukrainians forests with various levels of $^{134+137}$Cs pollution in regional state forest management's (condition on 01.01.93)

	Association	Area of forests, kha	Including on levels of pollution (Ci/км^2)							
			>1.0	1.1-2.0	2.1-5.0	5.1-10.0	10.1-15.0	15.1-40.0	40.1-80.0	>80.0
1	Vinnitsales	216.2	185.1	23.8	6.8	0.5				
2	Volynles	178.4	136.2	36.9	5.3					
3	Gitomyrles	732.3	292.4	185.2	158.3	50.3	16.4	27.0	4.8	0.6
4	Kievles	372.3	178.0	129.3	38.1	13.0	5.5	4.2	2.6	1.5
5	Rovnoles	671.5	293.6	215.3	151.6	10.7	0.3			
6	Sumyles	121.9	109.4	8.0	4.5					
7	Chercassyles	215.1	176.0	31.1	7.3	0.6	0.04			
8	Chernigovles	348.5	273.8	47.4	23.1	3.3	0.9	0.06		
9	Donetskles	16.0	13.1	2.9						
10	Dnepropetrovskles	19.5	19.5							
11	Kirovogradles	26.0	25.3	0.7						
12	Krymles	19.8	19.8							
13	Luganskles	26.3	25.3	0.9	0.1					
14	Nicolaevles	6.6	6.6							
15	Odesales	44.8	42.1	2.7						
16	Ternopolles	64.8	56.4	7.5	0.9					
17	Kharkovles	56.4	56.4							
18	Khmelnitskles	50.0	46.1	3.1	0.8					
	All of State committee of forestry of Ukraine	3186.4	1955.1	692.2	396.8	78.4	23.1	31.3	7.4	2.1
	Boiarskaya LOS	18.1	16.42	1.65	0.03					

Today our forestry enterprises rely on available maps, even though only one sample per 100 hectares was analyzed for Cesium-137 contents, one sample per 600 hectares correspondingly for Strontium-90, and one per 6000 hectares for Plutonium (Table 1).

The practice has been shown to not have enough spatial resolution even on Cesium for the radiation safety maintenance. The survey technique gives no more than one sample per forest quarter while the actual spatial variability of the contamination levels is much higher; an area of the forest management unit or, forest inventory parcel, is much smaller than one quarter. One forest quarter can contain up to 50 forest inventory parcels and as our research has demonstrated, different parcels of one quarter can be related to various zones in accordance with the forestry radiation safety regulations. It is evident that a more spatially detailed and more expensive survey is needed. Taking into account the extremely difficult economical situation for the Ukrainian forestry, now we can state that this is a real obstacle to accomplishing this job.

Another challenge is the prevention of radioactive releases. According to Ukrainian law "about the radiation safety regime of a territory, which has undergone radioactive contamination owing to the Chernobyl accident," all the production that is made in boundaries of the territory, with contamination density above 1 Curie per square kilometer, is subject to radiological control. Accordingly, within organizational structure of the Ministry of Forestry of Ukraine, the service of the radioactive control monitoring was established. Ten specialized radiological laboratories were created in the most affected regional forestry departments of the Ministry. Their main task is to carry out control monitoring of the whole spectrum of the forestry products. This is carried out at stages of forest cultivation, i.e. before harvesting, the stage of processing raw material and realization of ready products.

Analysis of the control monitoring data (Table 2) shows that the non-timber forest products, namely game fauna, mushrooms, berries, medicinal raw material, and hay, have high levels of radioactivity. These provide significant contribution to the collective effective dose of the population. The existing spectrometry equipment base enables only partial solution of the problem.

Scientific research supporting radiological forestry safety, is focused on development of survey techniques for radionuclide distribution mapping and investigation of radionuclide migration in the forest environment. Research on a representative set of permanent forest plots led us to ascertain that during the past 10 years since the accident, the main amount of radioactivity is concentrated in the topmost 2 centimeters layer of the forest soil. The values vary in range depending on forested land type and forest stand's age and species composition (Figure 1). The geographical feature of the Ukrainian Polessie is that the topmost layer of soil contains mainly the trees' root mass. It is easy to predict that we may not expect forest trees to clear themselves of radioactivity in the near future.

The change of radiocesium concentration in the forest vegetation can be arranged into the following decreasing sequence: shoots and leaves or needles of the current year, leaves or needles of the last year, branches, bark, and timber (Table 3). The timber has the least cesium content, however the process of radioactivity accumulation in the timber is not yet stabilized. As we can see (Fig. 2) Cesium content increases with the current year timber gain. We have to continue to improve our long-term research program to enable reliable predictions of future radioactive contamination of forestry products. In the new edition of the forestry management recommendations, prepared for publishing, last year's research results are summarized in the form of practical guide for radiation-safe forest management.

We have briefly outlined forestry management problems in Ukraine imposed by the legacy of the Chernobyl accident in 1986. The limited time for the presentation on the one hand and the complexity of radiation safety management in forestry on the other allows only a short review. However, we hope that it is appropriate to begin fruitful conversations at the symposium.

We would like to summarize the topics that have been discussed here.
First. We require better spatial resolution of map for radionuclide contamination in 55 forest product enterprises. These goals can be accomplished by funding for more measurement equipment to be used in extensive radionuclide monitoring surveys.
Second. To provide effective radiological control monitoring of non-timber forest's products we need improvements in our spectrometry equipment base.

TABLE 2. Generalized results of radiometric control a product of timber facilities on State Forestry Committee of Ukraine for 1997

Types to product	Amount samples		
	In all	Exceeding permissible level	%%
Softwood with bark	7460	3	0.04
Softwood	2291		
Bark	491	2	0.4
Timber products	2809		
Tinned foods	773		
Medicinal herbs	1943	332	17.0
Agricultural production	865	7	0.8
Hays. green provender	1640	230	14.0
Seed of trees & bushes sorts	147		
Coniferous paws	885	16	1.8
Coniferous-vitamin flour	12		
Product of wood manufacture	162		
Technological wood-chips	20		
Firewoods	992	25	2.5
Meat wild bestial and birds	126	36	28.6
Honeys	85		
Milks	53		
Juice a birch	294		
Fruits. Vegetables	118		
Fruits and berries	1608	255	15.8
Mushrooms fresh	2174	875	40.3
Mushrooms dried	91	71	^8.0
Mushrooms pickled	56	15	26.8
Fish	1		
Ashes of firewood	111		
Soil	3268		
Litter	318		
Wood shoot	576		
Juice a conifer	56		
Leaf. pine-needles	353		
Drinking water	40		
Beef. Pork	2		
Coniferous extracts	3		
Charcoals	1		
Buds of trees	3		
Mining upright	4		
New year's fir trees	247		
Wood sawdust	34		
Other product	148	8	5.4
THE WHOLE	30260	1875	

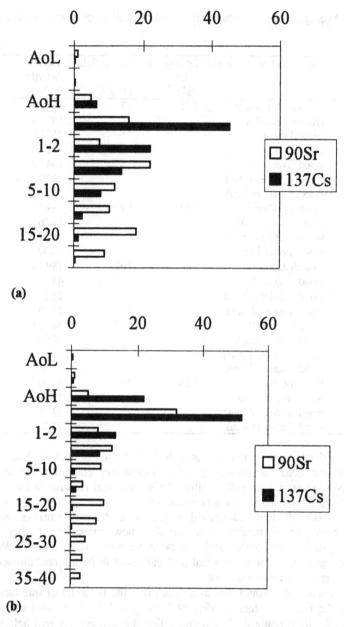

Fig. 1 Distribution of radionuclides into soil in mixed (a) and in pine (b) forests.

TABLE 3. Average factors of transition ^{137}Cs from soil in overhead phytomass wood plants

Species	Component of phytomass	Density of soil pollution ^{137}Cs Ci/km^2	kBq/m^2	Specific Activity Bq/kg
Pinus syl.	Annual pine-needles	2.25	83	898.8
	Old pine-needles			592.58
	Annual shoots			2582.64
	Old shoots			702.24
	wood. Top of trunk			104.73
	wood. middle of trunk			96.82
	wood. Lower of trunk			164.31
Betula pen.	Branch & shoots	2.63	97	144.07
	wood. Top of trunk			17.76
	wood. middle of trunk			24.22
	wood. Lower of trunk			19.49
Quercus rob.	Branch & shoots	2.3	102	189.15
	wood. Top of trunk			43.5
	wood. middle of trunk			23.14
	wood. Lower of trunk			21.48
Populus tre.	Branch & shoots	2.77	104	465.64
	wood. Top of trunk			74.09
	wood. middle of trunk			62.76
	wood. Lower of trunk			137.07
Alnus glu.	Branch & shoots	1.58	59	69.08
	wood. Top of trunk			19.73
	wood. middle of trunk			39.32
	wood. Lower of trunk			72.20

Third. We understand the value of research on behavior of radionuclides in forest environment and need for projecting future radioactive contamination of forestry products. We understand, as well as value, the international research and development cooperation in the field and welcome collaboration.

For more efficient radiological protection of forestry workers, we should reevaluate conventional methods and employ new management methods and silviculture systems that comply with the radiation safety requirements. We should improve management of forest fires and pest outbreaks in highly contaminated forest stands left without silviculture treatment.

In conclusion I would like to mention that the potential of international cooperation in the field of radiation safety of forestry and the Chernobyl consequences mitigation is far from realized. We do hope that this symposium will help us find a common interest and make positive moves toward research and design cooperation. We are also expecting better support from of the European Commission.

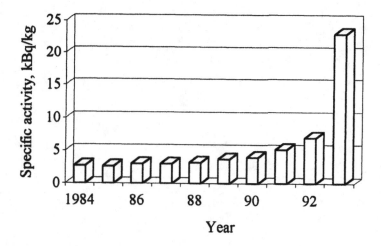

Fig.2. Specific activity of a current increment on height of Scotch pine in damp subor in a 10.01-15.00 Ci/km zone

BIOMASS-INTO-ENERGY OPTIONS FOR CONTAMINATED FOREST IN BELARUS AND RELEVANT RISK ISSUES

A. J. GREBENKOV, V. V. RYMKEVICH, A. P. YAKUSHAU
Institute of Power Engineering Problems, Sosny, Minsk, BELARUS

J. ROED, K. ANDERSEN
RISØ National Laboratory, Roskilde, DENMARK

T. M. SULLIVAN
Brookhaven National Laboratory, Upton, N.Y., USA

D. D. BREKKE
Sandia National Laboratories, Livermore, CA, USA

1. Introduction

The Chernobyl Accident caused serious impacts to the social well being of the population residing in contaminated regions of Belarus. Furthermore, the radioactive fallout has resulted in severe losses of economic values such as once productive forested and agricultural lands. In the aftermath of Chernobyl, these regions have suffered unemployment, inactive local industries, inefficiency of restrictions and a difficult demographic situation.

The greatest risk to human health is caused by the fact that the forest in the contaminated regions remains open to recreational and industrial uses by the general public and wood processing enterprises. Moreover, firewood from affected forest and woody waste after timber processing are extensively burnt in the municipal boilers and domestic stoves. This results in significant uncontrolled secondary contamination of "clean" regions and additional damage to the population's health.

Although the mitigation of harmful levels of radiation doses to the general population is the principal aim of implementing countermeasures; the possible options are largely dependent on economic issues. An additional economic benefit could be found on a basis of introduction of new profitable activities obtained from production of clean bioenergy using contaminated biomass. Consequently, the cost of averting the collective (population) dose could be offset by the economic return of the countermeasure option.

The energy potential of contaminated woody waste collected from Chernobyl forest by implementing conventional forestry is approximately 0.3 Mtce/year. However,

I. Linkov and W.R. Schell (eds.), Contaminated Forests, 289–301.

most of this biomass can not be utilized as a fuel without applying special environmentally sound technology. Positive economic justification is also required for such an activity. Thus, the goal of the proposed action is to provide a dual benefit in the remediation of the Chernobyl-affected area:

i.   In the near term, help reduce the health risks to the population. In the long term, the action will contribute to decontaminating and cleansing the affected land.

ii.   Provide sustainable power production and jobs to help stabilize the economic situation.

The proposed action for Belarus would establish a set of countermeasures to decontaminate the forest floor, introduce special forestry practice, and construct the appropriate biomass-fired power facilities. Ultimately, such power plants would be a reliable consumer of contaminated woody waste and forest litter, thus providing sustainable management of contaminated forest. In addition, the short rotation energy coppice technologies seem to be feasible for reuse of decontaminated areas that may add about 0.2 Mtce per year. This also would yield a certain decontamination effect of up to 1% of source terms removed during each harvesting period.

The present study evaluates the feasibility of energy production from identified contaminated biomass resources and potential resulting risks to forestry personnel, power plant workers, and the general public. The study is being conducted under concerted international projects involving research institutions from Belarus, Denmark and the USA.

2. Proposed Likely Remediation Strategies

There are several approaches to reclaim the land in contaminated forests. These can be grouped into two main options:

A. Do-nothing option. The implications of not instituting countermeasures are the following:

- No exposure to forestry personnel;
- Significant exposure to residents;
- Deterioration of forest biocenosis;
- Worse sanitary conditions for wood stand;
- High probability of fire;
- Loss of ripe commercial forests;
- Uncontrollable radionuclide expansion; and
- Long-term negative environmental impact.

B. Implement environmental countermeasures. This would consist of placing restrictions on the contaminated forest in the zone >40 Ci/km^2, and decontamination and proper use in the zone <40 Ci/km^2. This would lead to the following:

- Gradual mitigation of exposure to personnel and residents;
- Appropriate sanitary conditions;
- Minimum probability and consequence of fire;

- Reuse of forestland;
- Valuable commercial wood;
- Fuel and energy from biomass;
- Pulp & paper, other clean products;
- Employment and other social benefits;
- Radioactive waste treatment; and
- Radioactive waste disposal.

Feasible countermeasures and remediation strategies for operational forest are discussed below.

2.1. DECONTAMINATION OF FOREST FLOOR

In most cases, removing the top layers of soil (i.e. 4-5 cm of fresh and decomposed litter and a few centimeters of duff) results in removal of most of the radiological source terms (70 to 90%). According to modeling studies and actual experiments carried out in Gomel Province of Belarus, the decontamination factor could reach 3.5 [1]. After removing the top layer of forest floor material, reforestation or short rotation coppices must be provided to stabilize and prevent erosion of bare soil.

Conversely, it is a matter of common knowledge that forest litter contains a large fraction of the available nutrients for a wood stand. Therefore, this procedure requires a case-by-case evaluation in terms of the effects on forest health. The optimal scenario may be to remove the forest litter prior to harvesting the wood product (thinning or clear-cutting). At that, the felled trees would be left within the cutting area for the summer to dry. This is necessary in order to reduce the moisture content of wood, and causing the needles to fall. These needles and other debris from forestry operations are to re-establish a new litter layer, since most of the nutrients reside in the crown of trees.

This method would require an analysis in every case to assess efficiency, taking into account the amount of waste generated, thickness of humus, possibility of soil erosion, influence on bio-diversity, and availability of appropriate techniques. The utilization of the forest floor matter in a boiler could provide the appropriate solution for both reduction of waste volume and reduction of the relevant cost of decontamination.

2.2. CONTAMINATED FOREST MANAGEMENT [2]

Thinning in the zone of 15-40 Ci/km^2 should be used for such stands that can tolerate a very heavy thinning. The litter layer in these stands would be removed. The extent of thinning would be heavy enough to allow the stand to remain unattended for a period of 20 to 30 years. In that period, the contaminants will decline by natural radioactive decay and make the wood suitable for application to purposes other than energy.

Only on the clear cuts around the villages should the litter and duff layers be removed, if the radioactivity has not yet reached the mineral soil. For all other clear cuts, the area should be deep ploughed to bury the radionuclides under a layer of 50-60 cm of soil. In order to facilitate the deep ploughing, it may be necessary to remove tree stumps. These stumps would be crushed and used for energy as well.

Reforestation should be applied after clear-cutting with as low density of plantation as possible (1-3 thousand trees per hectare for pine). This will make the first thinning no sooner than 20-30 years after planting, allowing a significant portion of the ^{137}Cs and ^{90}Sr to decrease naturally.

2.3. CONTAMINATED LOG PRETREATMENT

If the contamination level of forest soil is less than 5 Ci/km^2, there is a minimal risk in harvesting the wood, and the resulting wood material may be allowed unrestricted use in any type of industry. In the zone of contamination with a level of 5-15 Ci/km^2, raw wood after felling requires bark stripping at a lumber processing plant or *in situ*, which would remove approximately 7% of the biomass and 60-70% of the radioactivity. At this point, valuable wood trunks received from this zone could be used without any limitation. In the zone of 15-40 Ci/km^2, the quality control of wood must be provided. At this level, even after stripping the bark, valuable wood trunks would not be recommended for direct use but could be sawed into the beams. Phloem layers of 2-3 cm thick would be stripped too, so the average size of square beam would not exceed 70% of stem diameter.

Nevertheless the wood processed would still contain not less than 30-40% of activity, which may not be in compliance with permissible limit of any wood processing industry. This wood would undergo yet another decontamination step (especially if wood is delivered to the processing plant with a high enrichment factor). Alternatively, this wood along with forestry debris, litter and duff from forest and woody waste from lumber processing industries is to be burned into energy through controlled combustion process. This could provide appropriate control of further fate of radionuclides and give the clean valuable product in form of heat and power.

3. Bioenergy option

At present, the Republic of Belarus suffers from an energy deficit due to a lack of indigenous resources. In addition, more than 60% of the power plants operated with natural gas and fuel oil have reached the end of their designed lifetime and should be rehabilitated. This power could be replaced with energy of bio-resources that would also provide a pathway for remediation of radiologically affected territories. Belarus is ideally suited for bioenergy production due to the large area of productive industrial forest, flat landscape, well-developed power distribution and district heating infrastructure. In contaminated areas, the energy potential derived from technically available woody waste, amounts to approximately 25 PJ/year. Our proposed action is construction and operation of biomass-fired power facilities that would provide sustainable power production from identified biomass resources.

The incineration with recuperation of thermal power (co-generation) is considered as one of the most feasible and economically sound technologies for radioactive wood treatment. Several traditional technologies of vaporization of biomass are under consideration, i.e. (i) grate-based systems; (ii) bubbling fluid beds; (iii) circulating fluid beds; (iv) suspension-fired systems; and (v) gasification.

Each option should be evaluated in order to provide the data on ecological parameters and economical efficiency. In 1997-1998 several pilot- and demonstration-scale tests were carried out by IPEP (Institute of Power Energy Problems) and SNLs (Sandia National Laboratories) to help address these issues.

3.1. REVIEW OF THE RESULTS OF TEST BURN [2, 3, 4]

Last year (1997) the equipment for gasification of radioactive woody waste with capacity of 100 kg (dw) per hour was developed at IPEP. Scientific research was conducted to study the isotopic composition of the initial raw materials, activity distribution during wood thermal treatment and the determination of activity fraction in ash and flue gas. The results show that the ^{137}Cs concentration in flue gas before passing through the emission release control system, at the temperature ranges up to 750 °C does not exceed 10%. Studies include the influence of temperature regime on remnant tar substances, degree of carbon content, characteristics and composition of gas produced and its purification, control and monitoring systems. At the inverse process, the gas was obtained with the parameters close to the gas used in the define ICE-cycle. The gas produced had caloricity of about 5.5 MJ/m^3 and traces of ^{137}Cs content.

In April and May 1997 the full scale test burn of litter and duff along with wood fuel was prepared and carried out at the Wheelabrator's Shasta Power Plant in Anderson, California, USA. During the test, boiler #1, with a maximum rating of 20 MW(e) equipped with separate electro-static precipitators (ESP), was fired. The primary objective of the test was to demonstrate that forest litter and duff could be burnt on a commercial scale. A secondary objective was to evaluate stack emissions and ash composition during the test.

About 70 tons of litter and duff were collected in Northern California. The composition of the combustion materials was similar to actual composition of litter, duff and wood sampled in the Chernobyl zone in Belarus. Two combinations, i.e. 20% and 50% of forest floor matter fed together with regular wood fuel, were burned. To simulate the ^{137}Cs and ^{90}Sr radionuclides, aqueous solutions of stable Cs and Sr (in form of cesium sulfate and strontium nitrate) were injected into the fuel feeder. Then the content of the above elements and total hydrocarbons were measured in bottom and fly ash and in the exhaust gas downstream of the ESP. A schematic view of the boiler and sampling points is presented in Figure 1. In addition, the pilot-scale tests have been performed at the Combustion Research Facility (CRF) at Sandia National Laboratories in Livermore, California. These laboratory studies were performed using a pilot-scale, down-fired facility of 100 kW(th). Preliminary results of the test burns at Shasta Power Plant and studies at CRF/SNLs are as follow [4]:

- A boiler of conventional type can effectively vaporize the materials investigated;
- The volatile radionuclides are condensed onto particles in the 0.3-0.5 μm range;
- The content of Cs in fly ash is 4-7 times as much compared to that in bottom ash;
- Cs distribution among different ash streams is defined by temperature in the combustion chamber and sulfur and chlorine content of the fuel;
- Capture efficiencies for Cs and Sr in a stoker-grate biomass-fired boiler with an ESP are 99.98 and 99.93%, respectively;
- A baghouse filter with advanced technologies for bag materials, e.g. Gore-Tex™ (or NOMEX™) may provide considerably improved efficiency for submicron-particle capture compared to ESP.

In the summer of 1998, it is planed to verify these conclusions during the real trials at an existing industrial boiler situated in the contaminated region of Belarus.

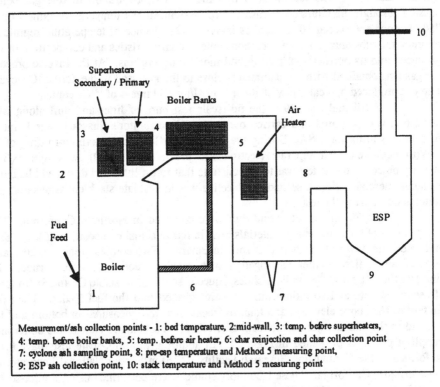

Measurement/ash collection points - 1: bed temperature, 2:mid-wall, 3: temp. before superheaters, 4: temp. before boiler banks, 5: temp. before air heater, 6: char reinjection and char collection point 7: cyclone ash sampling point, 8: pre-esp temperature and Method 5 measuring point, 9: ESP ash collection point, 10: stack temperature and Method 5 measuring point

Fig. 1. Scheme and sampling points at the Shasta Energy Plant

3.2. RESULTS OF ECONOMIC AND SYSTEM EVALUATION

The ENPEP/WASP-III code was used by IPEP to define a change of energy quantity and cost at different stages of power generation, conversion and consumption in the Belarus energy sector, with a focus on utilization of contaminated wood resources.

The country's local and regional energy systems are presented in diagram form, which shows the running of fuel and energy flows and their re-distribution among separate sectors of the economy. This database was used to calculate the fuel supply and energy demand on a mass balance approach. The calculation of energy flows and prices for all the elements of the diagram was carried out. Comparison of the different technologies for utilization of contaminated woody waste as a fuel for heat and electricity generation was performed. Results of the analysis show that the most feasible and cost effective option is related to power generation and co-generation, that would assume achievement of an internal rate of return of about 30-40% (Figures 2 and 3).

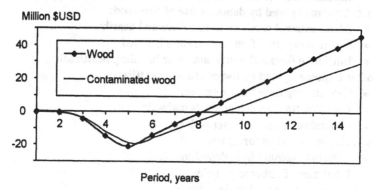

Fig. 2. Net present worth estimated for wood-fired steam turbine power plant

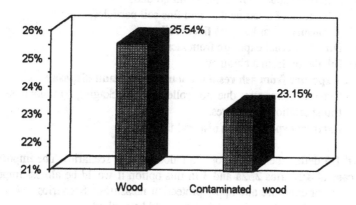

Fig. 3. Internal rate of return for wood-fired steam turbine power plant

4. The Relevant Doses to the Personnel and General Public

4.1. SCENARIOS AND EXPOSURE PATHWAYS

While investigating the contaminated forest utilization pathways, the exposure scenarios along with risk parameters for each scenario were evaluated, in order to formulate a quantitative basis for final decision on remediation options.

Six potential scenarios were investigated as the most probable exposure pathways:

Scenario 1: Exposure due to routine forestry and recreation:
- External exposure from trees and ground;
- Inhalation of airborne sawdust and fine soil particles;
- Ingestion from forest food pathway.

Scenario 2: Exposure posed by domestic use of firewood:
- Direct external exposure from a stove and hearth;
- External exposure from ash residues handled;
- Inhalation from a chimney and while handling hearth ash.

Scenario 2a: Exposure posed by use of ash as a fertilizer:
- External exposure from a garden;
- Ingestion from garden products pathway;
- Groundwater/surface water pathway;

Scenario 3: Exposure due to forest fire
- External exposure from deposition;
- Inhalation of airborne particles;
- Ingestion from food chain.

Scenario 4: Exposure due to active forestry:
- External exposure from trees and ground;
- Inhalation of airborne dust and fine soil particles;

Scenario 5: Exposure from biomass fired boiler:
- Direct external exposure from facility;
- Inhalation from a chimney.

Scenario 6: Exposure from ash residues management and disposal:
- External exposure due to collection, packaging (immobilization) and transportation procedures;
- External exposure from disposal facility.

Scenarios # 1-3 define the baseline option ("do-nothing" scenario). The important issue is that in case of scenarios 2, 2a and 3 in this option it would be almost impossible to provide control of exposure and apply protection measures. Scenarios # 4-6 represent the active option where the proposed action would be applied.

The main objective of the reported work was to develop a methodology for calculating the radiological doses from each of these scenarios. From assumptions of the local behavior patterns, which could easily be scaled up or down in order to reflect the conditions in a particular scenario, estimates were made of the typical doses that would be expected. The general approach of this study was to first estimate the maximum expected doses. Based on this calculation, together with a more realistic evaluation of critical parameters, the level of doses that were likely to generally be observed over workers/population was refined. These latter individual doses would then form the basis for an evaluation of collective doses while applying remedial action.

The following codes were used for dose and risk assessment:

FORESTLIFE (developed by A. Dvornik et al.[5]):
- Prognosis of radionuclides migration in forest ecosystem;
- Scenarios #1 and #4.

Universal Monte Carlo code MCNP4A (provided by LANL, modified by IPEP):
- External exposure dose rate;
- Scenarios #1 - #6.

LOCMIGR (developed by IPEP) and **COSIMA** (provided by RISØ):
- Transport of radionuclides released from furnace and forest fire;
- Scenarios #2, #3 and #5.

RESRAD / BUILD (provided by BNL):
- Inhalation dose and ingestion dose;
- Scenarios #1, #2a, #3 and #4.

DOZA (developed by IREP)
- Multi-cameral model for inhalation dose;
- Scenarios #1, #2, #3, #4, and #5

4.2. RESULTS OF BASELINE RISK ANALYSIS

The data for baseline dose assessment were collected from a number of materials published including our own investigations. These data are presented in [6] along with the relevant references.

From the present study it is possible to project the following overview of the relative importance of the different dose pathways:

1. External dose contribution from ground contamination is highly significant. In all cases, the external exposure from trees is less than that from the ground contamination. Handling of forestry material gives an additional external dose contribution to workers, but from a collective dose point of view does not have much importance. The inhalation dose from sawdust was found to be insignificant.

2. The internal individual dose contribution from consumption of forest foods is generally high depending on the actual contamination level and consumption rate, and the collective dose is very significant.

3. The direct external dose from domestic furnaces was found to be low, even if a furnace is fired with wood containing the maximum reported

radioactivity level. On the other hand, collective dose is relatively high since most of the rural population routinely uses the wood-fired domestic stoves. The dose received from inhalation of aerosols released from the domestic wood-fired furnaces was estimated to be negligible.

4. Due to the generally followed application procedure of using radioactive ash for fertilizing, the inhalation dose from this practice was found to be negligible. After the contaminated ash is applied for fertilizing a field, the additional amount of contamination would be small as compared to what was already in the field. On the other hand, the plant uptake of radionuclides was significantly much more influenced by the fertilizing effect, resulting in noticeable ingestion dose.

5. The dose received from a forest fire would only affect people within a small area. Further, the probability of a fire affecting a particular area was found to be extremely low. With the current frequency of forest fires, there was no significant collective dose. However, significant individual doses have, in some cases, been reported for fire-fighting personnel.

The results of actual measurements of average annual external doses correspond well to the baseline risk assessment. The range of calculated doses and results of direct measurements of external doses are given in Table 1. Again, it should be noted that assessments were made for maximally exposed individual.

TABLE 1. Annual Effective Doses to Exposed Individuals, mSv/year

Pathway	Annual Individual Doses			External Dose (measured in [7])
	External	Internal	Total	
Occupation: Forestry worker	1.4-5.5	0.002-0.16	1.4-5.7	0.21-4.2
Recreational use	0.02-1.15	9-22	9-23.2	0.05-1.6
Domestic use of firewood	0.04-1.0	< 0.001	0.04-1.0	
Use of ash as a fertilizer	0.01-0.13	0.086-1.23	0.1-1.4	
Forest fires:				
Firemen			8.0 per fire	
Residents			<0.001 per fire	

4.3. DOSES FROM ACTIVE OPTION

Three pathways were estimated within the "active" option: (i) external exposure from biomass-fired facilities and ash handled, (ii) external exposure from ash repository, and (iii) inhalation from radionuclide release.

To evaluate doses from the identified facilities, the Monte Carlo simulations of three radiation sources – fuel, furnace chamber media, and ash effluent – were used to provide the values of energy deposited per source particle. This was converted to absorbed dose rate using the specific activity of sources in each element of a facility. The facility under evaluation was a stoker-grate type boiler identical to the boiler we tested in Shasta. Construction materials used for Monte Carlo simulations were chosen from those normally applied in Belarus for a facility of similar design.

The maximal average activity of fuel was assumed to be 2850 Bq/kg, corresponding to wood supposed to be collected from the zone of 15-40 Ci/km². The radionuclide enrichment factor for bottom ash (approx. 15) and fly ash (approx. 70) was calculated based on the results of test burns at Shasta power plant [4]. The contaminants were distributed between the main compartments of a facility as shown in Table 2 (see also Figure 4).

TABLE 2. Mass Percent of Cs and Sr at samling points in Fig.1 in Effluent.

Sampling Point	1	7-9	10
Cs	87.81	12.18	<0.008
Sr	97.03	2.90	0.065

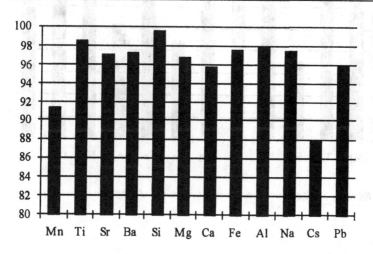

Fig.4. Percentage of different components of ash captured in bottom ash

The flue gas clean-up efficiency controlled by bag-house filter also was estimated based on the data [4] from the Shasta study (see Table 2 and Figure 5). The COSIMA code was used to determine radionuclide distribution in surrounding area.

The resulting doses to residents and personal were assessed. The calculated radiation doses to the maximal exposed individual from the pathways identified within active scenario were as follow:

- Worker: - is less than 1 mSv/year (external and inhalation exposure);
- Resident: - is less than 0.001 mSv/year (external exposure and due to inhalation).

Doses from the ash disposal site were evaluated assuming that this low-level radioactive waste repository would be simply a surface, trench-type waste disposal facility of 50000 m³ covered with 1- m thick ground layer. It was presupposed that after 50 years the state lost control of the facility and a certain family resides over the sealed repository. Hypothetical individuals residing in the repository area were assumed not to break the cover material, but they did consume the plants grown in the topsoil.

The total annual dose potentially obtained from such an intruder scenario was about 0.2 mSv/year for maximal exposed individual. The last was a person who spent all the time on site and ate the food produced from his own garden cultured on the ground covering a repository.

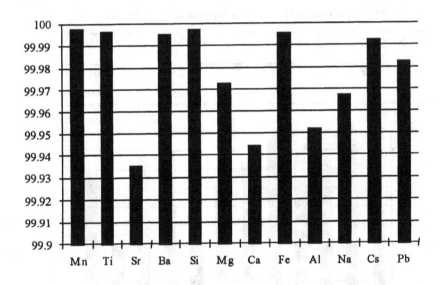

Fig. 5. Efficiency (in %) of electrostatic precipitator with regard to different components of ash

5. Conclusion

As a part of remediation strategy for contaminated forests in Belarus, we propose to construct and operate biomass-fired power facilities fuelled with wood and organic matter from the forest floor in the contaminated regions. These facilities would be capable of using both uncontaminated and contaminated fuels. The technologies involved (stoker-grate boiler, baghouse particle collection, etc.) are mature, reliable and comparatively simple for uncontaminated fuels. The radioactive species would be captured in the ash, which can be deposited in appropriately designed landfills. An economical analysis of this proposal shows that such an approach could provide feasible way for contaminated biomass utilisation.

Preliminary risk assessments of the proposed action indicate that contaminated biomass conversion is environmentally sound and does not contribute any significant risk to population.

More comprehensive risk evaluation, which include averted collective doses, must be conducted on a basis of a case study for the specific site of the identified biomass-fired facility. The demonstration scale power plant study is scheduled for 1999. The results obtained will be provided to decision-makers in the Belarus government to support a sound, risk-based decision on the proposed strategy.

Work is supported by the U.S Department of Energy under the Initiatives for Proliferation Prevention Program, Wheelabrator Environmental Systems Inc., and the government of the Republic of Belarus and the Danish government.

6. References

1. Anonymous (1996). "Strategies of decontamination. ECP-4 Final Report". EC-DGXII, EUR 16530 EN, Brussels: 174 pp.

2. Anonymous (1998). "Chernobyl Bioenergy Project". Progress Report. Elsamprojekt, IPEP, SNLs, FSL, RISØ. Elsamprojekt (publisher), Fredericia, Denmark: 138 pp.

3. I.A. Savoushkin, A.J. Grebenkov, V.N. Solovjov, and B.G. Louchkin (1996). "Radioactive Wood Waste Treatment in the Chernobyl Zone", In Proc. of Nuclear and Hazardous Waste Management International Topical Meeting "SPECTRUM '96", Seattle, WA, August 18-22, 1996. American Nuclear Society, Inc. (publisher), La Grange Park, IL: p. 25-31.

4. S.G. Buckley, M.M. Lunden, A.L. Robinson, D.C. Allen, A.J. Grebenkov, and L.L. Baxter (1998). "Fate of Sr and Cs in Biomass Combustion", In Proc. of AIChE Spring Meeting, New Orleans, LA, March 8-12, 1998: in press.

5. A.M. Dvornik and T.A. Zhuchenko (1995). "Behavior of ^{137}Cs in Pine Wood Stands in Belarus Polesje: Modeling and Prognosis", ANRI Scientific and Informational Journal (in Russian), No 3/4, "Doza" Information Center (publisher), Moscow: p. 59-66.

6. Anonymous (1997). "Data Collection Required for the Analysis", Progress Report, IPEP (publisher), Minsk: 48 pp.

7. Anonymous (1997). "Implementation of Individual Dosimetry Control of the Forestry Workers and External Exposure Dose Assessments", Technical Report (in Russian), RCIRME/GB, Gomel: 42 pp

CONTAMINATED TREE BIOMASS IN ENERGY PRODUCTION – POTENTIAL NEED FOR RADIATION PROTECTION

A. H. RANTAVAARA, K. M. MORING
STUK, Radiation and Nuclear Safety Authority, P. O. Box 14, FIN-00881 Helsinki, FINLAND

1. Introduction

1.1 OBJECTIVE

The radioactivity of the ashes from the wood-fired boilers of forest industry was studied in Finland in 1996-1997, with a view to collecting data for assessments of radiation exposure, and determining the need for continuos surveillance. The objective was to analyse the real exposure situation with representative sampling and to establish which radionuclides in wood ash contribute most to the dose received by workers in the industry and to the members of the public. The data collected in 1996-1997 provide insight into the doses due to ^{137}Cs related to the dynamics of radiocaesium in forest ecosystems. We also discuss the new assessments that will be needed if the ash is used for forest fertilisation in the future instead of being dumped as at present.

1.2 RELATED STUDIES

In Sweden, Hedvall [1] has been studying radiological issues related to the production of bioenergy since the 1980s. His emphasis has been on three fuel types - peat, wood chips and straw and several natural radionuclides. Swedish results show, among others, the importance of atmospheric pathways for doses due to ^{210}Pb.

In the USA the National Council of the Paper Industry for Air and Stream Improvement has analysed ash from wood-fired industrial boilers for ^{137}Cs [2]. Both the ingestion and inhalation pathways were assessed for ^{137}Cs from atmospheric nuclear tests. The ingestion pathway was the use of vegetables after the ash had been used to fertilise kitchen gardens. The influence of added nutrients on the uptake of caesium by plants was not considered.

Intensive experimental studies on the dynamics of radiocaesium in forests are under way in some projects of the 4th Framework Programme of the European Commission (LANDSCAPE, SEMINAT, EPORA). Surveys on radioactivity of timber have also been carried out for the industry and users of wood [3].

I. Linkov and W.R. Schell (eds.), Contaminated Forests, 303–310.
© *1999 Kluwer Academic Publishers. Printed in the Netherlands.*

2. Material and methods

2. 1. SAMPLES

The majority of the 22 wood fired boilers included in our study represented the pulp or mechanical forest industry; only a few district heating plants delivered samples. The fluidised bed was commonly used in combustion. Of the 87 ash samples, 51 were fly ash and 29 bottom ash; seven samples were of unknown type.

The composition of the fuel varied, the fuels contained also non-wood materials (Fig.1). The samples were mailed to the laboratory by the technical personnel of energy production plants. The sampling was repeated a few times, at most plants, and a total of 2 – 4 samples were taken during a few weeks.

Most of the industrial boilers and other plants received the wood from an area extending to about 100 km from the plant. Imported wood was used in some cases, but their fractions were not known exactly. The most common part of the tree used as fuel was bark.

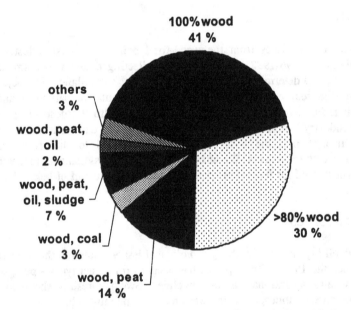

Fig. 1. *Composition of fuels in ash samples. In 71% of the ash samples the fuel had consisted of more than 80% wood and bark.*

2. 2. RADIONUCLIDE ANALYSIS

All samples were analysed for radiocaesium (^{134}Cs, ^{137}Cs), and also for ^{232}Th, ^{238}U, ^{235}U, ^{226}Ra and ^{40}K with a low-background gamma spectrometer (HPGe). Gamma lines of daughter nuclides where used to determine ^{232}Th, ^{238}U and ^{226}Ra. ^{226}Ra was also detected with the aid of its own gamma energy. In samples in which no ^{235}U was found, the results were in good agreement with those for the radon daughters, giving a low radon emanation coefficient for the ash and good reliability for the results of measurements of daughter nuclides. All measurements were made using Marinelli or cylindrical geometries, in accordance with the normal procedure at STUK for environmental sample measurements.

Some of the samples were analysed radiochemically for ^{210}Pb and ^{226}Ra after the ash had been dissolved in hydrochloric acid. Radium was separated by barium sulphate co-precipitation and lead by specific lead sulphate precipitation. The alpha activity of ^{226}Ra and the three alpha-emitting daughters were measured after an ingrowth period of about 4 weeks. The beta activity of ^{210}Pb + ^{210}Bi was counted after an ingrowth close to equilibrium.

^{90}Sr was determined in ash after the sample had been dissolved in a Na_2CO_3 melt. The method was based on nitrate precipitation and further separation of yttrium and strontium. Both alpha and beta activities were measured with a low-background liquid scintillation counter.

Solubility tests for the radionuclides were performed using acid ammonium-acetate (pH = 4.65) for extraction. Solutions and ash residues were analysed with gamma spectrometers for radiocaesium and ^{40}K, the only natural radionuclide detectable in the solutions by direct gamma measurement. The radiochemical separation procedures for ^{90}Sr, ^{210}Pb and ^{226}Ra were slightly modified for the analysis of NH$_4$Ac solutions.

2.3 DOSE ASSESSMENT

For some exposure pathways of the external radiation caused by wood ash, the dose was assessed by activity indices developed earlier for fuel peat ash. The indeces used derive from the Finnish safety requirements [4], and are based on annual doses of 0.1 mSv for members of the public and 1 mSv for workers in the relevant industry [5]. The activity indices include the real fractions of the dose caused by the measured concentration of each nuclide, typical background levels are also considered. Assumed exposure scenarios were used for different types of use or handling of the ash. The exposure pathways are related to use of the ash as a constituent of the materials used for the construction of streets and playgrounds, and for land filling. External exposure is the only dose pathway. Atmospheric releases are assumed to be insignificant. This is not necessarily so in all plant types, but the dose assessment methodology was developed for facilities provided with efficient filtering of fine particles. The exposure of workers, e.g. truck drivers, was assessed for handling of ash. The activity index for handling of ash, based on 1 mSv a^{-1}, is defined by Equation 1.

$$I = \frac{C_{Th}}{3000\,Bq/kg} + \frac{C_{Ra}}{4000\,Bq/kg} + \frac{C_K}{50000\,Bq/kg} + \frac{C_{Cs}}{10000\,Bq/kg} \qquad [1]$$

Separate activity indeces are used for public exposure (based on 0.1 mSv a^{-1}), due to landfilling etc.

306

Wood ash differs from peat ash in contents of nutrients and is a rather good substitute for PK fertilisers. Practical guidance for surveillance or control of the use of wood ash needs further analysis of the exposure conditions, especially if the ash is to be returned to the forest as a fertiliser.

3. Results

The total contents of radionuclides varied by nuclide, but rather little by type of ash (Fig. 2). ^{137}Cs, ^{40}K and ^{90}Sr were the dominant nuclides, whereas contents of natural radionuclides other than ^{40}K were rather small. However, when the fuel contained peat, the uranium abundance was higher than otherwise. In measurements of fly and bottom ash (samples taken from the same plant and during the same period), the levels of ^{137}Cs where higher in the fly ash by a factor of 1.7. No such trend was found for the other nuclides. For ^{40}K the high amounts of potassium in the sand of bottom ash interfered, and for ^{90}Sr the number of analysed bottom ash samples was too low to permit a conclusion to be drawn.

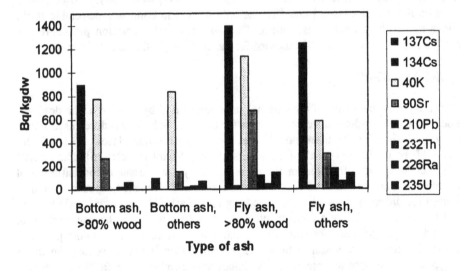

Fig 2. *Activity of radionuclides in ash samples. Geometrical means for samples of different types of ash and fuel*

The solubility of radionuclides in acid ammonium acetate varied considerably in most ash groups (Fig. 3, 4). ^{137}Cs, ^{40}K and ^{90}Sr were the most soluble nuclides. The solubilities of ^{137}Cs and ^{40}K showed a positive correlation with each other, whereas ^{90}Sr did not correlate with any of them. ^{226}Ra and ^{210}Pb dissolved less than the dominant nuclides, but the range of solubility was considerable in all ash groups. When the fuel contained sludge, the solubility of ^{137}Cs and ^{40}K in bottom ash was minimal.

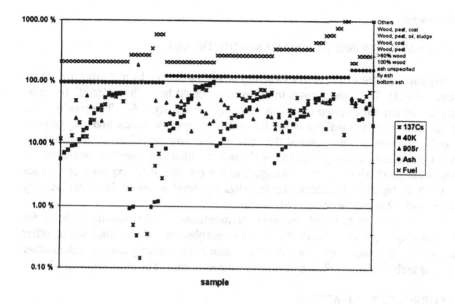

Fig. 3. Solubility of ^{137}Cs, ^{40}K and ^{90}Sr in NH_4Ac solution. Single analysed samples. The upper part of the graph shows the type of ash and fuel in the samples. Determination errors were small

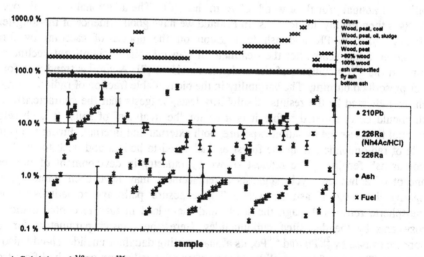

Fig. 4. Solubility of ^{210}Pb and ^{226}Ra in NH_4Ac solution compared with their total amounts and HCl soluble fraction (^{226}Ra). Single analysed samples. The upper part of the graph shows the type of ash and fuel in the samples. The bars indicate the error of determination

4. Discussion

4.1. RADIATION DOSES DUE TO HANDLING OF ASH

The exposure caused by the use or handling of ash varied, but by and large was of no concern (Fig.5). The annual doses to the workers could be of the order of 0.1 mSv based on the exposure scenario used for deriving the activity index. The results show that in some cases the activity index for materials used for streets and playgrounds (based on 0.1 mSv a^{-1}) may exceed the value 1. Vast amounts of such ash should not be used at sites continuously occupied by the public without site specific assessments of doses. Most wood ash is nowadays dumped, and in practice, then, exposure only arises when ash is handled. Reaching the investigation level of doses for other activity indices, was not therefore considered critical.

The contributions of different radionuclides to the activity indices for handling (Fig. 6) also indicate the relative contributions of the nuclides to other indices. ^{226}Ra dominates only the index for bottom ash containing less wood than other groups of fuels. ^{137}Cs contributes most to the exposure caused by other types of ash.

4.2. FOREST FERTILISATION

Recycling of the nutrients of trees by returning the ash to the forest calls for new assessments to be made. For peatlands, a single dose of ash may be 5000 kg ha^{-1}; for mineral soils it is usually smaller. Under optimised conditions this might give an additional annual growth of wood of 10 m^3 ha^{-1} [6]. The additional external dose in forests is then mainly due to ^{137}Cs. In Finland we have good evidence of the long-term influence of both PK and ash fertilisation on the uptake of caesium by forest vegetation. Some years after the treatment of mineral soils, a significant decline was found in the ^{137}Cs contents of the understorey plants [7]. A similar trend was found after prescribed burning. The variability in the bioavailable fraction of radionuclides of ash, as indicated by the results of solubility tests, suggests that the characteristics of ash should be considered before it is used for the treatment of soils. It is therefore essential to assess the radiation exposure, both external and internal, during the whole cycle of radionuclides, from the forest as fuel wood to boilers and back to the forest floor as ash fertiliser. The interest shown by industry in development of this new concept of a nutrient cycle suggests a replacement may have been found for the ongoing dumping of ash as waste. The exposure pathways sometimes include atmospheric releases through the stack, and these lead in turn to contamination of ecosystems by the deposited radionuclides. Here, the transfer parameters for and exposure caused by ^{210}Pb and ^{210}Po, its alpha emitting daughter nuclide, should also be estimated. The fate of radionuclides in soils after ash fertilisation certainly differs from that of their transfer in undisturbed soil, thus adding to the complexity of the radiological problem. Clearly, site-specific assessments are also needed; it is, for example important that the local consumption patterns should be known when the internal exposure is being estimated.

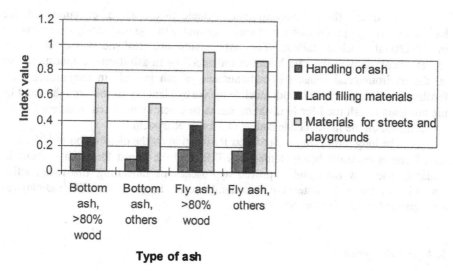

Fig. 5. *Geometrical means of activity indices for construction materials in the measured samples of different types of ash and fuel.*

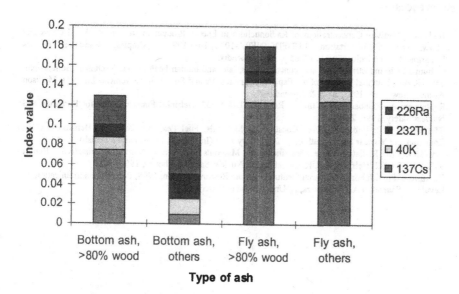

Fig. 6. *Radionuclide composition of activity index for handling of ash in the samples measured. Mean values for different types of ash and fuel.*

310

All in all, the radiological issues related to the use of contaminated tree biomass for energy production are of vital importance in regions heavily contaminated by accidental nuclear fallout. The assessments are realistic when based on representative data on the enrichment of radionuclides in ash during combustion, and on the environmental transfer of radionuclides of ash in soils in conjunction with fertilisation. The reduction in radionuclide uptake by plants from soil brought about by the nutrients in ash used for fertilisation has to be carefully analysed to determine the net influence on the radionuclide budget of a forest ecosystem.

The long-term influence of forest fertilisation on the distribution of ^{137}Cs in a pine forest is currently being studied by STUK and the Forest Research Institute in Finland. The new data and improved approaches in modelling the radionuclide dynamics in forest ecosystems will certainly result in more versatile radiological assessments for the use of wood ash.

5. Acknowledgments

The study was conducted with the aid of a grant from the Imatran Voima Foundation.

6. References

1. Hedvall R. Activity Concentrations of Radionuclides in Energy Production from Peat, Wood Chips and Straw. Doctoral dissertation. LUNFD6/(NFRA-1035)/1-65/1997. Nyköping: Radiation Physics Department, Lund University, 1997, 65 p. with Appendix.
2. Cesium-137 in industrial wood ash: concentrations, fate, and human health risks. Technical Bulletin No. 639. NCASI, National Council of the Paper Industry for Air and Stream Improvement, Inc., 260 Madison Avenue, New York. 1992, 25 p. & Appendices.
3. Rantavaara A., Radioactivity of timber. Report STUK-A 133, Helsinki: Finnish Centre for Radiation and Nuclear Safety, 1996, 29 p.
4. ST-GUIDE 12.2: Radioactivity of Construction Materials, Fuel Peat and Peat Ash Helsinki: STUK, Radiation and Nuclear Safety Authority. 2 February 1993 (In English, Finnish and Swedish).
5. Markkanen M. Radiation Dose Assessments for Materials of Elevated Natural Radioactivity. Report STUK-B-STO 32. Helsinki: STUK, Radiation and Nuclear Safety Authority, 1995, 40 p.
6. Kaunisto S. Finnish Forest Research Institute, Parkano Research Station, 1998, Personal communication.
7. Levula T., Saarsalmi A, Rantavaara, A. Unpublished manuscript.

FIRE AND RADIOACTIVITY IN CONTAMINATED FORESTS

B.D. AMIRO
Canadian Forest Service, Northern Forestry Centre
5320 - 122 Street, Edmonton, AB, T6H 3S5, CANADA

A. DVORNIK AND T. ZHUCHENKO
Forest Institute of the National Academy of Sciences of Belarus, 71
Proletarskaya Street
246654 Gomel, BELARUS

Abstract

Fire is an important recurrent disturbance to many forest ecosystems. In the forests contaminated by the 1986 Chernobyl accident, fire will inevitably occur and needs to be incorporated into models and plans for remediation. The fire regime is changing because of changes in land use, public access, and climate, and cannot be predicted easily based on historical information. We describe the Canadian Fire Behavior Prediction system as a possible tool to identify fire behaviour in the contaminated *Pinus sylvestris* forests in Belarus. However, our knowledge of fire behaviour is better than our knowledge of the ecological effects following fire. Presently, much of the ^{137}Cs inventory is in the top layers of the soil, and burning of the surface duff layers will affect radionuclide cycling, as well as alter the forest ecosystem. The effects of fire depend on burn severity, with light surface burns at cooler temperatures being much different than severe crown fires. Volatilisation of ^{137}Cs as a gas is expected to occur at fire temperatures greater than 400°C, but these temperatures would seldom occur in the mineral soil. Enhanced leaching of ^{137}Cs will likely occur following a fire, but this may only be to the rooting zone where surviving trees or new plants will access it. The role of charcoal sequestering ^{137}Cs is unknown. Fires are expected to change the radiation fields in the forest through the removal of the relatively uncontaminated recently formed surface litter layer that acts as shielding, and exposing the more contaminated soil layers (typically 5 to 10 cm). Although radiological doses to the public from exposure to smoke is of some concern, fire fighters and forestry workers are the critical groups that are most affected by fires.

1. Introduction

The 1986 Chernobyl accident contaminated a large extent of forest in Belarus, the Ukraine and Russia with sufficiently high concentrations of ^{137}Cs to impact forest use.

I. Linkov and W.R. Schell (eds.), Contaminated Forests, 311–324.
© 1999 *Kluwer Academic Publishers. Printed in the Netherlands.*

In Belarus alone, almost 2 million ha, or 24% of the national forest is contaminated with more than 3.7×10^{10} Bq ^{137}Cs km^{-2} (Belarus Ministry of Forests estimates). Throughout much of the contaminated areas, the commercial forests are plantations of *Pinus sylvestris* (Scots pine). However, there are many other forest types containing a wide range of species, as well as wetlands that are partially or totally forested. Soon after the accident, forest fires were identified as a potential concern in the contaminated areas of Belarus [1], and this was highlighted by a severe fire season in 1992 [2]. The concern about fires is mostly centred around potential resuspension of radioactivity into the atmosphere, with associated increase in radiological exposure to a wider population. Further, this resuspension could move ^{137}Cs from a trapped form in the forest to more mobile forms in surrounding agricultural areas.

Much work has been done on the radioecological aspects of contaminated forests [3,4] and models have been developed to evaluate radionuclide distributions [5,6]. Forests are looked on as a possible barrier to reduce the spread of contamination [7] and various countermeasures have been proposed [8,9]. However, almost all of the processes that have been examined are continuous ones, such as tree uptake and leaching of radionuclides. Fire is a periodic catastrophic process that can drastically change the landscape and supersede many of the slower, continuous mechanisms of radionuclide transfer.

The objective of our study is to review the features of fire that will impact the radioecology of contaminated forests. Here, we mostly focus on forests of Belarus, but the principles also apply to other forests. We also concentrate on ^{137}Cs, although other radionuclides such as ^{90}Sr and ^{239,240}Pu are present. We look at fire as a natural agent that will unavoidably affect at least some of the contaminated forests. As such, we need to assess the effects of fire in these ecosystems, and perhaps find methods to ameliorate the impact. Our prime questions are how does fire affect:
- the total radionuclide inventory in the forest ecosystem?
- radiation exposure received by humans in the forest?
- radionuclide concentrations in forest products?

2. The Role of Fire in Ecosystems

Fire is a key component of most northern forest ecosystems. The fire regime is the kind of fire activity or pattern of fires that generally characterise a given area [10]. This includes aspects of fire frequency, size, intensity, seasonality, type, and severity. Smith et al. [2] summarised some aspects of the fire regime in Belarus based on government officials' estimates. Peak fire years occur about every 33 years, with a recent severe situation in 1992. Most burned stands are relatively young and even aged, resulting in low to moderate intensity fires. Plantings are often closely spaced and ladder fuels are often present to offer the opportunity for crown fires. However, fuel loading of dead-and-down vegetation are generally low to moderate, and high windspeeds may be needed to achieve high fire intensities. Most fires are ignited by humans with few lightning-caused fires. The relatively humid climate of much of the area means that severe fire risk occurs infrequently. However, fire occurrence is controlled by the exceptional weather conditions, so that fires are sure to be an important part of the forest ecosystem. Fire may not be generally accepted as a main driving force in Belarus forests, but natural *P. sylvestris* forests in northern Europe are adapted to periodic wildfire [11,12]. Therefore, we must expect that fire will occur and that it is a natural part of ecosystem function.

Although it is possible to define the current fire regime in the forests contaminated by Chernobyl, it is likely that the fire regime is changing. Some of these changes result from the alteration of land use in contaminated areas where public access has become restricted since 1986. Reduced agriculture on contaminated land and return to forests is creating a different mosaic on the landscape where agricultural fires have decreased but forest fuels will increase. Fewer people in the contaminated forests may result in fewer ignition sources (most fires here are human caused), but still it is likely that there are sufficient sources of ignition. The age distribution of Belarus forests is also changing with older age classes being predicted over time [2]. In addition, recent assessments suggest that projected global warming will increase forest fire potential in both Canada and Russia [13], and we can expect that areas close to Chernobyl may also experience fire weather conditions that are different from historical ones.

Fire statistics for the 1980 to 1992 period in Belarus are reported by Smith et al. [2] based on the Republic of Belarus fire plan. Although in most years less than 1000 ha are burned, over 22,000 ha were burned in 1992. In this same year about half of the area burned in the Gomel Oblast (1390 ha) was in areas classified as radiologically contaminated. Based on statistics for the whole country, about 56% of the area burned was by surface fires, 37% by crown fires, and 7% as "underground" fires. From these data, it is clear that fire is still a major factor in Belarus forests, and that the contaminated areas will experience fire in the future.

Despite the best efforts at suppression, fire will continue to be a factor in contaminated forests near Chernobyl. Perhaps our goal should be to determine the effects of the fire regime, and to use this knowledge in landscape planning. Such planning, especially related to radioecology, requires enhanced knowledge of fire behaviour and how it will affect the radiation regime.

3. Fire Behaviour in the Contaminated Forests of Belarus

Fire behaviour is a complex science (and art) that involves the physics of fluid motions, the chemistry of ignitions, and the biology of organic material composing the forest. Fire behaviour has been modelled using various techniques, ranging from extensive mathematical relationships [14,15,16] to more empirical observations made over many years [17,18]. Here, we use the Canadian Forest Fire Behaviour Prediction System (FBP) as an empirically derived model that gives the main features of fire behaviour that are required to estimate the impact of fire on the forest radiation regime. A full description of FBP is given in Forestry Canada [18] and Hirsch [19].

314

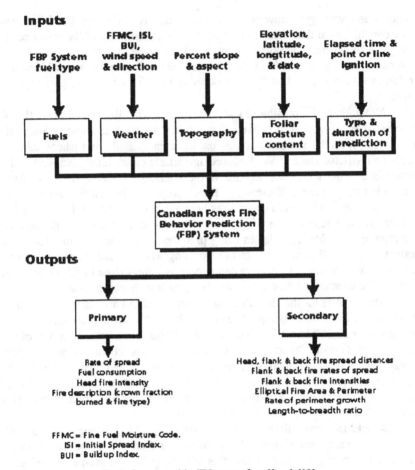

Inputs

| FBP System fuel type | FFMC, ISI, BUI, wind speed & direction | Percent slope & aspect | Elevation, latitude, longtitude, & date | Elapsed time & point or line ignition |

| Fuels | Weather | Topography | Foliar moisture content | Type & duration of prediction |

Canadian Forest Fire Behavior Prediction (FBP) System

Outputs

| Primary | Secondary |

Primary:
Rate of spread
Fuel consumption
Head fire intensity
Fire description (crown fraction burned & fire type)

Secondary:
Head, flank & back fire spread distances
Flank & back fire rates of spread
Flank & back fire intensities
Elliptical Fire Area & Perimeter
Rate of perimeter growth
Length-to-breadth ratio

FFMC = Fine Fuel Moisture Code.
ISI = Initial Spread Index.
BUI = Build up Index.

Fig. 1. Structure of the FBP system from Hirsch [19]

The main inputs to the FBP system are fuel type, weather, topography, foliar moisture content and type and duration of prediction (e.g., point or line source, and accelerating or equilibrium fire). These components are shown in Figure 1. Most of the area of concern in Belarus is relatively flat, so topographic influences will be ignored here. In addition we will be evaluating a fire that started from a point and has reached an equilibrium rate of spread (i.e., a fire that is progressing hours or days after ignition from a point).

There are many types of forest communities in the contaminated forests of Belarus, ranging from pine plantations to wetland forests. Here we will concentrate on pine forests that are commercially important for the Belarus forest industry. In the FBP system, a conifer plantation is classified as a fuel type (C6) where observations of fire behaviour have been made during 12 experimental fires in pine (*Pinus resinosa*) plantations. These plantations have similar characteristics to *P. sylvestris* plantations and are good analogues for the Belarus forest communities.

Weather conditions control most aspects of fire behaviour and are an integral part of FBP. The weather inputs have been experimentally related to fire behaviour, and are tied to the Canadian Forest Fire Weather Index system (FWI) [20,21]. The FWI uses data on temperature, relative humidity, wind and rain to derive indices for the Fine Fuel Moisture Code (FFMC), the Duff Moisture Code (DMC) and the Drought Code (DC). These three moisture codes estimate the moisture in:

- FFMC - litter and other fine cured fuels, indicating relative ease of ignition and flammability of fine fuel,
- DMC - loosely compacted organic layers of moderate depth, indicating fuel consumption in moderate duff layers and medium-sized woody material,
- DC - deep, compact organic layers, indicating seasonal drought effects on forest fuels and amount of smouldering in deep duff layers and large logs.

These moisture codes are then used to estimate the:

- Initial Spread Index (ISI) - numerical rate of the expected rate of fire spread, based on FFMC and wind speed,
- Buildup Index (BUI) - numerical rating of the total amount of fuel available for consumption, based on DMC and DC, and
- Fire Weather Index (FWI) - numerical rating of fire intensity combining ISI and BUI.

The FWI and FBP systems can use hourly weather data, but can also use daily noon values of temperature, relative humidity and wind speed, and 24-h total rainfall.

Our goal is to estimate ecological consequences of fires in contaminated forests, and many of the FBP outputs relate to fire behaviour that only indirectly controls these consequences. For example, the rate of fire spread in itself is not a direct indicator of ecological effects, but it is related to consumption of fuels, which are ecologically important. The FBP system indicators of interest are the surface fuel consumption (SFC) and crown fuel consumption (CFC). For the fuel type of interest:

$$SFC = 5 \cdot [1 - \exp(-0.0149 \cdot BUI)]^{2.48} \qquad (1)$$

where SFC is in units of kg m^{-2}, expressed as oven-dry weight. If the fire is sufficiently intense, the tree canopy will also burn in a crown fire, and crown fuel consumption (kg m^{-2} oven-dry weight) is estimated as:

$$CFC = CFB \cdot CFL \qquad (2)$$

where CFL is the crown fuel load (taken here as a typical value of 1.8 kg m^{-2}) and CFB is the crown fraction burned (unitless), as

$$CFB = 1 - \exp[-0.23 \cdot (ROS - RSO)]. \qquad (3)$$

ROS is the rate of spread (m/min) = RSI·BE, where

$$RSI = 30 \cdot [1 - \exp(-0.08 \cdot ISI)]^{3.0} \tag{4}$$

and

$$BE = \exp(0.18 - 11.16/BUI). \tag{5}$$

RSO is the critical surface fire spread rate that describes the transition from a surface fire to a crown fire, calculated as

$$RSO = CSI/(300 \cdot SFC) \tag{6}$$

where CSI is the critical surface fire intensity,

$$CSI = 0.001 \cdot CBH^{1.5} \cdot (460 + 25.9 \cdot FMC)^{1.5} \tag{7}$$

and CBH is the crown base height (m). For our example, we select CBH = 7 m, which was the typical crown base height for many of the experimental fires used to develop the relationships. In the forests of Belarus, this CBH varies depending on stand age and height, and will affect the ability of a fire to reach the crown. In Eq. (7), FMC is the foliar moisture content, which typically varies from 85% to 120% [18]. This value only affects CSI, and is only important in the case of a crown fire. For example, with the parameters outlined here, varying FMC from 85% to 120% varies CSI from 2540 to 3950 W m^{-1}. Although this transition can be important, it will have no effect during conditions when there is a low surface fire intensity without crowning, and also no effect when there is a high fire intensity (>4000 W m^{-1}) when crowning will commence irrespective of the value of FMC. For our analysis, we select an FMC value of 100% as an intermediate condition.

Using the above equations, we can estimate a range of SFC and CFC for a typical *P. sylvestris* forest for a range of fire weather conditions. Figure 2 shows estimates of SFC and CFC for various values of BUI. The SFC values increase with BUI only, whereas CFC values are dependent on both BUI and ISI. We have shown CFC values for ISI values of 5, 10 and 30 as a range of values that could be experienced during a fire episode. The higher values of ISI would most likely occur during periods of strong winds. We note that CFC quickly levels off at a given BUI value, such that the crown fire is mostly determined by the ISI. Figure 2 illustrates some of the principles that would need to be considered if a prescribed fire is used in contaminated forests. For example, if our goal was to remove surface duff layers that contain ^{137}Cs, we would best select conditions where the BUI is relatively high but the ISI is low. This would prevent crown fires and reduce scorching that kills trees; i.e., we would create a surface fire that removes large amounts of forest floor litter and the top organic layers of soil.

4. Inventory of Radioactivity in the Forest

Much of the ^{137}Cs is still present in the top layers of the soil, typically in the 5- to 10-cm layer [3,7]. Therefore, much of the recently formed litter layer is relatively uncontaminated. However, as ^{137}Cs is leached downward, more becomes available for root uptake and this uptake can be seen in the new wood [22]. Wetter landscapes tend to have a greater incorporation of Cs into plants, presumably caused by more leaching to the root zone [3,23]. This trend of progressive downward leaching means that a greater amount of the total inventory will be lost from the forest through export to the water table and subsequent runoff in a dilute form. However, this process is expected to be very slow with half-times on the order of a century [5]. The greatest impacts of progressive leaching is the movement of ^{137}Cs into the root zone where trees can access it and recycle it through the ecosystem.

Model predictions indicate that maximum tree concentrations will be experienced between 5 and 20 years after the deposition and that this will decrease very slowly, typically at a rate where the 30-year radiological decay becomes an important consideration [5]. Therefore, we have a situation where the trend in concentration is for the top soil litter and duff layers to decrease over time, the tree wood concentration to reach a peak and then decline, and the mineral soil to increase and then decrease through leaching. This trend needs to be considered when the effects of fire are assessed.

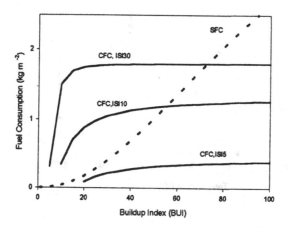

Fig. 2. Surface fuel consumption (SFC) and crown fuel consumption (CFC) vs. build-up index (BUI) for different values of initial spread index (ISI-5, ISI-10, ISI-30) for a C6 plantation type fuel (typical of *P. sylvestris* plantation) based on the Canadian Fire Behavior Prediction system. Model parameter estimates are given in the text.

5. Fire Effects on Radionuclide Cycling

The direct effects of fire on radionuclide cycling relate to mobilisation of ^{137}Cs in the forest. We can divide this into mobilisation that occurs during the fire, and that which occurs in the post-fire ecosystem. Many elements are lost during fires [24]. During the fire, ^{137}Cs can be resuspended into the atmosphere as particulate matter in smut and smoke, or as a gas if the temperatures are sufficiently high. The smut component is defined as particle sizes that have large settling velocities and will often be deposited close to the source, whereas smoke is truly suspended and is more widely dispersed [25]. Enhanced concentrations of ^{137}Cs have been measured in smoke [26,27], but we do not yet have a mechanistic description or reliable model of the suspension of ^{137}Cs in smoke.

The situation with possible gaseous releases of ^{137}Cs is also not totally clear. There appears to be a relationship between fire temperature and volatilisation of ^{137}Cs. For example, in a combination of controlled field and laboratory combustion, Amiro et al. [28] showed that higher temperatures cause a greater loss of Cs (Figure 3).

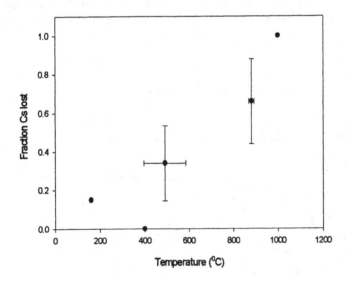

Fig. 3. Fraction of cesium lost in a fire vs. fire temperature from Amiro et al. [28]. Data at 200, 400 and 1000°C are from furnace combustion, data at about 500°C are from three straw fires and data at about 900°C are from two wood fires. The field data show mean values ± S.E. among burns.

Although Figure 3 shows that little Cs is lost at temperatures below 400°C, the question is whether there is a true threshold at which we can assume that almost no ^{137}Cs is volatilised.

Surface fire temperatures are not easily predicted because of the wide range of burning conditions that occur at the surface. For example, we may have quite hot flames in the air above the surface while there is a cold smouldering condition at some depth. Hungerford et al. [29] reviewed much of the literature related to soil heating and temperature during fires. When duff moisture content is high, only minimal soil heating results whereas a drier duff allows smouldering with prolonged heating. The duff can act as a barrier to heating the mineral soil. Experiments by Frandsen and Ryan [30] indicate that if both the duff and mineral soil were wet during a burn, temperatures did not exceed 90°C. They suggest that land managers could strive to burn when mineral soils are near saturation to minimise mortality to plants and loss of soil organic. Although extreme slash fires can produce surface temperatures of about 1000°C, many measurements of surface fire temperatures are in the range of 200 to 500°C [31]. Unfortunately, we do not have adequate methods or models to predict the amount of soil heating during fires, largely due to the complex nature of fires and the spatial variability in site conditions. However, it is reasonable to conclude that many surface fires have temperatures that are less than 400°C, which is the point when we start to get appreciable Cs volatilisation.

The situation with temperatures in the forest crown is different. Here, we experience higher temperatures during intense crown fires. The FBP system estimates crown fire temperature according to the relationship

$$\text{crown fire temperature (°C)} = 1227 - 2.75 \cdot FMC, \tag{8}$$

so our typical FMC value of 100% gives a crown fire temperature of about 950°C. The range of FMC from 85% to 120% gives a temperature range of 995°C to 900°C. In this case, temperatures are sufficiently great to allow for most of the ^{137}Cs to be volatilised. The combustion during these crown fires is mostly leaves and small twigs which contain relatively low concentrations of ^{137}Cs at present. However, if root uptake of ^{137}Cs becomes greater over time, then the crown may contain substantial amounts of radionuclides. This gaseous portion would be widely dispersed, and eventually condense and be deposited elsewhere. Radionuclides in the boles of trees would be largely unaffected during the passage of the fire because they are not consumed. However, dead trees will fall over and decompose. Therefore, the ^{137}Cs in the boles of trees will eventually be released back to the soil surface and be available for cycling through the ecosystem.

The remaining ash after a fire may contain enhanced concentrations of [137]Cs. For example, Amiro et al. [28] measured enrichment factors for Cs in ash ranging from 4 to 25 Bq kg[-1] ash per Bq kg[-1] wet biomass. Following natural fires in Canada, ^{137}Cs concentrations were higher in surface soils of burned stands compared to unburned stands [32]. However, the total ^{137}Cs load was less in burned stands; i.e., ^{137}Cs became enriched in soil although there was presumably some loss to the atmosphere. In the areas contaminated by Chernobyl, the enrichment of Cs in ash products following incineration and other waste-generating activities is a concern and various processes for immobilisation of the ash are being investigated [33]. The higher concentrations in the remaining ash may be of substantial concern, but their enhanced mobility is of greater ecological importance. Much of the original ^{137}Cs is tied up with organic matter, which slows leaching, and the newly formed ash is more mobile and soluble. Even for the primary deposition, Chernobyl particle size is important in the rate of leaching [23]. The expected effect after a fire is that ^{137}Cs will be more easily carried in water runoff and more quickly leached downward into the rooting zone where it will be recycled in plants.

6. The Implications of Fire to the Forest

Fire changes both the forest itself and the radioecological relationships. The direct forest effects vary greatly, depending on burn severity. In some instances, the outcome could be a light surface burn where some of the top soil organic material and fine fuels are removed, perhaps with some shrubs and trees being killed. Alternatively, a severe crown fire could kill a whole stand of trees, and burn the soil to a depth where mineral soil is exposed.

Depending on severity, fire will change plant communities. In much of Belarus, a severe stand-replacing fire would likely be followed by new site preparation and a new plantation being established. Such a situation would result in the new young trees having easier access to near surface ^{137}Cs, unless these surface layers were also burned and the ^{137}Cs inventory reduced. A potential loss in surface litter will also change germination and growing conditions for many plants. Unless dead and decomposing material is removed, there will be a greater substrate for decomposing organisms, and mushrooms (locally sought as forest produce) may flourish. Additional harvesting issues need to be addressed after fire, such as salvaging dead trees. Volunteer species that may flourish after a fire may need to be controlled so that they do not compete with the new plantations. The after-fire ecosystem will be different than the before-fire case.

7. The Implications of Fire to the Radiation Regime

Quantitatively, we know very little about the expected changes to the radiation regime following fire in the contaminated forests. Some of the uncertainty is caused by the wide range of burn severities that are possible. Burn severity can be estimated by applying a fire behaviour model such as the FBP model outlined previously. The FBP model makes quantitative estimates of the surface fuel consumption, which relates directly to the depth of the burn. Further, we can use the model to estimate conditions where we start to consume foliage in the tree crowns. However, the model does not directly predict endpoints such as tree death which, in the case of a surface fire of a given intensity, depends on tree species and size (e.g., bark thickness, height to live crown).

Even if we could accurately predict burn severity, we still have little field experimental evidence of what happens to Chernobyl-deposited ^{137}Cs following fires. For example, charcoal production can be important, and this charcoal can adsorb phenolic compounds [34]. Could charcoal be a method to adsorb and sequester ^{137}Cs following fires? And what will be the relative amount of leaching and root uptake of ^{137}Cs after a fire? These issues appear to be prime areas for field experimental research. Table 1 outlines qualitatively some potential effects that may occur depending on burn severity.

TABLE 1. Potential Effects of Fire on the Radiation Regime

Effects	Relative Burn Severity		
	Low	Moderate	High
Surface fuel consumption	some	some	much
Crown fuel consumption	none	some	much
Surface fire temperature	<400°C	300-600°C	>500°C
Atmosphere losses	low, smoke only	moderate, mostly smoke	high, smoke and gas
Ash enrichment	some	highest	lower
Biomass changes	low	moderate	high
Tree growth	little effect	some death	stand replaced
Ecosystem "function"	surface flora and fauna affected	species changes	large impact
Soil surface ^{137}Cs	some mobilized	enriched ash	lost
Mineral soil ^{137}Cs	little impact	some leaching	enhanced leaching
Deep leaching ^{137}Cs	little	some	more
Charcoal production	little	some	more

Measurements have verified that fires will increase ^{137}Cs concentration in the atmosphere [26,27] causing the radiological dose to the general population to increase after a fire. Further, fire fighters exposed to higher concentrations of smoke and gases will receive enhanced doses. The issue here is quantifying the dose amount, which depends on a combination of environmental concentration and exposure time. In addition to the direct exposure caused by inhalation and immersion in the plume, deposition to surrounding areas can result in ingested doses if agricultural products are contaminated. Further resuspension of deposited radionuclides can result in more widespread dispersion. However, these concentrations will quickly become dilute, and although the collective dose may increase, the incremental dose to an individual will likely be minimal in most cases. Calculations of the dose commitment for the most severe case of suspension by fire are needed to put into context with the magnitude of the original accidental release. Given that only a small portion of the forests are likely to burn in a given year, and only a small part of the inventory could be resuspended, the dose impact to the general population may be negligible.

Even if the radiological dose to the general population from fires is small, exposure to forest workers may still be a concern. Doses to forest workers in the more contaminated areas, even without fires, may limit harvesting [35]. Enrichment of ash with ^{137}Cs may be an issue if uncontaminated shielding materials are removed, so that the radiation field increases. For example, fire may remove the top litter and organic layers of the soil, which today are less contaminated than the lower layers. This could expose the 5- to 10-cm-deep soil layer where much of the ^{137}Cs has migrated, and therefore increase the local radiation field. However, a deeper burn could mineralise much of the ^{137}Cs and make it more easily leached to deeper layers, thereby decreasing the radiation field. In these cases, burn severity needs to be considered before any general predictions can be made.

Salvage harvesting may be practised following a fire where dead trees are harvested. In these instances, forest workers may be exposed to the freshly enriched ash with little shielding, and it is likely that the post-fire radiation field will be greater than the pre-fire one. Also, immersion exposure may increase because enriched particles will be deposited on skin and clothing. Aside from the fire fighters, these salvage workers may be the most exposed group that receive increased risk from fires.

8. Conclusions

Fire is a periodic, inevitable event occurring in northern forest ecosystems, including forests contaminated by the Chernobyl accident. Although we have information and models to predict many aspects of fire behaviour, the radioecological effects are uncertain. This makes it difficult to assess the true impacts of fires, especially on the changes to radionuclide cycling in the forest. It is most likely that fires will mobilise ^{137}Cs from the upper soil layers, although this depends on burn severity. It may be possible to use prescribed fire to manage forests to reduce radiological risk, but such a plan would need substantial supporting research. If this research demonstrates that risk will increase, irrespective of the type of burn that can be achieved, then a program of landscape management to limit fires may be warranted. In either case, the influence of fire needs to be incorporated into radionuclide cycling models and predictions of the future radiation regime in contaminated forests.

9. Acknowledgements

We thank Dr. Mike Weber for comments on a draft of this manuscript and Kelvin Hirsch for permission to use the FBP system diagram. The senior author also thanks Marty Alexander and other CFS Fire Network scientists for discussions of the FBP system, and Brad Hawkes for information on soil temperatures. Travel funds for the senior author was kindly provided by NATO.

10. References

1. Molodykh, V.G. (1993) Radiological consequences of forest fires. Preprint IREP-4, Academy of Science of Belarus, Inst. of Radioecological Problems, Minsk, 17 pp. (in Russian).
2. Smith, C.B., Amiro, B.D., Lewis, G., MacAulay, E., and Stauber, D. (1993) Assessing the state of forest fire protection in the Republic of Belarus, Final Report, Fire Management Mission. Canadian Interagency Forest Fire Centre, Winnipeg, Canada.
3. Ipatyev, V., Bulavik, I., Baginsky, V., Goncharenko, G., and Dvornik, A. (1998) Forest and Chernobyl. Forest ecosystems after the Chernobyl nuclear power plant accident, 1986-1994. *J. Environ. Radioactivity* (in press).
4. Ipatyev, V. (ed.) (1994) Forest and Chernobyl. Forest Institute of National Academy of Science of Belarus, Minsk, Belarus. (in Russian).
5. Schell, W.R., Linkov, I., Myttenaere, C., and Morel, B. (1996) A dynamic model for evaluating radionuclide distribution in forests from nuclear accidents. *Health Phys.* 70, 318-335.
6. Dvornik, A.M. and Zhuchenko, T.A. (1995) Behavior of ^{137}Cs in pine stands of Belarus Polesye: modelling and prediction. *ANRI (J. Radioecology)* 3/4, 59-66. Moscow (in Russian).
7. Tikhomirov, F.A., Shcheglov, A.I.,and Sidorov, V.P. (1993) Forests and forestry: radiation protection measures with special reference to the Chernobyl accident zone. *Sci. Total Environ.* 137, 289-305.
8. Guillitte, O. and Willdrodt, C. (1993) An assessment of experimental and potential countermeasures to reduce radionuclide transfers in forest ecosystems. *Sci. Total Environ.* 137, 273-288.
9. Guillitte, O., Tikhomirov, F.A., Shaw, G., Johanson, K., Dressler, A.J., and Melin, J. (1993) Decontamination methods for reducing radiation doses arising from radioactive contamination of forest ecosystems - a summary of available countermeasures. *Sci. Total Environ.* 137, 307-314.
10. Merrill, D.F. and Alexander, M.E. (1987) Glossary of forest fire management terms. National Research Council of Canada NRCC No. 26516, 4th ed., Ottawa, Canada.
11. Zackrisson, O. (1977) Influence of forest fires on the northern Swedish boreal forest. *Oikos* 29, 22-32.
12. Engelmark, O. (1984) Forest fires in the Muddus National Park (northern Sweden) during the past 600 years. *Can. J. Bot.* 62, 893-898.
13. Stocks, B.J., Fosberg, M.A., Lynham, T.J., Mearns, L., Wotton, B.M., Yang, Q., Jin, J-Z., Lawrence, K., Hartley, G.R., Mason, J.A., and McKenney, D.W. (1998) Climate change and forest fire potential in Russian and Canadian boreal forests. *Climatic Change* 38, 1-13.
14. Grishin, A.M. (1997) Mathematical modelling of forest fires and new methods of fighting them. Publishing house of Tomsk State University, Tomsk, Russia, 390 pp.
15. Grishin, A.M. (1992) Mathematical modelling of forest fires and new methods of fighting them. Nauka Pub. Novosibirsk Russia. 407 pp. (in Russian).
16. Albini, F.A. and Reinhardt, E.D. (1995) Modelling ignition and burning rate of large woody natural fuels. *Int. J. Wildland Fire* 5, 81-91.
17. Stocks, B.J., Lawson, B.D., Alexander, M.E., VanWagner, C.E., McAlpine, R.S., Lynham, T.J., and Dube, D.E. (1989) Canadian forest fire danger rating system: an overview. *For. Chron.* 65, 258-265.
18. Forestry Canada (1992) Development and structure of the Canadian forest fire behavior prediction system. Forestry Canada Inf. Rep. ST-X-3. Ottawa, Canada.
19. Hirsch, K.G. (1996) Canadian forest fire behavior prediction (FBP) system; user's guide. Natural Resources Canada, Canadian Forest Service, Spec. Rep. 7, Edmonton, Canada.
20. Van Wagner, C.E. and Pickett, T.L. (1985) Equations and FORTRAN program for the Canadian forest fire weather index system.. Canadian Forestry Service, For. Tech. Rep. 33, Ottawa, Canada, 18 pp.
21. CFS (Canadian Forestry Service) (1987). Tables for the Canadian forest fire weather index system. Can. For. Serv., For. Tech Rep. 25, Ottawa, Canada, 48 pp.
22. Greben'kov, A.J., Jouve, A., Rolevich, I.V., and Savushkin, I.A. (1994) Possible technologies for Belarus forest site remediation after Chernobyl accident. Proc. Nucl. Haz. Waste Manage. Topical meeting Spectrum '94, Atlanta, U.S.A. Aug. Vol. 3, pp 1640-1644, Amer. Nucl. Soc.
23. Mamikhin, S.V., Tikhomirov, F.A., and Shcheglov, A.I. (1997) Dynamics of ^{137}Cs in the forests of the 30-km zone around the Chernobyl nuclear power plant. *Sci. Total Environ.* 193, 169-177.
24. Evans, C.C. and Allen, S.E. (1971) Nutrient losses in smoke produced during heather burning. *Oikos* 22, 149-154.
25. Lee, M.J. (1988) The modelling of smut and smoke emissions from straw fires. *Agric. For. Meteorol.* 42, 321-337

26. Dusha-Gudym, S.I. (1996) The effects of forest fires on the concentration and transport of radionuclides. in J.G. Goldammer and V.V. Furyaev (eds.), *Fire in Ecosystems of Boreal Eurasia*, Kluwer Academic Publishers, Dordrecht, pp. 476-480.

27. Dusha-Gudym, S.I. (1993) Forest fires on radionuclide contaminated territories. Guarding and protection of forest, mechanization, using of forest. Revue Information, 9, Federal Serv. Russian Forestry, Moscow, 50 pp. (in Russian).

28. Amiro, B.D., Sheppard, S.C., Johnston, F.L., Evenden, W.G., and Harris, D.R. (1996) Burning radionuclide question: what happens to iodine, cesium and chlorine in biomass fires? *Sci. Total Environ.* 187, 93-103.

29. Hungerford, R.D., Harrington, M.G., Frandsen, W.H., Ryan, K.C., and Niehoff, G.J. (1990) Influence of fire on factors that affect site productivity. Symp. Management and Productivity of Western-Montane Forest Soils, Boise, ID, U.S.A.

30. Frandsen, W.H. and Ryan, K.C. (1986) Soil moisture reduces belowground heat flux and soil temperatures under a burning fuel pile. *Can. J. For. Res.* 16, 244-248.

31. Smith, D.W. and Sparling, J.H. (1966) The temperature of surface fire in jack pine barren. I. The variation in temperature with time. *Can. J. Bot.* 44, 1285-1291.

32. Paliouris, G., Taylor, H.W., Wein, R.W., Svoboda, J., and Mierzynski, B. (1995) Fire as an agent in redistributing fallout ^{137}Cs in the Canadian boreal forest. *Sci. Total Environ.* 160/161, 153-166.

33. Greben'kov, A.J., Trubnikov, V.P., Nikolaev, V.A., and Surzhikov, G.T. (1993) Radioactive waste immobilization technology for Belarus site remediation after Chernobyl accident. Proc. 1993 Int. Conf. Nuclear Waste Management and Environmental Remediation, Prague, Czech Republic, Sept., Amer. Soc. Mech. Eng.

34. Zackrisson, O., Nilsson, M.-C., and Wardle, D.A. (1996) Key ecological function of charcoal from wildfire in the boreal forest. *Oikos* 77, 10-19.

35. Szekely, J.G., Amiro, B.D., Rasmussen, L.R. and Ford, B. (1994) Environmental assessment of radiological consequences for forestry in contaminated areas of the Republic of Belarus. Consultant report to the World Bank, Washington, U.S.A. January, 57 pp. + appendices.

CONTRIBUTION OF FOREST ECOSYSTEM TO DOSE FORMATION FOR INHABITANTS OF THE UKRAINIAN POLESYE

V. GIRIY, I. YASKOVETS, Yu. KUTLAKHMEDOV, V. ZAITOV,
A. KUPRIYANCHUK, V. ONISCHUK
Institute of Radioecology, L.Tolstoy St. 14, Kiev 252033 UKRAINE

R. HILLE and F. ROLLOFF
Research Center Julich, D-52425, Julich, GERMANY

1. Introduction

After the Chernobyl accident large areas of forests in Ukraine received a significant fallout of radionuclides. During the next 50-70 years the most significant problem for the general population of Ukraine is that of ^{137}Cs (see Table 1 for ^{137}Cs data). Despite the fact that the soil's contamination levels in the western part of Ukrainian Polesye (the Volyn' and Rivne Regions) are relatively low, the problem of forest component of the total dose from internal exposure is of main concern, mainly because of the soil's and climatic conditions resulting in persistent high levels of ^{137}Cs in food products in contaminated areas of those Regions [1-4].

TABLE 1. Areas of radionuclide contaminated forests in the northern Ukraine

Region	Total Fore Areas	Areas of forests (km^2) With contamination density of $Ci\ km^{-2}$ ($kBq\ m^{-2}$)						Total amount of Cs-137 activity, Ci (Bq)
		<1,0 (3 (37-74	1-2 (74-185	2-5 (185-370	5-10 (370-555)	10-15	>15 (55	
Zhytomyr	7322	2924	1825	1583	503	164	324	21083 (7,8E+14)
Rivne	6715	2936	2153	1516	107	3	0	11315 (4.2E+14)
Kiev	3723	1780	1293	382	130	55	93	7494 (2.8E+14)
Chernihiv	3486	2738	474	231	33	9	0,6	4402 (1.6E+14)
Volyn'	1784	1362	369	53	0	0	0	1918 (7.1E+13)

The contribution of forests to the total dose to the public from radiocaesium exposure can be split into external and internal components.

The external dose arises from direct irradiation due to radionuclides present in forests or incorporated into construction materials. Generally, in the presently populated forested territories in Ukraine, the external forest component can be critical for the general public in the areas that are the most contaminated (mostly in the Zhytomyr and Kiev Regions) and for residents of the evacuated zone. Also, there is occupational

I. Linkov and W.R. Schell (eds.), Contaminated Forests, 325-332.

exposure that can be received through working in contaminated forests and through irradiation from timber cut for industrial purposes and for fuel. Now this component can be practically important only for forestry workers in the most contaminated areas and for residents of the evacuated zone [1, 5].

After the initial phase following a nuclear accident such as Chernobyl, ingestion pathways become increasingly important in formation of the exposure dose to humans. Large forest areas in the Ukrainian Polesye (UP) are traditionally used for growing private fodder and as pastures for cattle. Generally privately-owned livestock graze on marginal land (mainly areas of forest landscapes). This privately used land is of little value to collective farms. That causes higher intake of radionuclides from the soil. On the other hand, in families of the resident population in the UP, especially in remote villages of its western part, virtually all food is home-produced or gathered in forests; the consumption of mushrooms and berries contribute sufficiently to the population's diet in the UP [6-8]. Also, countermeasures implemented by the authorities after the Chernobyl accident were known to have been mainly oriented to collective farms rather than private ones [6, 7]. Therefore rural populations in the radionuclide contaminated zone may be at risk of high internal exposure, mainly because of the high levels of radionuclides in home-produced animal products (milk, meat) and so-called 'forest gifts' (mushrooms, berries as well as herbs, nuts, occasionally game, birds).

The forest-derived component of the total dose to the resident population from internal and external exposure is studied herein.

2. Materials and methods

Sets of data have been collected in 1991-1998 for radioecological, natural and socio-economic conditions, internal exposure doses (body burden measurements using whole body counters) for numerous settlements situated throughout the Chernobyl fallout contaminated zone located in three Regions. Detailed data are available for twelve villages in Manevychi District of the Volyn' Region; Velyky Cheremel, Vezhytsia, Stare Selo, Drozdyn' in the Rivne Region; Narodychi, Bazar, Chyhyri, Olevs'k in Zhytomyr Region. Those settlements have been chosen mainly because of the fact that they are partially surrounded by forests (mainly pine) (see Table 2, more details are in [1]). The levels of soil contamination in order of 1000 $kBq\ m^{-2}$ are detected only outside Narodychi and Bazar. The most contaminated forest areas are closed to the general public, but this prohibition is sometimes violated.

The comprehensive description of the data used can be found elsewhere [1-3, 6-9]. The studies in the Volyn' Region involving a mobile laboratory and local specialists are currently continuing.

The data include: contents of radionuclides (mainly ^{137}Cs) in soils, fodder, locally produced foodstuffs (milk, meat, potatoes, fresh and dried mushrooms, berries, and some other items), forest products (mainly mushrooms, berries), ^{137}Cs and ^{134}Cs content in human body and the survey-determined diet with respect to main domestic and forest products. Each datum in the database is mainly accompanied by a name of resident (if appropriate), by a particular area used for growing private fodder or collecting forest products (i.e. a location where grassy fodder/hay or a forest product is originated from). In general, all main natural pastures and hay-making areas in the settlements under study are covered by data collected.

TABLE 2. The ratio of forest-covered area to the total area within the 6 km circle around the settlements under study (per cent)

Settlement	Region	Ratio	Settlement	Region	The ratio
Galuziya	Volyn'	72	Bazar	Zhytomyr	13
Velyky Cheremel	Rivne	62	Narodychi	Zhytomyr	25
Vezhytsia	Rivne	65	Chyhyri	Zhytomyr	16
Stare Selo	Rivne	75	Olevs'k	Zhytomyr	48
Drozdyn'	Rivne	84			

Annual effective doses for internal exposure from ^{137}Cs for adults have been assessed by calculations based on the ECOMODEL, a compartment model describing the radionuclide transport through food chains from the soil to the human body, and compared with direct estimation (according to [10]) based on the annual ingestion intake (AII) of ^{137}Cs [2].

The model has been validated using the measured data of ^{137}Cs concentration in foodstuffs and the collected whole body counters (WBC) data [1- 3, 8, 9].

The direct estimation for internal exposure dose caused by the consumption of a forest food product was performed as:

$$D = k I r P, \qquad (1)$$

where D is the above-mentioned dose, mSv year^{-1}, k is the dose coefficient according to [10], $mSv Bq^{-1}$; I is the ingestion rate for this product in a settlement, kg day^{-1}; r is the food processing retention factor; P is the weighted-mean specific activity in this product, $Bq kg^{-1}$.

3. Results and discussion

While analyzing the ^{137}Cs contamination levels for foodstuffs it turned out that the greatest ^{137}Cs content variations are for 1) animal foodstuffs produced using forest areas (milk as well as meat); 2) mushrooms and forest berries. A transfer factor (TF) which is equal to the ratio of radionuclide specific activity in a dry product to its surface activity on soil ($m^2 kg^{-1}$) and an accumulation factor which is equal to the ratio of radionuclide specific activity in a dry product to its specific activity in soil (non-dimensional) are used as the parameters characterizing migration of ^{137}Cs from the soil to the forest products (Table 3). For unimproved forest pastures up to 60-80 per cent of the ^{137}Cs content is still kept in the upper (0-5 cm) layer. Therefore TFs 'soil - grassy part of plants' are one order higher than ones for improved pastures [1-3].

Also, the turf of unimproved pastures usually contains some grass grown during previous seasons which is also partially eaten by cattle while grazing. It causes the most contamination in milk [3]. The intake of ^{137}Cs by mushrooms is particularly high in some areas of the Volyn' and Rivne Regions and relatively low in those areas of the eastern part of the UP where gray forest soils and black soils predominate [1, 4].

TABLE 3. The range of the measured values of the accumulation and transfer factors for forest pastures in the settlements under study (1996-1997) [1-3]

Region	Type of plants	Type of soil	Transfer factor min./aver./max. $10^{-3} \, m^2 \, kg^{-1}$	Accumulation factor min./aver./max.
Zhytomyr	grass	Podzol sandy loamy	0.7/ 1.5 /2.7	0.15/ 0.19 /0.25
Zhytomyr	grass	Podzol sandy	0.2/ 1.25 /2.2	0.05/ 0.16 /0.4
Zhytomyr	grass	Peat	0.5/ 25.1 /113.4	0.05/ 2.9 /12
Rivne	grass	Podzol sandy loamy	0.6/ 18.2 /67.6	0.16/ 3.0 /16.95
Rivne	grass	Turf sandy	1.4/ 6.9 /19.8	0.2/ 0.93 /1.65
Rivne	grass	Podzol sandy	1.1/ 6.1 /14.7	0.35/ 0.9 /1.5
Rivne	grass	Peat	3.1/ 31.2 /71.7	0.5/ 3.9 /6.5
Rivne	mushrooms	Podzol sandy	101.7/ 124.1 /146.4	7.4/ 19.1 /30.8
Rivne	berries	Podzol sandy loamy	4.2/ 4.3 /4.6	0.4/ 0.8 /1.2
Rivne	berries	Podzol sandy	5.8/ 7.1 /8.3	0.4/ 1.1 /1.75
Volyn'	grass	Podzol sandy	1.9/ 38.2 /147.1	0.3/ 2.9 /16
Volyn'	grass	Peat	90.7/ 118.0 /264.2	0.6/ 4.6 /18.7

Some of model based calculations are shown in Table 4. All the data presented in the Table are given for ^{137}Cs.

The calculated results for the majority of our case studies are in good or satisfactory agreement with the measured data of ^{137}Cs concentration in foodstuffs and the collected WBC data. The soil factor (column 2), introduced while validating the ECOMODEL, integrates the effects of the main physical-chemical factors influencing the radiocaesium intake by plants (see [1] for details). Since the greatest variations of both ^{137}Cs content and consumption rate occur for mushrooms, we decided to use the ranges of individual values for mushrooms for each settlement. To assess the contribution of mushrooms to the annual exposure dose from ^{137}Cs the equation is derived on the basis of calculations using the ECOMODEL:

$$d = 4.3\text{E-}6FC, \tag{2}$$

where d is the exposure dose from ^{137}Cs, mSv year^{-1}; F is the ingestion rate for mushrooms, g day^{-1}; C is the ^{137}Cs concentration in mushrooms, $Bq \, kg^{-1}$.

In Table 5 the relative contributions of the main foodstuffs to the exposure dose from ingestion of ^{137}Cs in two settlements that are typical for the eastern and western parts of the UP respectively are presented.

The differences between those parts arise from the landscape conditions and peculiarities of private farming systems.

In the radionuclide contaminated areas of the western part of UP (the Volyn' and Rivne Regions) remote settlements are virtually surrounded by forests. In private farms a significant part of natural pastures and hay-making areas are located on forest lands and fluxes of fodder come mainly from forest areas; mushrooms and berries contribute sufficiently to the population's diet [7, 8].

TABLE 4. The ^{137}Cs contamination data and calculations using the ECOMODEL versus the measured WBC data for several settlements under study (for ^{137}Cs, 1996) [1]

Settlement	Soil factor	Soil contamination density (aver./max.) $kBq\ m^2$	Specific activity in (aver./max.)		Dose of internal exposure, $mSv\ year^{-1}$		Body burden, Bq		Dose of external irradiation, $mSv\ year^{-1}$		WBC data Bq	WBC dose, $mSv\ year^{-1}$
			milk, $Bq\ L$	potatoes, Bq kg	aver.	max.	aver.	max.	aver.	max.	aver.	aver.
Narodychi*	6.5	351/5698	353/5700	36/570	0.56	7.7	1.5E+4	2.3E+5	0.52	8.4	8.9E+3	0.36
Bazar	7.8	362/1380	138/522	14/52	0.27	0.78	6.5E+3	2.2E+4	0.53	2.04	3.0E+3	0.12
Olevs'k	6.0	63/207	98/320	10/32	0.22	0.52	4.9E+3	1.4E+4	0.09	0.31	-	-
Chyhyri	7.0	371/1270	252/870	25/80	0.42	1.23	1.1E+4	3.5E+4	0.55	1.87	-	-
Velyky Cheremel	4.3	144/352	1300/3270	135/322	1.85	4.4	5.3E+4	1.3E+5	0.3	0.74	4.4E+4	1.85
Vezhytsia	4.1	89/172	1060/2050	107/200	1.49	2.8	4.3E+4	8.2E+4	0.19	0.36	9.8E+4	4.03
Stare Selo	3.85	51/115	851/1930	86/192	1.22	2.63	3.5E+4	7.6E+4	0.11	0.24	4.9E+4	2.04
Drozdyn'	3.9	53/148	826/2320	84/230	1.18	3.14	3.4E+4	9.2E+4	0.11	0.31	5.1E+4	2.09
Galuziya	3.8	34/56	90/410	42/470	1.26	3.99	3.3E+4	7.5E+4	0.04	0.14	8.5E+3	0.37

* Narodychi is the settlement with nourishment for the main part of the residents from the centralized governmental purchases made in non-contaminated areas

- no data

TABLE 5. Contribution of the main foodstuffs (per cent) to the internal exposure dose from ingestion of Cs-137 (for the diets with maximum, minimum and averaged activity respectively). Food processing retention factors are taken from [11]

Type of products	Bazar			Velyky Cheremel		
	min.	max.	aver.	min.	max.	aver.
Mushrooms and berries	2	46	24	4	39	21.5
Milk and milk product	71	33	52	79	45	62
Meat	18	9	13.5	3	6	4.5
Potatoes and vegetables	7	9	8	13	8	10.5
Other	2	3	2.5	1	2	1.5

In the radionuclide contaminated areas of the western part of UP, the forest component exceeds 90 per cent of the internal exposure dose for remote villages without good transport connection; the remaining dose can be assigned mainly from potatoes and vegetables grown in private gardens [1, 7, 8]. In the eastern part of the UP, there are many more non-forest areas for private pastures and hay-making where the dose can be lowered. Therefore, the contributions by milk and meat in many cases are smaller, in terms of absolute and relative values, and can be assigned to forests only partially. However, mushrooms and berries can contribute even more to the internal dose due to high levels of ^{137}Cs contamination in these forest products [1]. The absolute values of the forest-derived component of internal exposure dose vary for the settlements under study in the eastern and western parts of the UP from 0.1 to 7.7 and from 0.2 to 5.0 mSv per year respectively, depending on minimal and maximal values of the AII for [137]Cs [1, 2, 7]. Generally, our estimates agree with the data and results obtained for some villages in the Rivne Region in [12].

For the main part of the presently populated forested territories, the significant part of the external dose (from 13 to 70 per cent) for the average individual external dose depending on fractions of time spent in the different locations including house, yard, garden, street, and forest can be assigned to forests. As a rule, radioactive contamination of settlements' territories decreased due to cultivation of the land and decontamination measures implemented after the Chernobyl accident in many locations throughout the radionuclide contaminated zone, mainly in the eastern part of the UP. The external irradiation dose in forest is 1.5 - 2 times higher than the dose in a neighboring non-forested area having the same contamination density [1, 5].

The internal exposure doses to the resident population of the UP have been shown to be strongly depend on the combination of a number of economic, social and demographic factors, including the following: a) low productivity of private and collective farms; b) large number of children in families; c) export of the significant part of foodstuffs collected in forests and produced by private farms to non-contaminated regions; d) selective consumption of locally produced foodstuffs with lower contamination; e) state-funded nourishment for children; f) migrations of residents for temporary work [6, 7]. To study these dependent factors the generalized criteria of assessment have been defined and some practical estimates have been performed [7].

4. Conclusions

The contribution of forests to the internal exposure doses for the resident population in the main part of the Ukrainian Polesye cannot be limited to consumption of forest products. Taking into account a more realistic broader number of agricultural products that are produced using forest areas, the contribution of forests to the internal exposure dose is estimated from a few per cent in the settlements of the Zhytomyr Region to 90 per cent in the remote villages of the Volyn' and Rivne Regions and tends to increase because of the extremely slow rate of decrease of radioactive contamination of forest products and unfavorable socio-economic conditions in the Ukraine. Diets based on privately produced milk and mushrooms can result in dose estimates of up to 7 to 8 mSv year^{-1}.

These conclusions should be taken into account in managing the radionuclide contaminated zone and in planning appropriate countermeasures. In particular, the standard diet used in dose assessment, in which the main contribution to the internal exposure is caused by milk and potatoes, needs to be revised (as shown in [8]).

Comparing results of dose assessment using the ECOMODEL in 1996-1997 and those on the basis of newly adopted National Radiation Safety Standards of Ukraine (NRBU-97) [10] using the same AII, we concluded that the dose assessment based on NRBU-97 and the AII is 1.4-1.6 times more conservative (more details in [2]).

5. Acknowledgments

The main part of the work presented herein has been performed within the framework of the governmental research project "Comprehensive assessment of radioecological situation, validation of radionuclide migration models" (Contract No. 01/6N-96) funded by the Ukrainian Ministry of Emergencies, formerly the Ukrainian Ministry of Chernobyl. Portions of this work were performed under the auspices of the German Federal Ministry of Environment (the measuring program coordinated by the Research Center Julich). Additional support has been provided by the Ukrainian Academy of Agricultural Sciences under a budget program.

The authors are grateful to the local authorities and specialists in the radionuclide contaminated zone (particularly Mr. Grigory Karpchuk, a radiologist) for their invaluable support.

Special gratitude is expressed to Mr. John Karwatsky, NATO Center in Kiev, for his support and assistance in preparing the presentation and this article.

6. References

1. Giriy, V., Yaskovets, I., Zaitov, V., Kupriyanchuk, A., Onischuk, V., and Akinfiyev, G. (1996) Comprehensive assessment of radiological situation and validation of radionuclide migration models. Report of the Institute of Radioecology. State Registration No. 0196U003796. The Ukrainian Ministry of Emergencies (formerly the Ukrainian Ministry of Chernobyl), Kiev.*

2. Kutlakhmedov, Yu., Yaskovets, I., Giriy, V., Zaitov, V., Kupriyanchuk, A., Kutlakhmetov, V., and Sklabinsky, E. (1997) Comprehensive assessment of radiological situation and validation of radionuclide migration models. Report of the Institute of Radioecology. The Ukrainian Ministry of Emergencies, Kiev.*

3. Zaitov, V., Kupriyanchuk, A., Kutlakhmedov, Yu., and Smirnov, A. (in press) Radionuclide pollution of food products in the settlements under study in the Volyn' Region, Ukraine: general features, seasonal changes, predictive assessment. *Reports of the National Conference "Science. Chernobyl-97", February 11-12, 1998, Kiev.* The Ukrainian Ministry of Emergencies, Kiev.*

4. Krasnov, V. (1998) Radioecology of forests of the Ukrainian Polesye. " Volyn' ", Zhytomyr.

5. Hille, R., Hill, P., Heinemann, K., and Heinzelmann, M. (1996) The impact of the Chernobyl accident - an evaluation from the German perspective. Final report of a three-years investigation. Berichte des Forschungszentrums Julich; 3186. ISBN 0944-2952. Julich.

6. Kupriyanchuk, A. (1997) Present state, socio-economic problems and perspectives of the Unconditional Resettlement Zone. *Proc. of the National Conference "Science. Chernobyl-96", February 11-12, 1997, Kiev.* ISBN 966-95131-0-3. The Ukrainian Ministry of Emergencies, Kiev.*

7. Kupriyanchuk, A. (in press) Effects of socio-economic conditions on individual and collective doses to the population of the Volyn' and Rivne Regions, Ukraine, from ingestion of ^{137}Cs: generalized criteria of assessment and practical results. Paper accepted for presentation at the First International Symposium *Issues in Environmental Pollution: The State and Use of Science and Predictive Models* to be held in Denver, Colorado on August 23-26, 1998. *Environmental Pollution.*

8. Kupriyanchuk, A., Zaitov, V., Kutlakhmedov, Yu., and Yaskovets, I. (1998) The population's diet in the radionuclide contaminated zone (upon the results of the questionnaire survey in the Volyn' Region). *Proc. of the National Conference "Science. Chernobyl-97", February 11-12, 1998, Kiev.* ISBN 966-95131-2-X. The Ukrainian Ministry of Emergencies, Kiev.*

9. Akinfiyev, G. and Kupriyanchuk, A. (1997) Irradiation doses to the population in the radionuclide contaminated zone in Ukraine: distribution on professional and age groups. *Proc. of the National Conference "Science. Chernobyl-96", February 11-12, 1997, Kiev.* ISBN 966-95131-0-3. The Ukrainian Ministry of Emergencies, Kiev.*

10. *National Radiation Safety Standards of Ukraine (NRBU-97). National Hygienic Standards* (1997), Kiev.

11. Rantavara, A. (1987) *Radioactivity of vegetable and mushrooms in Finland after the Chernobyl accident in 1986.* STUK-A5. Finnish Center for Radiation and Nuclear Safety, Helsinki.

12. Strand, P., Howard, B., and Averin, V. (1996) *Transfer of radionuclides to animals, their comparative importance under different agricultural ecosystems and appropriate countermeasures. The final report of ECP-9.* CEC, EUR 16539EN. Brussels, Luxembourg.

---

* in Ukrainian and Russian. English versions can be obtained from the Institute of Radioecology

GAMMA RAY EXPOSURE DUE TO SOURCES IN THE CONTAMINATED FOREST

V.GOLIKOV, A.BARKOVSKI, V.KULIKOV, M.BALONOV.
Institute of Radiation Hygiene, Mira Street 8, 197101 St.Petersburg, RUSSIA

A.RANTAVAARA, V.VETIKKO
Finnish Centre for Radiation and Nuclear Safety, P.O.Box 14, FIN-00881, Helsinki, FINLAND

1. Introduction

A contaminated forest can contribute significantly to human exposure for a few decades after the initial contamination. The general strategy for modelling external exposure of people in a contaminated forest includes the solution of two main tasks:

- the development of an ecological model for the migration of the relevant radionuclides in evaluating of their content in various components of the forest ecosystems. This applies at times after the radioactive fallout which occurred according to the given scenario of the accident;
- the development of a calculational model for evaluating the kerma rate in air and the effective dose rate to people, at various tames from the external gamma radiation field in the contaminated forest. This applies to the given distribution of activity in the various compartments.

When these two tasks are complete, the general goal of determing the dynamics of exposure to the people in a forest, which has undergone radioactive contamination, will be achieved.

The calculation of the effective dose from external exposure involves the following major steps:

- computation of energy and angular distributions of gamma-radiation flux of monoenergetic sources distributed in the environmental media;
- calculation of kerma rate in free air;
- calculation of the effective dose of the incident gamma-radiation, characterized above in terms of energy and angular distribution, for each of the initial energies considered; and
- calculation of the effective dose for specific radionuclides, considering the gamma-quanta energies and yields per decay for the nuclides present.

I. Linkov and W.R. Schell (eds.), Contaminated Forests, 333–341.
© 1999 *Kluwer Academic Publishers. Printed in the Netherlands.*

The goal of the present work was to complete the first two of the above tasks for a series of radionuclides, which potentially contribute significantly to the external exposure rate from deposited radioactivity in using of the information collected from different forest types.

2. Materials and methods

2.1 MODEL OF CONTAMINATED FOREST

The dose coefficients for external irradiation to the human body from radionuclides distributed in the environment have been tabulated in a number of reports [1,2]. The authors of these works considered the following situations of human exposure and the corresponding source geometries, respectively:

- a semi-infinite source is assumed for submersion in contaminated air and an infinite source is assumed for immersion in contaminated water;
- the two-layered system soil-air is considered for calculations of dose coefficients for human exposure from contaminated soil.

In the case of human exposure in the forest, the number of radiation sources should correspond to the number of compartments included in the ecological model of radionuclide migration in the forest ecosystem. Besides that, the different components of the forest ecosystem are not only the emitting media, but also the absorbing media, with the absorption magnitude between air and water.

As a first approximation, the vegetation will be represented by trees only, without consideration of the understorey in connection with the primary fallout. In calculating the external exposure, this compartment will be used to evaluate the attenuation coefficients for gamma-radiation. To consider the vertical distribution of radionuclides deposited on the trees, we divide the biomass compartment into parts by a horizontal plane with limits at the height of the crown.

The basic source for the dose calculations is taken to be an infinite isotropic planar source of monoenergetic photons, located at a specified depth in the soil, or at specified heights in the forest biomass. The kerma rate due to a distributed source in the soil or in the forest biomass may be readily computed from the kerma rates due to a series of planar sources.

The forest ecosystem is represented by a three-layered compartments. The two upper layers (crowns and trunks of trees) have the same elemental composition but different physical density, and the third layer is the soil (See Fig.1).

Here A_1 and A_2 are the specific activities of the layers (homogeneous over the volume) in Bq/g; ρ_1 and ρ_2 are the layer densities in kg/m^3; h_1 and h_2 are the heights in m of the trunk layer and crown layer, respectively.

In the calculations, we used the following atomic compositions (mass percents) of forest ecosystem model components and their physical densities:

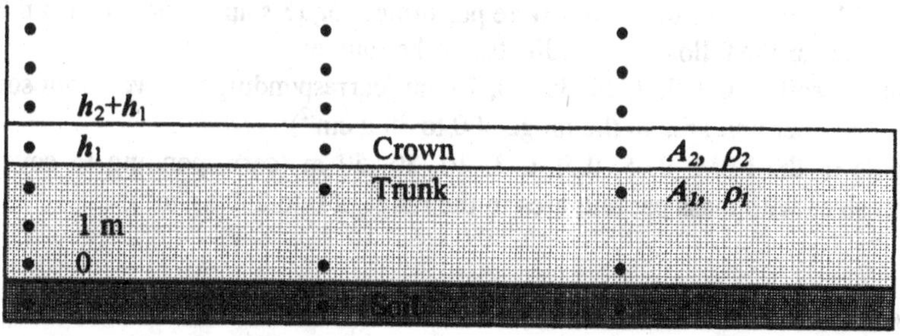

FIG. 1. Spatial structure of the forest ecosystem model for external dose calculations.

- Air: N - 75.5, O - 23.2, Ar - 1.3; the composition is for the condition of 40% relative humidity, a pressure of 760 mm Hg, a temperature of 20 °C, and density of 1.2 kg·m^{-3}.
- The atomic composition of the forest biomass was determined from data in the literature [3]: C - 27, O - 63.9, H - 8.6, N - 0.19, Ca - 0.13 and 0.18 due to the trace amounts of other elements (K, P, Mg, Fe, Zn, S). The physical density of the forest biomass depends on its type, using as a basis the literature data [3]. For calculations we have chosen a range from the set values of the forest biomass density of from 0 to 5 kg/m^3 (the first value corresponds to the free air). We also considered a set of values of the forest height: from 1 to 50 m. The data presented above belong only to the forest biomass and do not include the air that is also present in this medium, and in our calculations we used the sum of this two components of forest-equivalent medium (FEM).
- The atomic composition for soil was taken from paper [2]: Si - 26.2, Al - 8.5, Fe - 5.6, H - 2.2, O - 57.5. The soil density was taken to be 10^3 kg/m^3.

2.2 THE BRIEF DESCRIPTION OF A METHOD OF CALCULATION

To develop a sufficiently simple model for formation of human external exposure doses from all elements of the forest ecosystem subjected to radioactive contamination, we performed a series of calculations of the gamma-radiation dose rate from planar isotropic sources in FEM, using the computer code VICAR-2 for the numeric solution of the transport equation [4]. This code uses a multi-group approximation of the method of integral equations [5]. We performed the calculations using the 23-group system of constants, which was created especially for this kind of problem [6].

We calculated numeric values for the kerma rate at the height of 1 m above the ground created by gamma radiation of planar isotropic sources located at different depths in the soil and at different heights in the FEM. This conditional medium has an average mass density and weighted atomic mass of uniformly mixed biomass of trees and air in the corresponding layer of the forest ecosystem model (see Fig. 1). We used 5 densities of forest biomass: 0, 0.5, 1, 2, and 5 kg/m^3. In the calculations, air mass was added to the forest biomass, so that the density of FEM composition was 1.2, 1.7, 2.2,

3.2, and 6.2 kg/m³. The calculations were performed for 18 source energies from 20 keV to 3 MeV and for the following coordinates of the source:

- depths in soil: 0, 0.3, 1, 3, 10, 30, 50, 70 cm (corresponding to cover from soil slabs with mass per unit area in the range of 0 to 70 g·cm⁻²);
- heights in the FEM: 0.05, 0.5, 1, 3, 10, 30, 50 m (corresponding to cover from forest slabs with mass per unit area in the range of 0 to 250 kg·m⁻²).

3. Presentation of calculation results for radionuclides

In this report, we present the results of kerma rate calculations for the following radionuclides: Cs-137→Ba-137m, Cs-134, I-131 which potentially contribute significantly to the external exposure rate from deposited activity. Calculations of kerma rate for radionuclides was performed on the basis of the energies and intensities of the radiations emitted by radionuclides reported in ICRP Publication 38 [7]. Calculation of kerma rate was performed for a point source located at the height of 1 m above the ground. The sources located in the soil and in the forest biomass were considered separately.

3.1 KERMA RATE FROM CONTAMINATED SOIL

At the first step, kerma rate values were calculated at 1 meter above the soil surface from the flat isotropic sources situated on 8 depths ζ (g·cm⁻²) in soil with gamma- spectra corresponding to the listed radionuclides. The complete set of the results includes data obtained for the five values of forest biomass density ρ_l specified above.

A further task was to describe the results obtained by simple analytical expressions. On the basis of the experience in previous research [8], we chose the two-component exponential function:

$$K_a(\zeta, \rho_l) = p_1 \cdot exp(-p_2 \cdot \zeta) + p_3 \cdot exp(-p_4 \cdot \zeta), \tag{1}$$

as the approximating one The parameters p_1, p_2, p_3 and p_4 were determined within the nonlinear method of least squares. Preliminary analysis of the results showed that the parameter values p_3 and p_4 are not influenced by the forest biomass density ρ_l, and parameters p_1 and p_2 depends on it as exponents. Therefore let us rewrite equation (1) as the following:

$$K_a^{sol}(\zeta, \rho_l) = a_1 \cdot exp(-a_2 \cdot \rho_l) \cdot exp(-(a_3 \cdot exp(-a_4 \cdot \rho_l) \cdot \zeta)) + a_5 \cdot exp(-a_6 \cdot \zeta) \tag{2}$$

where: $K_a^{sol}(\zeta, \rho_l)$ is the air kerma rate $[10^{-12} \cdot (Gy \cdot s^{-1})$ per $(Bq \cdot cm^{-2})]$ at the height of 1 m above the ground in the forest with biomass density ρ_l (kg·m⁻³) from a planar isotropic source located in the soil at depth ζ (g·cm⁻²). Parameter values a_1 - a_6 for separate radionuclides are presented in the Table 1.

TABLE 1. Values of the parameters in the equation (2), (4) and (5)

Radionuclide	a_1	a_2	a_3	a_4	a_5	a_6
Cs-137→Ba-137m	3.39	0.094	0.492	0.051	2.38	0.0723
Cs-134	9.46	0.093	0.470	0.055	6.31	0.0699
I-131	2.29	0.091	0.540	0.048	1.72	0.0845

Given an arbitrary distribution in soil of specific activity $A_m(\zeta)$ of the radionuclide, the kerma rate in air at the height of 1 m above the ground in the forest with biomass density ρ_1 can be written as follows:

$$K_a^{soil}(\rho_1) = \int_0^\infty A_m(\zeta) \cdot K_a^{soil}(\zeta, \rho_1) \cdot d\zeta, \tag{3}$$

For the most frequently used distributions of activity in soil, (exponential or homogeneous in a soil layer between depths ζ_1 and ζ_2) we may derive the explicit form of the expression (3).

For the exponential distribution in the form $A_m(\zeta) = A_a \cdot \beta \cdot exp(-\beta \cdot \zeta)$, where $A_m(\zeta)$ is the specific activity of soil (Bq·g⁻¹) at the depth ζ (g·cm⁻²), A_a is the activity per unit area (Bq·cm⁻²) and β is the depth distribution parameter, which is the reciprocal of the relaxation length (cm²·g⁻¹):

$$K_a^{soil}(\beta, \rho_1) = 3.6 \cdot A_a \cdot \left\{ \frac{a_1 \cdot exp(-a_2 \cdot \rho_1)}{1 + \dfrac{a_3 \cdot exp(-a_4 \cdot \rho_1)}{\beta}} + \frac{a_5}{1 + \dfrac{a_6}{\beta}} \right\}, \tag{4}$$

where $K_a^{soil}(\beta, \rho_1)$ is the kerma rate (nGy·h⁻¹); A_a is the activity per unit area (Bq·cm⁻²); and ρ_1 (kg·m⁻³) is forest biomass density.

For homogeneous distribution of radionuclide activity in soil layer between ζ_1 and ζ_2 (g·cm⁻², $\zeta_1 < \zeta_2$) with specific activity A_m (Bq·g⁻¹):

$$K_a^{soil}(\zeta_1, \zeta_2, \rho_1) = 3.6 \cdot A_m \cdot \left\{ \begin{array}{l} \dfrac{a_1}{a_3} \cdot exp((a_4 - a_2) \cdot \rho_1) \cdot exp(-a_4 \cdot exp(-a_4 \cdot \rho_1) \cdot \zeta_1) \cdot \\ [1 - exp(-a_3 \cdot exp(-a_4 \cdot \rho_1) \cdot (\zeta_2 - \zeta_1))] \\ + \dfrac{a_5}{a_6} \cdot exp(-a_6 \cdot \zeta_1) \cdot [1 - exp(-a_6 \cdot (\zeta_2 - \zeta_1))] \end{array} \right\}, \tag{5}$$

where $K_a^{soil}(\zeta_2, \zeta_1, \rho_1)$ is the kerma rate (nGy·h⁻¹), and ρ_1 (kg·m⁻³) is the forest biomass density. Equation (5) is directly applied to the calculation of the kerma rate from the experimentally taken soil samples in the form of layers with measured specific activity of radionuclides.

The results of the calculation of air kerma according to equations (4) and (5) for different radionuclides were compared with data obtained by the method of Monte

Carlo for the flat isotropic source at a depth of 0.5 g·cm^{-2} [9], for the exponential (β = 0.5... 5 cm^2·g^{-1}) distribution of the activity in soil [10], and also for uniform distribution of activity in the layers with thickness of 1 and 15 cm [1]. The results of the comparisons do not differ by more than by 10%.

3.2 KERMA RATE FROM CONTAMINATED CROWNS AND TRUNKS

As in the case above, air kerma rates for the flat isotropic monoenergetic source situated at the different heights in the FEM were calculated. The analysis of the results of the calculations showed that by taking into consideration area changings of FEM parameters of kerma rate values in a point of 1 m above the soil surface, it could be described by following equation:

$$K_a^{FEM}(E,x) = b_1 + b_2 \cdot x - b_3 \cdot ln(x),\qquad(6)$$

where $K_a^{FEM}(E,x)$ is the air kerma rate from the contaminated forest biomass at the height of 1 m above the ground from a plane isotropic source with energy E located in FEM at height x (in units of mean free paths). The parameters b_1, b_2, and b_3 were determined using the nonlinear method of least squares fit and the analytical expressions for any energy were obtained. The additional error caused by using of this equation is not more than 10%.

In the present version of the model, we assume that the distribution of activity over the volume of the specified forest zones is constant within each zone (trunks and crown). Then, using equation (6), the energies and intensities of the radiations of radionuclides and the fulfilling integration in areas of trunks and crown, the equations for kerma rate at 1 m height above the soil surface from contaminated trunks and crown are obtained:

from trunks

$$K_a^{tr}(h_1,\rho_1) = 0.36 \cdot A_m \cdot \omega_1 \cdot \left[R_1 \cdot (x_0 + x_1) + R_2 \cdot (x_0^2 + x_1^2) - R_3 \cdot (x_0 \cdot ln(x_0) + x_1 \cdot ln(x_1))\right],\qquad(7)$$

where x_o=0.12+0.11·ρ_1; x_1=(0.12+0.11·ρ_1)·(h_1-1); ω_1=ρ_1/(0.12+0.11·ρ_1). $K_a^{tr}(h_1,\rho_1)$ is the air kerma rate (nGy·h^{-1}) from contaminated trunks with specific activity A_m (Bq·g^{-1}) at the height of 1 m above the ground; ρ_1 (kg·m^{-3}) and h_1 (m) are density and height of trunk, respectively;

from crown

$$K_a^{cr}(h_2,\rho_2) = 0.36 \cdot A_m \cdot \omega_2 \cdot \left[R_1 \cdot x_2 + R_2 \cdot (x_2^2 + 2 \cdot x_1 \cdot x_2) - R_3 \cdot ((x_1 + x_2) \cdot ln(x_1 + x_2) - x_1 \cdot ln(x_1))\right],\qquad(8)$$

where x_2 = (0.12 + 0.11·ρ_2)·h_2; x_1=(0.12+0.11·ρ_1)·(h_1-1); ω_2=ρ_2/(0.12+0.11·ρ_2). $K_a^{cr}(h_1,\rho_1)$ is the air kerma rate (nGy·h^{-1}) from contaminated crown with specific activity A_m (Bq·g^{-1}) at the height of 1 m above the ground; ρ_2 (kg·m^{-3}) and h_2 (m) are density and height of forest crown. The values of the parameters R_1, R_2 and R_3 are presented in the Table 2.

TABLE 2 The values of the parameters R_1, R_2 è R_3 in the equation (7) and (8).

Radionuclide	R_1	R_2	R_3
Cs-137→Ba-137m	5.60	0	1.28
Cs-134	15.3	0	3.49
I-131	3.86	0.000525	0.887

4. Examples of calculations and discussion.

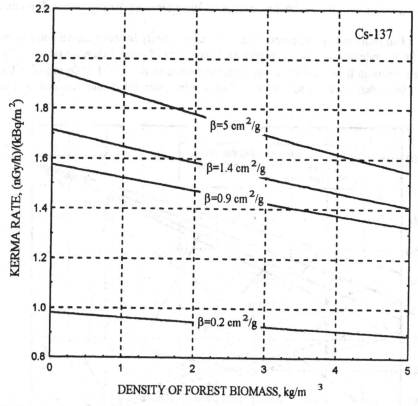

FIG. 2. Air kerma rate at the point 1 m over the soil in the forest with different densities of biomass for various exponential distributions of activity in soil.

In the Figure 2 the results of the calculation of the air kerma rate from Cs-137→Ba-137m for various exponential distributions in soil and for various forest biomass densities are presented. The values of the index (β) of the exponential distribution of activity that correspond to the following naturally observed data were chosen:

• dry fallout (β = 5 cm²·g⁻¹),

340

- wet fallout ($\beta = 1.4$ cm$^2 \cdot$g^{-1}),
- distribution of activity in a pine forest of the Bryansk region in 1993 ($\beta = 0.9$ cm$^2 \cdot$g^{-1}),
- average activity distribution on the open plots of the virgin soil in the Bryansk region in 1993 ($\beta = 0.2$ cm$^2 \cdot$g^{-1}).

It can be seen from the figure that the kerma rate at 1 m above soil surface does not depend significantly on the forest biomass but does depend on the location of the radiation source. The deeper that the radionuclide is in the soil, the smaller is the dependence of the kerma rate from the forest biomass density. It is also necessary to point out the kerma rate above the open plots of virgin soil and under the forest cover differ by a factor of 1.3-1.6 with dependence of forest biomass density (Bryansk region, 1993)

The influence of relocation of the deposited activity between crown and soil on the air kerma rate at 1 m in a forest with various height of trees is demonstrated in Fig. 3. In this scenario it was assumed that from the total deposition of 1 MBq·m^{-2} of ^{137}Cs, p% has been deposited on soil, and (100-p)% were intercepted and retained in tree crown.

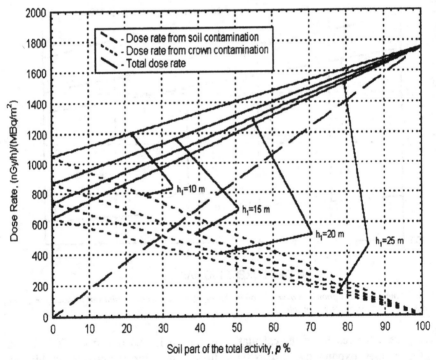

Fig 3. Gamma-radiation dose rate at the point 1 m over the soil in the forest at various activity distributions between crown and soil

For calculation, the following parameters of a forest and its compartment contamination were selected:

<u>crown:</u> h_2=5 m; ρ_2=0.5 kg·m^{-3}; $A_2=\frac{100-p}{100} \cdot 400$ Bq·g^{-1};

<u>trunks:</u> h_1=10 m, 15 m, 20 m, 25 m; ρ_1=1.5 kg·m^{-3}; A_1=0;

<u>soil:</u> exponential depth distribution of activity with β=3 cm^2·g^{-1}; $A_a=\frac{p}{100} \cdot 100$ Bq·cm^{-2}.

In a hypothetical case where there was complete deposition of activity on the soil, the greatest value of the dose rate is observed. Where there was complete deposition of activity on the crown, the dose rate is by 2-3 times less and depends considerably on the height of the forest crown. Thus, it is possible to envisage a protective effect of the forest on the external exposure of a man in comparison with the exposure in a open field. This occurs at an early time after the fallout. At a later time the situation changes to the opposite direction because of the migration of radionuclide in the forest ecosystem.

5. Acknowledegments

This work was partially supported by the European Commission within the project RODOS-C.

References

1. K.F. Eckerman, J.C. Rymon. External exposure to radionuclides in Air, Water, and Soil. Federal Guidance Report No.12. Office of Radiation and Indoor Air, U.S. EPA, Washington, DC 20460 (1993).
2. Saito and P. Jacob. Gamma ray fields in the air due to sources in the ground. Radiation Protection Dosimetry Vol. 58, No. 1 pp. 29-45 (1995). Peter Jacob and Herwig G. Paretzke Gamma-Ray exposure from contaminated soil. Nuclear Science and Engineering 93, pp. 248-261 (1986).
3. STUK-RTC "Protection" Subcontract, 2-nd Interim RTC Report, St.-Petersburg, November 1997.
4. M.N.Nikolaev, B.G.Rjazanov, M.M.Savoskin and A.M.Tsybulya. "Multi-group approximation in the theory of neutrons transfer". Moscow: Energoatomizdat, 1984, 256 pp. (In Russian).
5. R.Bergelson, A.P.Suvorov, B.Z.Torlin. Multi-group methods for calculation of protection from neutrons. Moscow: Atomizdat, 1970. (In Russian).
6. Golikov V.Yu., Barkovski A.N. and Likhtarev I.A. Parameters of the photonic radiation field of plane sources in tissue-equivalent plates. Atomnaya energiya, 1989, V. 67, No 5, pp. 341-346. (In Russian)
7. ICRP (1983) International Commission on Radiological Protection, "Radionuclide Transformations: Energy and Intensity of Emissions," ICRP Publication 38, Annals of the ICRP, Vols. 11-13.
8. Golikov V.Yu., Balonov M.I., and Ponomarev A.V. Estimation of external gamma radiation doses to the population after the Chernobyl accident. In: Chernobyl Papers, vol. 1, 1993, pp. 247-288.
9. Jacob P., Rosenbaum H., Petoussi N., and Zankl M. Calculation of organ doses from environmental gamma rays using human phantoms and Monte Carlo methods. Part II: Radionuclides disributed in the air or deposited on the ground. GSF- Bericht 12/90. 1990.
10. International Commission on Radiation Units and Measurements, Gamma-Ray Spectrometry in the Environment, ICRU Report 53.

THE ROLE OF THE FOREST PRODUCTS IN THE FORMATION OF INTERNAL EXPOSURE DOSES TO THE POPULATION OF RUSSIA AFTER THE CHERNOBYL ACCIDENT

G.J. BRUK, V.N. SHUTOV, I.G. TRAVNIKOVA, M.I. BALONOV,
M.V. KADUKA, L.N. BASALAEVA
*Institute of Radiation Hygiene, 8 Mira Str., 197101, S.-Petersburg,
RUSSIA*

Abstract

It was shown that radioactive decontamination of mushrooms, berries and game during 11 years after the Chernobyl accident was very slow in contrast to agricultural food products (where natural decontamination took place quickly). Use is made of aggregated radioecological data together with a broad assumption about the consumption of forest products to predict doses that might arise from consumption of such foodstuffs. These predictions are then compared with values derived from whole body measurements.

1. Introduction

The present work is devoted to an analysis of the data on the dynamics of ^{137}Cs transfer factors in the current most critical (for the population of Russia) links of food chain: "soil-mushrooms", "soil-berries" and "soil-game". These data were obtained during 1986-1997 in the most contaminated region of Russia - the Bryansk region - and then were used for assessing internal exposure doses of the population as well as our data about the contribution of forest products consumed by local inhabitants [1-3]. It was shown that radioactive decontamination of mushrooms, berries and game during 11 years after the Chernobyl accident was very slow in contrast to agricultural food products (where natural decontamination took place quickly). This paper makes use of aggregated radioecological data together with a broad assumption about the consumption of forest products to predict doses that might arise from consumption of such foodstuffs. These predictions are then compared with values derived from whole body measurements.

2. Materials and methods

Southwest districts of the Bryansk region were chosen as the zone of study (about 200 km to the north-east from the Chernobyl nuclear power plant). These districts belong to

I. Linkov and W.R. Schell (eds.), Contaminated Forests, 343–352.
© *1999 Kluwer Academic Publishers. Printed in the Netherlands.*

the most contaminated areas of the former USSR after the accident. In this zone surface activity on soil (σ) reaches 3700 kBq/m^2 for ^{137}Cs and 110 kBq/m^2 for ^{90}Sr. About 110,000 people live in the so called "controlled area" (σ ^{137}Cs > 555 kBq/m^2) in this zone.

The turf-podzol sandy, sandy-loam and peat-bog soils predominate in the top soil of the investigated area. The main part of forest tracts in the given region contains mixed woods, where birch, aspen, pine, and spruce prevail. The characteristics of forest soils (organic horizon) varied in the limits: pH$_{KCl}$ - 3.0-4.7; concentration of exchangeable potassium K$_2$O - 19-280 mg/kg; humus C - 0.3-5.9%; exchangeable bases S - 1-19 meq/100 g; cation exchange capacity (CEC) - 6-30 meq/100 g; clay content - 1.4-44%.

We used transfer factor (TF) which is equal to the ratio of radionuclide specific activity in raw product to its soil surface activity (m^2/kg), as the parameter characterising migration of caesium radionuclides from soil to forest products.

In this work we present data on transfer coefficients only for ^{137}Cs. The ratio of caesium radioactive isotopes in samples of soil and forest products was the same, because contribution of ^{137}Cs of global origin is negligibly small for such high levels of accidental contamination in the Bryansk region.

The species of mushrooms and berries growing in the Bryansk region are typical for the central and northern European part of Russia. We analysed the ^{137}Cs content only in the species of mushrooms and berries consumed actively by the local population. The main part of samples was obtained from the local population. Values of the ^{137}Cs transfer factor were calculated as the ratio of the radionuclide specific activity in a sample (wet weight) to the average density of contamination of the forest tract where mushrooms, berries and game were gathered. Periodically, sampling of "fruits of the forests" was accompanied by soil sampling, in these cases TF values were determined more accurately.

During the period 1987-1997 dietary surveys were carried out in the Bryansk region using questionnaire focused on the qualitatively consumption of natural food products [4, 5]. All the respondents replied to the question: «How often do you consume mushroom?» They had to answer: no, seldom, often or very often. In latter years several surveys focused on the estimation of the average consumption rate (quantitative data) of the «forest gifts» by the population were added. These studies were accompanied by measurements of ^{137}Cs content in the human body.

Measurements of whole body content within the survey of inhabitants were made using Robotron (Model 20046) portable radiometers, produced in the former German Democratic Republic and SKIF - portable spectrometer (made in Russia). These instruments consist of a scintillation spectrometer detector with a 6.3 cm x 6.3 cm NaI(Tl) crystal. In the first case signals were registered within the energy range of 0.5 to 1.0 MeV, which includes the ^{134}Cs and ^{137}Cs photopeaks, in another one we could register ^{137}Cs separately. The measurements were made with the person either in a sitting position with the detector positioned near the lower part of the abdomen or with the person standing with the detector positioned near the loins. The measurements time was 100 s, and the practical detection limits were 2 to 0,4 kBq, respectively. Calibration of the system was performed previously by administering a standard solution containing ^{137}Cs and ^{134}Cs to six volunteers who were employees of the Institute of Radiation Hygiene. The measurement technique was certified by the metrological service of Russia.

Nearly 10,000 measurements were taken to characterise the quantity of Cs radionuclides in food products of the agricultural and natural origin consumed by local inhabitants.

The ^{134,137}Cs concentrations in samples were measured by scintillation gamma-spectrometer with NaI detector (110 mm in diameter), with the sample well 200 cm^3 in volume. The minimum detected activity (MDA) of caesium radionuclides was about 2 Bq for measurement time 30 min and with a relative error 30 %. Content of caesium radionuclides in samples below MDA was determined within antimonium-iodine radiochemical technique.

3. Results

The analysis of the dynamics of radioactive contamination of food chain links traditionally giving the major contribution to the formation of the internal exposure dose in population (natural grasses, vegetation, milk and meat of cattle) has shown that during the first five-six years after the Chernobyl accident natural decontamination of agricultural food products took place with relatively high rate with a half-period of decontamination (T) about 1-1.5 years [6, 7].

The ^{137}Cs content in the bodies of inhabitants of the Bryansk region decreased in 1986-1991 with the half-period of decontamination about 15 months [6-8], close to the T value of decontamination in agricultural food products from this radionuclide. However, in 1992-1993 we found the fact of considerable increase of ^{137}Cs content in bodies of inhabitants of settlements in the Bryansk region [1, 8]. The analysis of the situation, including observation of the behavior of the local population and polls of inhabitants showed that this growth is connected with the growth of consumption of mushrooms and berries in the indicated period (in spite of the fact that the population is well informed that the ^{137}Cs content in mushrooms, and often in berries, growing in the controlled areas exceeded permissible standards almost everywhere).

In contrast to agricultural food products, natural decontamination of which took place sufficiently quickly, we either did not find statistically significant decrease of radioactive contamination of mushrooms and berries during 11 years after the Chernobyl accident, or their decontamination was very slow (Figures 1, 2).

Moreover, for almost all mushroom species, we found an increase of ^{137}Cs specific activity in the first 2-3 years after the accident. The reason is, on the one hand, in the penetration of the radionuclide from the upper forest fall to the layer of mushroom mycelium, and, on the other hand, by some increase of ^{137}Cs content in soil due to its additional ingress with fallen leaves in autumn 1986 and with needles of coniferous trees during several years after the accident.

Statistically significant natural decontamination from radioactive cesium has been found for berries of *Fragaria vesca* and *Rubus idaeuc*. The half-period of their decontamination from ^{137}Cs is 3.5-4 years. ^{137}Cs TF from soil to other species of berries did not change significantly with time.

346

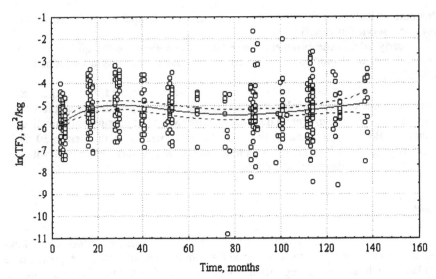

Fig. 1. ^{137}Cs transfer factor from soil into Boletus edulis

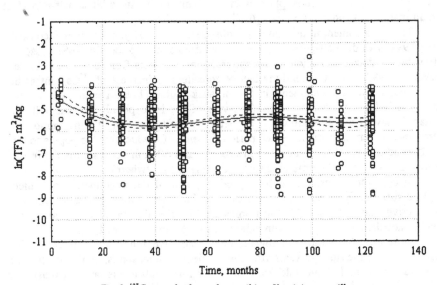

Fig. 2. ^{137}Cs transfer factor from soil into Vaccinium myrtillus

Unfortunately, there are not enough data for the first 2-3 years of observation to make a correct conclusion on the dynamics of ^{137}Cs TF from soil into game during the period of time elapsed after the Chernobyl accident. Statistically significant decontamination from ^{137}Cs has been found for duck and hare meat. The half-period of their decontamination from ^{137}Cs is about 2.2 and 3.5 years, respectively.

Values of ^{137}Cs specific activity in mushrooms, berries and game varied in wide limits in dependence on surface activity of this radionuclide in soil, reaching maximum values of about 500 kBq/kg(mushrooms), 180 kBq/kg (cranberry) and 230 kBq/kg (wildboar) respectively. The data on ^{137}Cs TF's for forest products sampled in 1986-1997 in the Bryansk region are shown in Table 1 (in descending order of TF).

For comparison, we present numeric values for ^{137}Cs TF's for some species of mushrooms that we gathered in contaminated districts of the Oryol and Tula regions, where black and gray forest soils predominate: *Boletus edulis* - $(0.45\pm0.19)*10^{-3}$, *Leccinium scabrium* - $(0.56\pm0.15)*10^{-3}$, *Russula cyanoxantha* - $(0.23\pm0.10)*10^{-3}$ m^2/kg. This is more than by an order of magnitude lower than the TF characteristics for districts where sandy soils predominate.

Statistical significance of the correlation between the ^{137}Cs TF's from soil into mushrooms and soil properties was checked by the method of pair correlation of logarithms of these values (Table 2). The soil characteristics arrange in the row according to the descending influence on ^{137}Cs TF: K$_2$O-Clay-S-CEC-pH-C.

Quantitative connection of Cs TF's and soil properties can be established in the form of equations of the one-dimensional regression with respect to the most closely correlating parameters. The equation of the multiple regression for the transfer factor with respect to the most closely correlating with it relatively independent parameters, give the highest accuracy of the prognosis for the content of ^{137}Cs in mushrooms - multiple correlation coefficient R=0.72. Hence, followed that more than 50% of variation of ^{137}Cs TF's is caused by the influence of taken into account soil characteristics.

4. Discussion

The population of contaminated regions of Russia, which consumed almost no mushrooms and berries before 1991-1992, in the last years, in spite of limitations, began to gather and use them intensively. This has caused a noticeable increase of ^{137}Cs content in the bodies of inhabitants and, significant correlation between "forest gifts" consumption and whole body content of ^{137}Cs were recorded. Every autumn we observed the increase of the ^{137}Cs content in human bodies, which was determined by intensified gathering and consuming of natural food products by local inhabitants in this period. Besides the contribution of forest products into the internal dose in the latter period after the accident has much increased also due to the extremely slow dynamics of decontamination of "fruits of the forest" from ^{137}Cs, in contrast to agricultural foodstuffs (see, for example Figure 3, where the dynamics of ^{137}Cs TF from soil into milk, potato, mushrooms, berries and the bodies of adult inhabitants of the Bryansk region are compared). The specific activity of ^{137}Cs in agricultural food products have decreased from 1987 to 1997 by one to two order of magnitude, while ^{137}Cs content in mushrooms, some species of berries and game remained practically constant.

Assuming that the contribution of each species of mushrooms and berries to the food ration of inhabitants is approximately equal, one can assess the average ^{137}Cs TF from soil to mushrooms and berries over the Bryansk region as $18*10^{-3}$ and $5.6*10^{-3}$ m^2/kg. These values are the same in 1987 and 1994-1997, respectively.

TABLE 1. ^{137}Cs transfer factors from soil to mushrooms (1986-1997),
berries and game (1990-1997), $*10^{-3}$ m^2/kg

Species	Number of samples	Average ± St.error	Median
Mushrooms			
Agaricus emeticus	2	80	-
Lactarius rufus	16	78±18	69
Xerocomus badius	24	66±21	32
Paxillus	66	54±9	28
Bupleurum	91	47±7	22
Xerocomus	45	47±12	22
Leccinum aurantiacum	56	36±18	4.9
Suillus luteus	355	33±2	21
Guski	71	26±7	9.8
Kurochki	51	24±5	11
Leccinum scabrum	261	16±1	7.6
Lactarius deliciosus	25	16±3	11
Russula	299	15±1	6.5
Lactarius necator	187	14±1	9.2
Redovka	58	12±7	3.7
Morchella	31	12±3	3.1
Cantharellus cibarius	579	11±3	3.7
Tricholoma flavovirens	152	10±1	6.5
Boletus edulis	446	9.2±0.7	5.0
Armillaria mellea	368	6.6±2	1.3
Dozdevik	7	4.8±1.2	-
Agaricus campestris	8	0.7±0.2	-
Berries			
Vaccinium oxicoccos	55	14±2	10
Vaccinium vitis-idaea	24	8.1±1	7.3
Vaccinium myrtillus	1043	5.8±0.2	4.7
Rubus idaeuc	567	5.3±0.2	3.7
Fragaria vesca	197	2.5±0.02	1.5
Ezjevica	10	1.0±0.3	-
Game			
Wildboar	115	20.4±3.6	7.9
Elk	141	5.3±0.8	2.2
Duck	42	5.0±1.0	1.8
Hare	39	3.1±1.6	0.55

Assuming that an adult inhabitant from the critical population group, i.e., actively gathering forest products for home use, consumes 10 kg of mushrooms (in raw weight), 5 kg of berries and 7 kg of game per year, one can assess the average daily intake of ^{137}Cs to the body of standard man from these products: $1.1*10^{-3}$ in 1987 and $0.54*10^{-3}$ in 1994-1997 m^2/day. In these calculations we assumed that the food processing retention factor for the ^{137}Cs specific activity in the cooked product as compared with the raw one is 0.5 for mushrooms and 1.0 for berries (Tables 3, 4).

According to the data of Tables mentioned above, the contribution of the natural foodstuffs into internal exposure of the critical groups of the population increased from 20 % in 1987 up to 80 % in 1994-1997 (Figures 4, 5). For the all population, the real data of our surveys of different food products by rural inhabitants living on the controlled area are used.

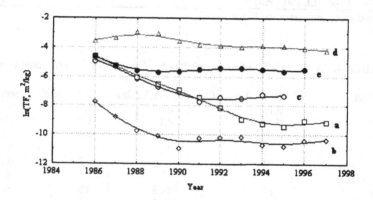

Fig. 3. ^{137}Cs transfer factor from soil into milk (a), potato (b), bodies of adults (c), Suillus luteus (d) and Vaccinium myrtillus (e)

TABLE 2. Pair correlation coefficients between logarithm of ^{137}Cs transfer factors from soil to Boletus edulis and logarithm of parameters of soil properties

	TF$_{Cs}$	pH	K$_2$O	C	S	E	Clay
TF$_{Cs}$	1						
PH	-0.36	1					
K$_2$O	-0.60	0.62	1				
C	-0.12	0.07	0.57	1			
S	-0.54	0.78	0.89	0.52	1		
E	-0.39	0.36	0.77	0.81	0.72	1	
Clay	-0.55	0.41	0.58	0.26	0.65	0.57	1

TABLE 3. Average daily intake by adults of ^{137}Cs in 1987 with local food products

Product	TF, $*10^{-3}$ m²/kg	Consumption, kg/day	K$^{*)}$	Daily intake, $*10^{-3}$ m²/day
Milk	5.2	0.7	1.0	3.6
Pork	4.5	0.1	1.0	0.45
Beef	18	0.06	1.0	1.1
Potato	0.15	0.3	0.8	0.036
Root crops	0.18	0.033	0.8	0.005
SUBTOTAL				5.2
Fish	0.62	0.018	0.9	0.06
Mushrooms	18	0.028	0.5	0.25
Berries	5.6	0.014	1.0	0.08
Game	40	0.02	1.0	0.80
TOTAL FOR ALL PRODUCTS				6.4

*) food processing retention factor

TABLE 4. Average daily intake by adults of ^{137}Cs in 1994-97 with local food products

Product	TF, $*10^{-3}$ m²/kg	Consumption, kg/day	K$^{*)}$	Daily intake, $*10^{-3}$ m²/day
Milk	0.12	0.7	1.0	0.084
Pork	0.09	0.1	1.0	0.009
Beef	0.45	0.06	1.0	0.028
Potato	0.02	0.3	0.8	0.005
Root crops	0.03	0.033	0.8	0.008
SUBTOTAL				0.13
Fish	0.62	0.018	0.9	0.01
Mushrooms	18	0.028	0.5	0.25
Berries	5.6	0.014	1.0	0.08
Game	10,2	0.02	1.0	0.20
TOTAL FOR ALL PRODUCTS				0.67

*) food processing retention factor

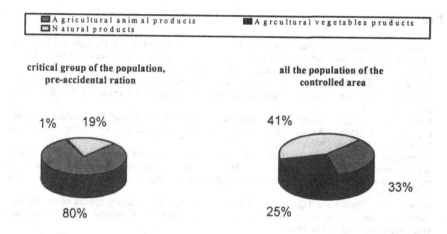

Fig. 4. *Contribution of different foodstuffs into internal exposure of rural inhabitants of the Bryansk region in 1987*

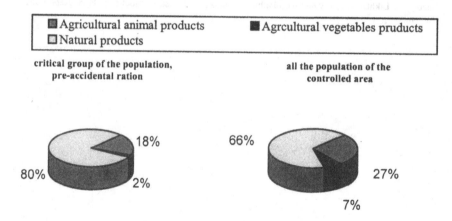

Fig. 5. *Contribution of different foodstuffs into internal exposure of rural inhabitants of the Bryansk region in 1994-1997*

352

5. References

1. Shutov, V.N., Bruk, G.Ya., Basalaeva, L.N., Vasilevitskiy, V.A., Ivanova, N.P., Kaplun, I.S. (1996). «The role of mushrooms and berries in the formation of internal exposure doses to the population of Russia after the Chernobyl accident». Radiat. Prot. Dosimetry, Vol. 67, No. 1,: 55-64.
2. Strand, P. et.al (1996). «Transfer of radionuclides to animals, their comparative importance under different agricultural ecosystems and appropriate countermeasures». Final report. ISBN 92-827-5201-1: 249.
3. Kenigsberg, J., Belli, M., Tikhomirov, F., Buglova, E., Shevchuk, V., Renaud, Ph., Maubert, H., Bruk, G., Shutov, V. (1996). «Exposure from consumption of forest produce». Proceeding of the First International Conference. Minsk, Belorus: 271-282.
4. Balonov, M.I., Travnikova, I.G. (1989). «The role of agricultural and natural ecosistems in internal dose formation in inhabitants of the controlled area». Proceedings of a symposium: «Transfer of radionulides in natural and semi-natural environments». Udine, Italy, September 15-17, ISBN 1-85166-539: 419-430.
5. Balonov, M.I., Travnikova, I.G. (1993). «Importance of diet and protective actions of internal dose from Cs radionuclides in inhabitants of the Chernobyl region». The Chernobyl Papers 1,: Research enterprises, Washington: 127-166.
6. Shutov, V.N., Bruk, G.Ya., Balonov, M.I., Parhomenko, V.I., Pavlov, I.J. (1993). «Caesium and strontium radionuclide migration in the agricultural ecosystem and estimation doses to the population». The Chernobyl Papers 1, Research enterprises, Washington: 167-218.
7. Bruk, G.Ya., Shutov, V.N., Balonov, M.I., Basalaeva, L.N., Kislov, M.V. (1998). «Dynamics of ^{137}Cs content in agricultural food products produced in regions of Russia contaminated after the Chernobyl accident». Radiat. Prot. Dosimetry, Vol. 76, No. 3: 169-178.
8. Jacob, P., Likhtarev, I. (1996-YII). «Pathway analysis and dose distribution». Final Report JSP-5, Luxembourg: 130 .

MODEL FORESTDOSE AND EVALUATION OF EXPOSURE DOSES OF POPULATION FROM FOREST FOOD PRODUCTS

T. ZHUCHENKO and A. DVORNIK
Forest Institute of the National Academy of Sciences of Belarus, 71 Proletarskaya Street, 246654 Gomel, BELARUS

1. Introduction

The radiation situation in the contaminated regions of Belarus is becoming stabilized. For 10 years different countermeasures were undertaken while producing agricultural products. In spite of that, forest and forest food products affect the population contributing to higher radiation doses. Belarus is the most forested country in Europe, so its population traditionally uses many forest food products in the diet. In the first period after the accident the forests role was protective limiting the contamination of agricultural lands and further radionuclide redistribution. At present, the radionuclides are included in the long-standing life cycle of forest ecosystems, where they now provide a significant radiation dose to the population.

The external radiation dose in the forests is 1.5 - 2 times higher than the dose in the surrounding areas. The half time for external radiation dose reduction in the forest is comparable to the half-life of 137 Cs, i. e. 30 years. Fixation of most radionuclides occurs in the upper horizon of the typical forest soils. About 6% of the total deposited radoinuclides contribute to the external dose from root uptake into the above ground layer. The contribution of the internal dose to the population has been increasing constantly and, at present, forms half of the total dose. This dose to the population does not include that portion from contaminated milk from cows feeding on hay grown in the contaminated forests. To evaluate the influence of the forest on the formation of internal radiation dose to the population from the contaminated regions of Belarus, especially where the annual radiation dose to man exceeds 1 mSv/y, an independent model for estimating the internal dose, caused by the use of forest food products, was developed. This model, FORESTDOSE, was developed in the Forest Institute of the National Academy of Sciences of Belarus. The model is based on past determinations of regularities of the growth of forest stands and on the experimental data on the contamination of the forest products.

It should be noted that at present, despite prohibition measures on forest food products, the population on the whole, reverted to the traditional use before the accident. According to the continuing analysis of the contamination levels of the forest food products, there is no tendency to the levels decrease.

I. Linkov and W.R. Schell (eds.), Contaminated Forests, 353–358.

2. General description of the model

The FORESTDOSE model was created for the estimating internal radiation dose, caused by the injection of forest food as an independent model products (mushrooms and berries).

The model parameters are determined using data from standard documents of the regional forest administration and analysis of literature values without the need for field measurements. Model estimates were carried out for 39 settlements of the Bragin and Vetka areas of the Gomel province and 5 contaminated areas of the Mogilev province in the framework of the State Rehabilitation Programme.

Information and assumption for FORESTDOSE are:

- an area within the radius of 5 km around a settlement is available for the collection of mushrooms and berries,
- a direct dependence exists between the consumption of mushrooms and berries and the percent forest cover of the settlement, i.e. the ratio of forest lands area to the total area of the 5 km circle around the settlement,
- the percent forest cover used by the settlement is determined only by the area of productive forests, i.e. only such types and ages of the forests, where food products grow,
- the type of the forest and its age determine the species of mushrooms and berries, their crop capacity, and the transfer factor of radionuclides into different species,
- the crop capacity of each mushroom species depends on weather conditions of the year. The concepts of low-, mean- and high-yielding years are used,
- there is a direct dependence between the collection of each mushroom species and its crop capacity,
- the exportable portion of the harvest and the culinary treatment of the produce are taken into account while estimating the dose,
- there is a connection between the consumption of the forest food produce and the percent forest cover of the settlement,
- only exploitable stock of produce is taken into account while estimating the crop capacity.

The scheme of the model is presented in Fig.1.

3. Model formalization

The internal dose caused by the consumption of forest food products was calculated as:

$$D = k*C*Q_{eff},$$

where
D = the internal dose, mSv/y,
k = dose coefficient, $1.4*10-8$ Sv/Bq,
C = consumption of food forest product in concrete settlement,

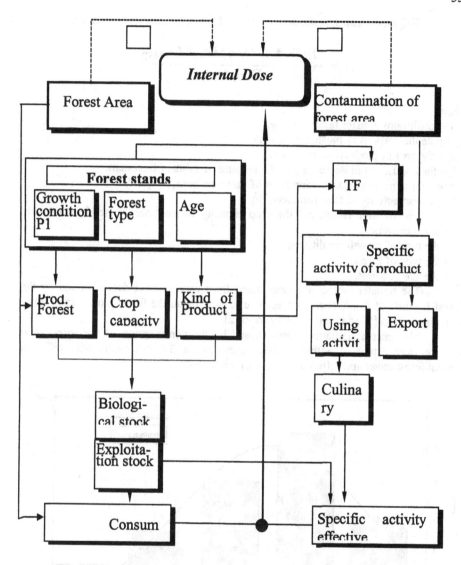

Fig.1. The scheme of the model FORESTDOSE

Q_{eff} = weighted-mean specific activity of food forest products, taking into account the culinary method of preparation.

The consumption of the forest food products is assumed to have linear dependence on percent forest cover area of the settlement. According to long-term statistical data including pre-accident observations, the average consumption of mushrooms in rural settlements of Belarus is 2.3 kg/y, and bilberries - 4 kg/y. The average percent forest cover area of Belarus is 36%.

Weighted-mean specific activity of food forest products is calculate as:

$$Q_{eff} = \frac{\sum(TF_i * H * Spr_i * Kd_i) * Z - A_{exp}}{\sum(H_i * Spr_i)}$$

where
TF_i = mushroom transfer factor, m2/kg,
H_i = potential harvest of mushroom, kg/ha,
Spr_i = the area of productive forests, ha,
Kd_i = the coefficient of decreasing of TF because of culinary treatment,
Z = weighted-mean contamination level of forest sections, kBq/m2,
A_{exp} = export activity of food products, Bq.

The transfer factor and the crop capacity of mushrooms are dependent on several parameters:
$P1$ = the type of growth conditions,
$P2$ = the type of forest,
$P3$ = the age of forest.

The transfer factors are determined from experimental data obtained by the Forest Institute of the Academy of Science of Belarus on the forest plots during long-term observation [1] and crop capacity from [2].

The internal radiation doses caused by the consumption of the forest food produce were estimated with the above methods for 39 settlements with different percent forest cover and different contamination levels.

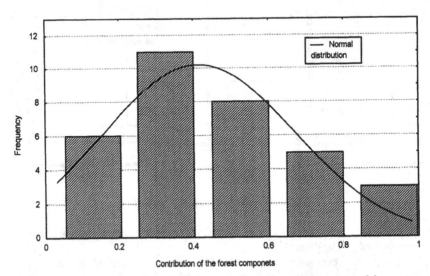

Fig.2 The distribution of the forest components contribution to internal dose

The relationship of the radiation dose from the forest component to the values of direct measurements of the internal dose with the whole body counter were estimated.

The correlation between the dose value taking into account all parameters was estimated to evaluate the most important factors affecting the formation of the internal radiation, caused by the forest products.

The results are presented in the Table 1.

TABLE 1. The correlation coefficient between significant parameters

	Parameters	1	2	3	4	5	6	7
1	Settlement deposition	1.00	0.57	0.43	0.64	0.41	0.72	0.64
2	Forest deposition	0.57	1.00	0.23	0.37	-0.05	0.79	0.95
3	Percent productive forest cover area	0.43	0.23	1.00	0.89	0.19	0.46	0.45
4	Percent forest cover area	0.64	0.37	0.89	1.00	0.21	0.66	0.55
5	Whole body counter	0.41	-0.05	0.19	0.21	1.00	0.17	0.02
6	Calculated dose by model	0.72	0.79	0.46	0.66	0.17	1.00	0.85
7	Estimation of dose	0.64	0.95	0.45	0.55	0.02	0.85	1.00

Table 1 shows the correlation between the internal radiation dose and the productive forest cover percent of the settlements and the level of the radionuclide contamination. The dependence internal dose from forest product from forest cover and deposition is described by formula:

$$D_{for} = -0.32 + 0.016*Z + 1.14*Y - 0.72*Y^2$$

where

Z = the level deposition, Ci/km2,

Y = -percent productive forest area

Fig.3 shows the approximation of this dependencies for 39 settlements.

For the estimation of the dose value the two-parameters dependence on the above values is suggested.

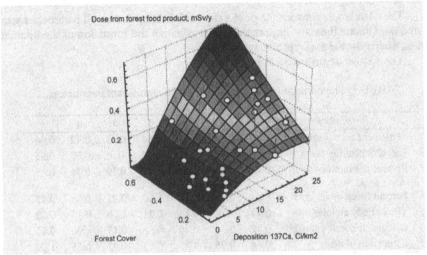

Fig.3. The dependence internal dose from forest product from forest cover and deposition

4. References

1. Dvornik A.M., Zhuchenko T.A. Model of the formation of internal irradiation dose of population from food forest products // Chernobyl: Ecology and Health - Gomel, 1998. - v1(5). - p.11-15.(in Russian)
2. Grimashevich V.V. Guidance for implementation of productivity of edible forest mushrooms and estimation their storage / Forest institute - Gomel, 1992. - 44p.

REDUCING THE CONSUMPTION OF [137]CS VIA FOREST FUNGI - PROVISION OF 'SELF-HELP' ADVICE

N.A. BERESFORD, B.J. HOWARD, and C.L. BARNETT
Institute of Terrestrial Ecology, Grange-Over-Sands, LA11 6JU, UK

G. VOIGT and N. SEMIOCHKINA
GSF-Institute of Radiation Protection, Neuherberg, GERMANY

A. RATNIKOV
Research Institute of Agricultural Radioecology, Obninsk, RUSSIA

I TRAVNIKOVA
Institute of Radiation Hygiene, St. Petersburg, RUSSIA

A.G. GILLETT
Environmental Science Division, School of Biological Sciences, University of Nottingham, UK

H. MEHLI and L. SKUTERUD
Norwegian Radiation Protection Agency, Østerås, NORWAY

1. Introduction

Since the Chernobyl accident it has been recognised that agriculturally produced foods are not the only important food types which contribute to the radiation dose received by man. The transfer of radiocaesium to 'wild food' products (e.g. edible fungi, freshwater fish, game animals) is often much greater than that to agriculturally produced foodstuffs. The accumulation of radiocaesium by certain species of edible fungi for instance, results in radiocaesium levels that are far in excess of most other foods consumed by man.

Within the Commonwealth of Independent States (CIS) there is a common tradition of collecting fungi (and berries) from forests (often termed *forest gifts*). In both rural and urban settlements in Russia approximately 40 % of the population surveyed consumed wild fungi [1]. Mean annual intake rates of fungi ranged from 1.2 to 14 kg fresh weight across the four Russian settlements [2]. Shutov et al. [3] estimated that fungi could contribute up to 60-70 % of dietary ^{137}Cs intake of those adults collecting fungi and berries from forests within Russia. Indeed, within the rural population in

I. Linkov and W.R. Schell (eds.), Contaminated Forests, 359–368.
© *1999 Kluwer Academic Publishers. Printed in the Netherlands.*

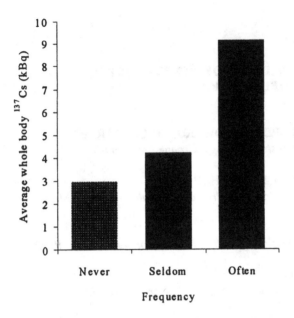

Fig. 1. A comparison of the consumption rate of fungi (expressed qualitatively as frequency eaten) and the whole body ^{137}Cs burden determine in people (n=102) living in an urban area of Russia (town of Klincy, population = 70 – 80 thousand) during July 1997 (from [1]).

Russia, a mean increase in the whole body radiocaesium activity of 60-70 %, in autumn as a result of fungi consumption has been noted [4]. More recently, fungi have been found to contribute significantly to the radiocaesium intake of urban population as well as rural communities (Figure 1).

Fungi, therefore, can contribute significantly to the ingested dose of some individuals whilst nutritionally representing an insignificant component of the diet. This provides the possibility to significantly reduce the intake of radiocaesium by individuals by trying to change their consumption habits with regard to contaminated fungi. Frank et al. [5] recommended that restrictions on the consumption of forest fungi would significantly reduce the internal dose in 18 of their 36 study settlements within the CIS where they estimated a total annual dose of >1 mSv. However, as already mentioned, the collection of fungi within Russia, Ukraine and Belarus is a strong tradition and many people have ignored bans placed by national governments on the collection of fungi from contaminated regions. Partially, this lack of willingness to adhere to national government instructions may be the result of a distrust of authorities within these countries and also the method of communication. Additionally, there is the possibility that simply banning something without simplified reasoning may have a detrimental psychological effect on populations already coping with the stress of living in contaminated areas [6-7]. Provision of clearly explained dietary advice has been very effective in other countries. For instance, in Norway the Directorate of Health published a brochure aimed at groups such as hunters, fishermen and the Sami reindeer breeders, who were most likely to consume large quantities of comparatively highly contaminated foods such as reindeer meat and freshwater fish. The brochure gave examples of the best ways to prepare food to limit radiocaesium intake and presented advise on intake rates in easily understandable units (e.g. 'meals per week'). Follow-up surveys estimated that up to 80 % of individuals within the target groups changed their diet as a consequence of the advice [8]. It was estimated that the radiocaesium intake of the Sami group was reduced by 400-700 % whilst that of hunters by up to 50 % as a result of these dietary changes [8].

In this paper, we present the rational behind the provision of self-help advice which we have produced in the form of posters and leaflets during the course of two European Commission funded projects [1,9]. These provide advice on how individuals can reduce their ^{137}Cs intake, giving recommendations on:

- the maximum quantity of fungi that can be consumed each year;
- preparation methods;
- species to collect which are the least contaminated;
- species which should be avoided.

The possibility of extending this approach to other food products collected from or produced within forests is discussed.

2. Selection of Criteria on Which to Base Advice

2.1 ANNUAL [137]CS INTAKE VIA FUNGI

Recognising that in the regions of the CIS contaminated by the Chernobyl accident a number of food products could contribute significantly to the total intake rate of ^{137}Cs, we adopted a maximum advised annual intake of ^{137}Cs due to fungi consumption of 35 kBq y^{-1} in our calculations. This equates to an annual committed effective dose of 0.5 mSv y^{-1} (according to ICRP 1993 [10]). We have recommended that children, and lactating or pregnant females ingest no more than 17.5 kBq y^{-1} of ^{137}Cs via fungi consumption.

2.2 IDENTIFICATION OF CONSUMPTION AND COOKING HABITS WITHIN THE CIS

In order to advise which species to select, and which to avoid, we initially had to identify the commonly eaten species within the areas of concern. Similarly, information on how each species is traditionally cooked was required. Colleagues from CIS institutes identified 33 commonly eaten species, 29 of which are mycorrizal species (i.e. their mode of nutrition is mutualistic) and are found within forests. A number of preservation techniques are used: salting (after picking the fungi are boiled 2-3 times and then stored in brine); pickling (boiled then stored in vinegar); drying (dried after collection and then soaked before cooking). The most common cooking method for fresh fungi was to boil them before frying; only the *Agaricus* species were recorded as being cooked by frying alone.

To make recommendations on the effect of cooking on the radiocaesium content of fungi we have used the data presented by Rantavara [11] and Jacob et al. [12] which are summarised in Table 1. All of the reductions shown within the table are only achieved if the cooking liquid, which will contain the radiocaesium lost from the fungi, is discarded. This aspect has been stressed within the leaflet as this will sometimes represent a change to the usual cooking procedure. On the basis of these data we have assumed that the reduction in radiocaesium due to multiple stage cooking processes is compounded (e.g. for boiling and then frying we have assumed a reduction of 75 %).

TABLE 1. The reduction achieved in the radiocaesium content of fungi following commonly used cooking procedures (adapted from [11-12]).

Cooking Procedure	Percentage radiocaesium lost compared to freshly collected fungi
Wash	20 %
Fry	50 %
Boil	50 %
Salt	50 %
Dry then soak	75 %
Parboil, salt then soak	90 %

2.3 DETERMINATION OF THE ^{137}CS ACTIVITY CONCENTRATION IN FUNGI

Within the current projects we have been collating available aggregated transfer factor (Tag) values for different species of edible fungi: Tag is defined as the ratio of the dry matter activity concentration in the fungal fruit body (Bq kg^{-1}) to the radiocaesium deposition (Bq m^{-2}). The Tag data base now contains over 850 values representing approximately 140 different edible species. The values originate from the literature review of Gillett & Crout [13], data presented by Barnett et al. [14] and new data supplied by the Ukrainian Institute of Agricultural Radiology, Research Institute of Agricultural Radioecology and Institute of Radiation Hygiene within the course of the current programme.

There is considerable variability in the transfer of radiocaesium to fungi, with the collated Tag values varying over nearly 6 orders of magnitude (7.12×10^{-5} - 1.08×10^{1} m^{2} kg^{-1}); even within species variation can be great, for instance Tag values for *Boletus edulis* vary from 2.73×10^{-3} to 9.98×10^{0} m^{2} kg^{-1}. However, there are clear differences in the transfer of radiocaesium to fungi exhibiting different modes of nutrition (i.e. mycorrizal, parasitic or saprotrophic). Figure 2 presents a summary of Tag values on the basis of nutritional type; the transfer to mycorrizal species is significantly higher than

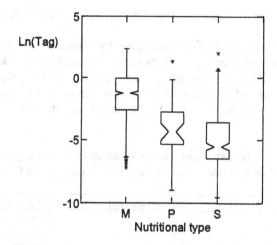

Fig. 2. Box whisker plot comparing variability in the transfer of radiocaesium to edible fungi of different nutritional types (M = mycorrizal; P = parasitic; S = saprotrophic).

that to parasitic or saprotrophic species. Since most of the fungi favoured for collection from forests within the CIS are mycorrizal species we have adopted the mean Tag value of 7.87×10^{-1} m^2 kg^{-1} for this nutritional type to enable us to calculate maximum fungi intake rates. Adopting the median Tag value of 2.88×10^{-1} m^2 kg^{-1} for mycorrizal species would better predict the activity concentration of fungi collected by the 'average consumer'. However, given the large degree of variability in transfer and the uncertainty which will be associated with estimated deposition values, we have adopted the mean value as the more cautious option. Obviously, we could have used the maximum Tag value (i.e. 1.08×10^1) on the basis that this was the most cautious. However, this would have resulted in the recommendation that fungi are not consumed at all. Whilst this may well be the most desirable option from the point of view of radiation protection, it is unlikely that such advice would be respected by the local population and it may give rise to undue concern by those people who have been consuming fungi over the last 12 years.

To convert estimates of the ^{137}Cs activity concentration in dried fungi to fresh weight values a dry matter content in fungi of 10 % has been used.

3. Recommended Fungi Consumption Rates

Within the leaflet we have assumed that the reader will know the local ^{137}Cs deposition and our example calculations have been based upon a ^{137}Cs deposition of 5 Ci km^{-2} (185 kBq m^{-2}); there are 28000 km^2 of land within the CIS which received a ^{137}Cs deposition of 5 Ci km^{-2} or more [15].

Assuming a deposit of 5 Ci km^{-2} the recommended maximum amounts of fresh fungi which could be consumed annually by adults without exceeding the intake limit of 35 kBq are presented in Table 2; amounts assuming different cooking regimes are presented. As well as recommending amounts in kilograms we have also expressed them in terms of baskets collected (1 basket = 6 kg fresh weight) as this is the traditional method of collection.

TABLE 2. Recommended maximum annual intake by adults of forest fungi assuming collected from within an area which had received a ^{137}Cs deposit of 5 Ci km^{-2}.

Preparation method	Maximum fresh weight (kg)	Amount per person
As picked	2	$^1/_3$ basket
Washed	3	$^1/_2$ basket
Fried or boiled or pickled	6	1 basket
Dry and soak	12	2 baskets
Parboil and salt and soak	30	5 baskets

For the posters, the table is presented in the form of a map showing the maximum annual consumption rates of fungi in different areas of the contaminated regions (Figure 3). The map is based on deposition data presented by IAEA [15]. The areas (3057 km^2 in total) where no fungi consumption is recommended are those which received over 1480 kBq m^{-2} (40 Ci km^{-2}) of ^{137}Cs.

4. Recommendation of Which Species to Select and Which to Avoid

We have restricted the advice on intake rates to mushrooms collected from the forest only. Although we have made recommendations that *Agaricus* species and edible parasitic species such as *Armillariella mellea* are likely to be much less contaminated. Unfortunately, we have received contradictory information as to whether these species are actually consumed within the CIS. For instance, when the content of the leaflet were discussed with inhabitants in a Belarussian village it became apparent that although the collection of mycorrizal species was common, no parasitic or saprotrophic species were routinely collected for eating. This is perhaps a reflection of local differences in abundance and consumption habits.

As discussed above there is considerable variability in the transfer of radiocaesium and to simplify the advice we chose to use a Tag value representative of all mycorrizal species. However, it was possible to suggest some species which should, if possible, be avoided (e.g. *Boletus badius*, *Lactarius rufus*, *Paxillus involutus* and *Russula vesca*).

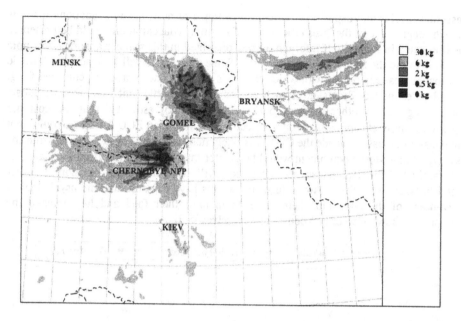

Fig. 3. Maximum recommended annual consumption rates by adults of fungi within the contaminated regions of the CIS. Values are fresh weight as collected but it has been assumed that the fungi will be washed and then fried or boiled before consumption (i.e. the initial ^{137}Cs activity will be reduced to 40 %). The map has been adapted from the IAEA [15] ^{137}Cs deposition map, the maximum deposition for each contour interval has been used in the estimation of the recommended intake rates.

5. Problems and Alternatives

In the approach described above we have assumed that people will know the amount of ^{137}Cs deposited in the area in which they live. We acknowledge that this may not always be the case and to an extent, the maps showing advised fungi intake rates (e.g. Figure 3) get over this problem. However, both deposition and the transfer of radiocaesium to fungi are highly variable. This variability is so high as to result in the probability of people ingesting more than 35 kBq y^{-1} ^{137}Cs even if they follow the guidance we have presented. It could of course be argued that if they had not followed the guidance their intake of ^{137}Cs would have been even higher.

It could be that advising that no fungi be consumed in areas with the highest levels of ^{137}Cs deposition may have a psychologically detrimental effect for those reasons already discussed with regard to national government bans on fungi collection in all of the contaminated areas. However, it is not possible to suggest an acceptable intake rate of forest fungi in areas with more than 1480 kBq m^{-2} ^{137}Cs using the approaches described above.

An alternative approach which should be considered is to base the recommendations on ^{137}Cs activity concentrations in collected fungi rather than trying to assume transfer and deposition values. An example of how, recommendations for how much fungi to eat on the basis of known ^{137}Cs activity concentrations could be presented is illustrated in Figure 4. If information was presented in this manner a worked example on how to use the figure would be required. Some settlements will have access to simple radioanalytical equipment with which the ^{137}Cs activity concentration in collected fungi could be estimated. Provision of such equipment on a settlement basis has previously been suggested by other workers [5,7]. Frank et al. [5] estimated that the cost per averted man Sv of supplying monitoring equipment was 2746 ECU (based on equipment made and purchased outside the CIS). It is likely that the averted dose resulting from the supply of monitoring equipment would be greater than that resulting from avoidance of contaminated fungi alone as the population will have the ability to monitor other foodstuffs. Pilot studies in two villages in Belarus have demonstrated the psychological advantage of the population being able to monitor their food and hence make the decision on how much of it to consume themselves [5,7].

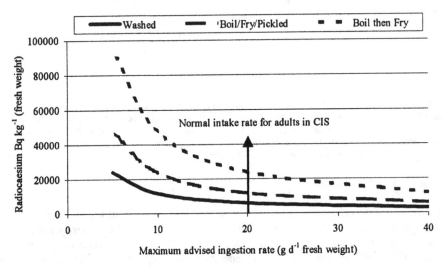

Fig. 4. Maximum advised adult daily ingestion rates of fungi using different preparation techniques (washing; boil or frying or pickling; boil then fry) on the basis of a know ^{137}Cs activity concentration in the fungi as collected. It is assumed that all cooking or pickling fluids are discarded and not consumed. The 'normal intake rate' is the mean of the intake rate by adults in the four Russian settlements presented by Mehli [2].

6. Benefits of the Approach

If successful, the provision of advice on how individuals can restrict their own ^{137}Cs intake via fungi consumption represents a very cost effective method by which their radiation dose can be considerably reduced. This reduction in dose would be achieved with a negligible effect on the nutritional content of the diet. There is also the considerable additional benefit that such approaches would enable local populations to understand their situation more clearly [6-7].

There are a number of villages and towns in Russia and Ukraine where estimates of the intake of fungi by the population have recently been made [2]. The 'control' data available for these locations provide an opportunity to test the effectiveness of the leaflet and poster before trying to distribute it to a wider population. The leaflet was discussed with a limited number of inhabitants in a Belarussian village during the autumn of 1997. The general impression given was that they felt that the provision of the advice it contained was useful and that they would consider using it.

7. Application to Other Food Products

The consumption of contaminated milk, especially from privately owned cattle, is the major 'agricultural' source of ^{137}Cs in the diet [e.g. Frank et al. 1998]. The numbers of private cattle are also increasing within the CIS. The production of contaminated milk is therefore one of the most important problems which needs to be addressed. It is apparent that there is some scope for 'self-help' within affected populations if they can be effectively communicated with. For instance, radiocaesium binders based on hexacyanoferrate compounds are generally available to the population. However, these are sometimes being ineffectively used, if used at all. In Belarus, ferrocyn (5% potassium and 95% iron hexacyanoferrate) is supplied to private cattle owners at the rate of 50 kg per cow every winter. During the winter many cows are not producing milk for 2 months, however, we know of owners still administering ferrocyn during this period. Similarly, private cattle owners often have a combination of relatively uncontaminated (supplied by the collective) and contaminated (collected from natural meadows/forest clearing) hay available for the winter feeding period. The optimum use of the uncontaminated hay would be to feed it to lactating dairy cattle; giving the more contaminated hay to dry cows and calves. It is, therefore, our intention to produce guidelines on the use of hexacyanoferrate compounds and winter feedstuffs to enable rural populations to minimise the ^{137}Cs content of milk.

A similar approach could be used for a number of other food products which can significantly contribute to the ^{137}Cs intake of individuals within the population. Food sources being considered include freshwater fish and forest berries.

8. Acknowledgements

The work described is being conducted as part of European Union funded projects F14-CT-95-0021 (RESTORE) and ERBIC15-CT96-0209 (RECLAIM). We are grateful for this support and also the help of the following colleagues: Katerina Kalchenko, Boris Prister (UIAR), Michael Balonov (IRH), Juliet Frankland (ITE), Prof. L.M. Levkina (Moscow Univ.), Simon Wright (ITE), Marcel van der Perk (Utrecht Univ.) and Per Strand (NRPA). The draft leaflet (in English and Russian) will be available on the programmes web site (www. nottingham.ac.uk/~sbzag/mirrorsindex.html) and we would welcome comments. Note this site is currently undergoing re-development but should be available by mid-July.

9. References

1. Mehli, H. & P. Strand (Eds.). 1998. *RECLAIM - Time-dependent optimalisation of strategies for countermeasures use to reduce population radiation dose and reclaim abandoned land.* First twelve months progress report for the period 1.2.97 - 31.1.98. EU contract no. ERBIC15-CT96-0209. Norwegian Radiation Protection Authority: Oslo.

2. Mehli, H. (Ed.) 1998. *Handbook on consumption habits in Russia and Ukraine with special focus on food consumption in the areas contaminated by the Chernobyl accident.* Deliverable of the EU projects RESTORE and RECLAIM. Contract numbers F14-CT95-0021 and ERBIC15-CT96-0209. Norwegian Radiation Protection Authority: Oslo.

3. Shutov, V.N., G.Ya Bruk, L.N. Basalaeva, V.A. Vasilevitskiy, N.P. Ivanova & I.S. Kaplun. 1996. The role of mushrooms and berries in the formulation of internal exposure doses to the population of Russia after the Chernobyl accident. *Radiation Protection Dosimetry*, 67, 55-64.

4. Skuterud, L., M. Balanov, I. Travnikova, P. Strand, & B.J. Howard. 1997. Contribution of fungi to radiocaesium intake of rural populations in Russia. *Science of the Total Environment*, 193, 237-242.

5. Frank , G., P. Jacob, G. Pröhl, J.L. Smith-Briggs, F.J. Sandalls, P.L. Holden, S.K. Firsakova, Y.M. Zhuchenko, A. Jouve, F.A. Tikhomirov, S.V. Mamikhin, T.V. Rusina, R.M. Alexakhin, S.V. Fesenko, N.I. Sanzharova, E. Nicaise, M. Medzadourian, I.A. Likhtarev, L. Kovgan & M.M. Kaletnyk. 1998. *Optimal management routes for the restoration of territories contaminated during and after the Chernobyl accident.* Final report COSU-CT94-0101, COSU-CT94-0102, B7-6340/001064/MAR/C3 and B7-5340/96/000178/MAR/C3. EUR 17627 EN. ISBN 92-828-2237-0. European Commission Directorate-General Environment, Nuclear Safety and Civil Protection: Brussels

6. Tønnessen, A., L. Skuterud, J. Panova, I.G. Travnikova, P. Strand & M.I. Balonov. 1996. Personal use of countermeasures seen in a coping perspective. Could the development of expedient countermeasures as a repertoire in the population, optimise coping and promote positive outcome expectancies, when exposed to a contamination threat? *Radiation Protection Dosimetry*, 68, 261-266.

7. Heriard-Dubreuil, G.F. (Ed.) 1998. Ethos mid-term report. Report to European Commission Directorate General XII. Mutadis Consultants: Paris.

8. Strand, P., T.D. Selnæs, E. Bøe, O. Harbitz & A. Andersson-Sørlie. 1992. Chernobyl fallout: Internal doses to the Norwegian population and the effect of dietary advice. *Health Physics*, 63, 385-392.

9. Voigt, G., B.J. Howard, U. Sansone, P. Strand, G. Rauret, P. Burrough & N. Crout. 1998. *Restoration strategies for radioactive contaminated ecosystems (RESTORE).* Nuclear Fission Safety Programme E.2. Association contract No F14-CT95-0021. Midterm report 1996-1997. GSF-Institute of Radiation Protection: Neuherberg.

10. International Commission on Radiological Protection (ICRP) 1993. Age-dependent doses to members of the public from intake of radionuclides: Part 2 Ingestion dose coefficients. *Annals of the ICRP*, Vol 23 No3/4, ISSN 0146-6453.

11. Rantavara, A. 1987. *Radioactivity of vegetable and mushrooms in Finland after the Chernobyl accident in 1986.* STUK-A5. Finnish Centre for Radiation and Nuclear Safety: Helsinki.

12. Jacob, P., G. Pröhl, L. Likhtarev, L. Kovgan, R. Gluvchinsky, O. Perevoznikov, M.I. Balonov, A. Golikov, A. Ponomarev, V. Erkin, A. Vlasov, V.N. Shutov, G.I. Bruk, Y.E. Kenigsberg, E.E. Buglova, V.E. Shevchuk, M. Morrey, S.L. Prosser, K.A. Jones, P.A. Colgan, R. Guardans, A. Suañez, P. Renaud, H. Maubert, A.M. Skryabin, N. Vlasova, I. Linge, V. Epifanov, I. Osipyants, A. Skorobogotov. 1995. *Pathway analysis and dose distributions.* Joint study project No 5. Final report EUR 16541 EN. ISBN 92-827-5207-0. European Commission Directorate General XII: Brussels.

13. Gillett, A.G. & Crout, N.M.J. (in preparation). A review of ^{137}Cs transfer to fungi: importance of species, spatial scale and time.

14. Barnett, C.L., B.A. Dodd, P. Eccles, N.A. Beresford, B.J. Howard & J.C. Frankland. 1997. *Radionuclide contamination of fruitbodies of macrofungi in England & Wales: A survey of contamination levels and dietary intake habits.* Project No: T07051w1, MAFF Contract no: RP0425. Final report to Ministry of Agriculture, Fisheries and Food. Institute of Terrestrial Ecology: Grange-over-Sands.

15. International Atomic Energy Agency (IAEA). 1991. *The International Chernobyl Project surface contamination maps.* ISBN 92-0-129291-0. IAEA: Vienna.

RADIOACTIVELY CONTAMINATED FORESTS: GIS APPLICATON FOR THE REMEDIAL POLICY DEVELOPMENT AND ENVIRONMENTAL RISK ASSESSMENT

V. DAVYDCHUK
Institute of Geography, National Academy of Sciences
44, Volodymyrs'ka St. Kyiv 252034 UKRAINE

1. Introduction

Environmental risk assessment of the radioactive contaminated forests includes definition, localization and evaluation of the criteria that determines formation and evolution of exterior and interior doses.

Under certain conditions of surface radioactive contamination, the main factors for exterior dose formation are secondary redistribution of the radionuclides in the environment, including horizontal or vertical migration and by biogenic accumulation. At the same time the biogenic accumulation of the radionuclides is the most important factor for internal dose deposition.

Because of the spatial character of the problem, the necessary component for evaluation of forest ecosystems contamination, remedial policy development and environmental risk assessment is cartographic modeling and GIS application of the radioactive polluted territories. This method creates the basis to estimate intensity and to define areas of redistribution and biogenic accumulation of the radionuclides. Radioactive contamination of the forests can be identified systematically, approaches to the forest management and wood utilization can be developed, and environmental risk factors evaluated.

2. Results Obtained

2.1. LANDSCAPE BASED RADIOECOLOGICAL GIS

The extent of contamination of the forest is now determined mostly by radionuclides in plant roots which depends on the local soil and other environmental conditions. The following information layers (basic maps) are needed for radionuclide cartographic modeling:

- map of soil radionuclide contamination, which reflects the extent of initial radioactive fallout;

I. Linkov and W.R. Schell (eds.), Contaminated Forests, 369–376.

- map of the forest ecosystems, including the types and age structure of the wood stands, and land use
- map (or maps) of the environmental factors which determine the extent of radionuclides in plant roots.

To effectively manipulate these cartographic information layers a special radioecological GIS has been created on the basis of the landscape approach [1].

These developments used programming tools of SPANS GIS, an ITERA TYDAC products, with a SUN SPARC 10 Workstation. This hardware and software configuration was installed in Kyiv as part of the Chernobyl GIS Pilot Project. This project represents part of Canada's assistance to Ukraine in dealing with the impact of the Chernobyl disaster[2].

The complex landscape (geosystem) approach was proposed as the structural base of the radioecological GIS. Landscape is the natural terrestrial system consisting of natural components as well as relief, surface deposits, soil and vegetation. The landscape map reflects also drainage conditions and geochemical peculiarities of the territory, which determine soil-to-plant transfer of the radionuclides.

The landscape maps of the Polessky district and of the Kyiv province have been based on the topographic map and remote sensing data. To create the landscape map, information which characterized geology, geomorphology, forests, soils and land use structure were integrated with data of our complex field survey and investigations [3].

The Polesian region is characterized by mainly low and plain relief (110-145 m above sea level) with a combination of different natural landscapes that have certain combinations of relief structures, surface deposits, soils and vegetation. These include river flood plains and terraces, extreme-moraine ridges, moraine-fluvioglacial and limno-glacial plains. Typical for the region are landscapes of moraine-fluvioglacial ripply plains, which consist of dusty sand, with thin (1,5-5 cm) interlayers of loamy sand at the depth 0,8-1,5 m, bedded by moraine boulder loam at the depth 2,5-3,5 m, with soddy-podzolic dusty-sandy soil (pH 4,5-4,9, humus 1-2%), covered by different types of the pine forests.

Total area of *Pinetum* in the accident zone covers more than 30,5% of the territory, or about two thirds of its forested lands. Forest ecosystems of the region are presented by *Pinetum cladinosum* and *Pinetum phodococco-dicranosum*, with the ground cover of *Koeleria glauca (Sorend.) DC, Festuca sulcata (Hack.) Num.p.p., Antennaria dioica (L.) Gaertn., Phodococcum vitis-idaea (L.) Avror., Dicranum scoparium Hedv., Cladonia silvatica (L.) Hoffm* and *C. rangiferina (L.) Webb.*

Under influence of the regular forest farming, the age of the plant life in the territory is mainly young to middle age.

Polesie is notable for the considerable swamping of the river flood plains and terraces. Bog zones account for 22,5% of the territory covered by radionuclides from the accident. Mostly typical swamps occupy the rear lowered flat parts of the river terraces and flood plains, composed by eutrophic peat (thickness 0,5-2,5 m), with peat bog soil (pH 5,0-5,5, organic matter up to 75%), covered with alder forests and sedge-reedous bog coenoses. Two thirds of the bogs were drained during last decades before the accident and were turned into farmlands, which account for 40% of the arable lands.

2.2. CARTOGRAPHIC MODELLING CONTAMINATION THE FORESTS

The landscape based radioecological GIS provide the possibility for more extensive modeling of the contamination of the forest ecosystems by ^{137}Cs. It allows the use of both the procedures: reclassification (creation of maps of the separate elements and components from complex maps) and overlay of the different information layers.

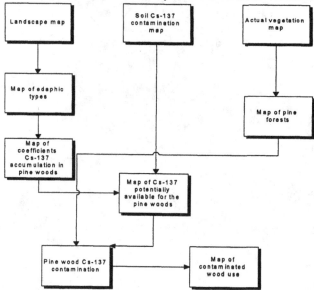

Fig. 1. Cartographic modeling of the pine wood contamination by ^{137}Cs

According to the cartographic modeling of the contamination of the forest ecosystems by ^{137}Cs, these procedures consist of the following steps, presented in Figure 1.

Originally, GIS, using a reclassification procedure, derived the map of edaphotopes from the basic landscape map. The definition of edaphotope, or type of texture or drainage conditions [4] consisting of areas homogenous by the fertility and humidity of soil. As the concept of the edaphotopes is accepted by Ukrainian forestry, the forests of the country are well documented by this definition. At the same time characteristic of the edaphic conditions is a necessary element of the landscape map legend.

Sampling data from the Ukrainian Forestry Ministry on the ^{137}Cs contamination of soils and pine wood, which have been well localized by edaphic conditions, were used to calculate the « soil-to-pine wood » transfer coefficients for each edaphotope [5]. It has provide opportunity to derive a map of the transfer coefficients from the map of edaphotopes.

By overlaying this map with the ^{137}Cs soil contamination map, the map of ^{137}Cs available to pollute pinewood was created (Figure 2). The maps of the ^{137}Cs biologically

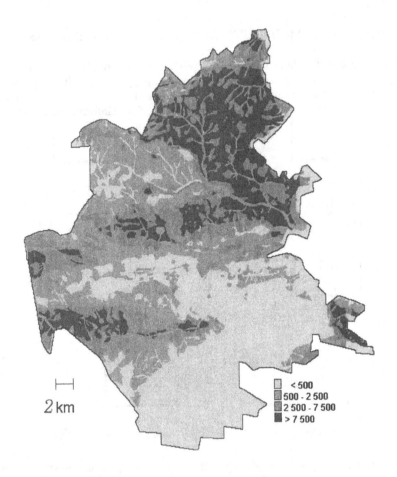

Figure 2. [137]Cs available to pollute pine wood

available, which determine contamination of both wood and other forest products, can be used in the environmental risk assessment.

Overlaying the «[137]Cs available» and «pine forests» maps, the map of the pine wood contamination by ^{137}Cs has been created. According to the Ukrainian national standards for wood utilization, forest areas can be classified into three categories, depending of their wood contamination levels:

1. Areas suitable for all kinds of utilization, both firewood and industrial – with contamination less than 750 Bq/kg.
2. Areas suitable for only industrial use –750 – 3700 Bq/kg.
3. Areas not suitable for any use – above 3700 Bq/kg.

The cartographic interpretation of this evaluation is given at Figure 3. This evaluation is related to 1992-93, according to the period of sampling of the data used. Of

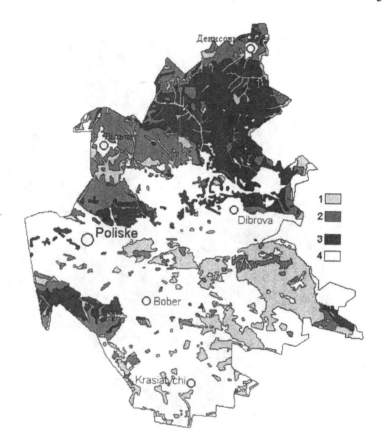

Fig. 3. Contamination levels of the forest pine wood (see text)

course, accuracy and completeness of these qualitative evaluations correspond to the experimental data used.

After that, from the basic map of the forest ecosystems, by its overlay with the map of the age groups of the wood stands, the map of distribution of the forest types by age groups was created. This map is important in the evaluation of the total forest phytomass stock and its fractions, including wood, and to calculate the volume of the wood for every contamination level. It is also useful in the creation of corresponding maps.

2.3. EVALUATION OF THE LOCAL BALANCES OF THE RADIONUCLIDES

The environmental risk assessment has to take into consideration the direction and intensity of the migration processes of the radionuclides at forested and potentially forested areas that determine future evolution of the dose rates. The processes of

Fig. 4. Balances of ^{137}Cs in the landscape aerials: 1.Negative max. 2.Negative moderate.
3.Negative min. 4. Neutral. 5.Positive min. 6.Positive moderate. 7.Positive max.

horizontal redistribution of the radionuclides vary depending of such stable components of the landscape as relief, lithology and soils, and variable ones, such as vegetation cover. These factors and processes result in the formation of the local radionuclide balances and therefore self-decontamination (or self-contamination) of the natural areas.

The balance of radionuclides in the landscapes, which reflects the ability of the natural systems to evacuate and/or accumulate the pollutants, can be negative, positive or neutral (Davydchuk, Arapis, 1995). The ability of the landscapes to evacuate ^{137}Cs can be evaluated mainly through their predisposition to wash-off processes.

Landscapes that occupy high levels of relief normally are characterized by negative balance. They are geochemically autonomous, and their balance depends only on the intensity of natural evacuation (washing) of the pollutants. The landscapes with positive balance (depressions, valleys etc.) are under geochemical influence of the surrounding territories, which belong to the same watershed basin. Neutral balance of the radionuclides by horizontal migration is obtained at the areas of intensive filtration, like forested dry sandy river terraces and fluvioglacial planes.

When evaluating perspective balances of the radionuclides, attention should also be paid to the natural evolution and self-restoration of the landscapes and ecosystems at the abandoned territories. This started after the evacuation of population and sufficient modification and limitation of traditional human activity occurred. This evacuation caused intensive development of plant succession processes, which modified initially cultivated forests, ploughed lands and meadows into relatively stable forest ecosystems according to edaphical conditions of the territory. Self-restoration of vegetation, accumulation of phytomass, deconsolidation and self-restoration of soil influence a decrease in wash-off and an increase in infiltration.

The map of expected balances of ^{137}Cs in the landscape aerials of Polessky district (Figure 4) was created by reclassification of the map of natural landscapes into some separate layers (relief, lithology, soils), and evaluating their role in the washing off/accumulation. After that, overlaying with the maps of vegetation/land use and in predicting plant successions.

The balance of ^{137}Cs was evaluated taking into account the plant succession processes, which are obtained in the evacuation zone. According to the succession model elaborated, these processes, in collaboration with human activity, shall create dense forest cover during the next 50-80 years. At present there exist long fallow grasslands (former arable lands), dry meadows and abandoned settlements.

The proposed above evaluation of the local balances of the ^{137}Cs is a qualitative one. Using experimental data its quantitative interpretation can be made, and the physical decay of the radionuclides can be taken into account.

3. Discussion

The maps of the ^{137}Cs biologically available, forest ecosystems contamination by ^{137}Cs and the ^{137}Cs balances in the landscape aerials, generalize and combine the main environmental factors which determine the dose rates and evolution of the forested territories. They are proposed as the background for the visualization, analysis and presentation of spatial aspects and factors for the environmental risk assessment. As GIS prepared these maps, they can be useful to analyze and predict very different scenarios in the remedial policy development for the radioactively contaminated forests.

In addition to the applications mentioned, some other radioecological tasks can be solved using this approach, such as the evaluation of the ^{137}Cs immobilized in the phytomass, the ecosystems radiocapasity, evaluation of rehabilitation strategies, etc.

4. References

1. Davydchuk, V.S., Palko, S. and Glieca, M. 1997. An Application of Radionuclude Migration Monitoring in the Polessie District, Ukraine. *Proceedings. Eleventh Annual Symposium on Geographic Information Systems.* Vancouver, British Columbia, Canada

2. Palko, S., Glieca, M., Dombrovski, A. and Kéna-Cohen, S. 1995. Belarus-Canada-Ukraine Project on Chernobyl. *CD-ROM Proceedings. The 7th International Conference on Geomatics,* June 11-15, 1995, Ottawa

3. Davydchuk, V.S., Zarudnaya, R.F., Sorokina, L.Yu., Mikheli, S.V., Petrov, M.F. and Tkachenko A.N. 1995. *Landshafty Chernobyl'skoy Zony i Ikh Otsenka po Ousloviyam Migracii Radionuclidov [Landscapes of the Chernobyl Accident Zone and Their Evaluation According to Radionuclide Migration Conditions].* Naukova Dumka. Kyiv

4. Pogrebniak, P.S. 1960. *Foundation of Forest Typology.* Russian translation by M. Anderson. University on Edinburgh

5. Sorokina L.Yu. 1996. Pro Nakopychennia ^{137}Cs Phytokomponentamy Lisovykh PTK Zalezhno Vid Edaphichnykh Umov [Accumulation of ^{137}Cs by Phytocomponents of Forest Ecosystems Depending of Edaphic Conditions]. *Ukrainian Geographical Journal,* №1, pp. 44-48

6. Davydchuk, V., Arapis, G. 1995. Evaluation of ^{137}Cs in Chernobyl Landscapes: Mapping Surface Migration Balance as Background for Application of Rehabilitation Technologies. *Journal of Radioecology,* №1, pp. 7-13

EVALUATION OF SHORT ROTATION COPPICE AS REMEDIATION OPTION FOR CONTAMINATED FARMLAND

H. VANDENHOVE, A. GOMMERS and Y. THIRY
Belgian Nuclear Research Centre (SCK•CEN), Radioecology, Boeretang 200, B-2400 Mol, BELGIUM

F. GOOR and J.M. JOSSART
Catholic University Louvain (UCL), ECOP, Louvain-la-Neuve, BELGIUM

T. GÄVFERT and E. HOLM
University of LUND (U.LUND), Radiation Physics, Lund, SWEDEN

J. ROED
RISOE National Laboratory, Roskilde, DENMARK

A. GREBENKOV
Institute of Power Engineering Problems (IPEP), Minsk, BELARUS

S. TIMOFEYEV and S. FIRSAKOVA
Research Institute of Radiology (RIR), Gomel, BELARUS

1. Introduction

Following the Chernobyl accident, many thousands of square kilometres were severely contaminated in the CIS and the application of corrective measures in contaminated areas remains a key issue in large territories of Ukraine, Belarus and Russia. When agricultural perspectives must be abandoned in these territories because of high levels in food products or because of economically or technically non-realistic corrective options, an increasing interest arises in developing more integrated and ecologically based approaches. In this context, industrial crops, not used for food production, may be an alternative remediation option.

In *Short Rotation Forestry* (SRF, coppice cultivation, cutover area), fast growing tree species [e.g. willow (*Salix spp.*)] are intensively managed and harvested for biomass in a 3-5 year cutting cycle and a 22-25 year rotation. Fast growing willows are among the highest biomass producers (10-12 t ha^{-1}, [1]). The harvested biomass is converted into heat and power. This non-food production, agricultural-industrial crop is a potential candidate for the valorisation of contaminated land with restrictive use.

I. Linkov and W.R. Schell (eds.), Contaminated Forests, 377–384.
© *1999 Kluwer Academic Publishers. Printed in the Netherlands.*

When dealing with radioactive contamination, apart from radioecology and dosimetry, economic and social aspects also should be considered in the optimisation approach to determine the most suitable corrective action. Under the EC-RECOVER project (RElevancy of short rotation COppice VEgetation for the Remediation of contaminated waste farmland), following domains are dealt with:

- Flux of Cs (and its analogue K) during willow cultivation and wood combustion
- Dose received during coppice culturing, subsequent handling and conversion
- Comparison of SRF at a technico-agricultural level with other energy crops
- Energy balance and economic cost benefit of SRF and other energy crops

From the radioecological, dosimetric, technical and economic analyses, parameters will be identified which affect the application of the SRF concept for site remediation. The ranking of these parameters will provide us with the necessary tool to validate this remediation option critically.

Present paper summarises the main results obtained under the RECOVER project with emphasis on the radiocaesium cycling during the second year of a SRF culture on two different soil types.

2. Radioecological aspects: radiocaesium flux and transfer in a coppice ecosystem

Since the coppice concept was not yet studied as a possible valorisation tool for radioactively contaminated land, the flux of radionuclides is studied to obtain information on the radionuclide levels in the wood, the exploitable plant part and in the ashes after combustion. Radiocaesium behaviour is studied in a coppice ecosystem by investigating the soil-to-plant transfer on lysimeter scale (detailed study), on farmers sites in Sweden, on test sites in Belarus and by studying the radiocaesium flux during conversion.

2.1. LYSIMETER STUDY (MOL, BELGIUM)

Materials and methods
Cuttings of *Salix Viminalis* L. (cv Orm) were planted in May 1996 on two different soil types, an orthic luvisol and an orthic podzol (hereafter called loamy and sandy soil). The experimental plots were 2*2 m^2 and replicated fourfold. The soil was contaminated with 10 MBq ^{134}Cs m^{-2} over a depth of 25 cm. Final contamination levels were 24 $\pm$ 6 10^3 and 30 $\pm$ 4 10^3 Bq kg^{-1} dry soil for the loamy and sandy soil, respectively. Initial soil characteristics of the two soils are summarised in table 1.

Results and discussion
The S-shaped pattern of the growth curve is consistent with findings for all types of vegetation growing in areas with pronounced seasonal variations in climate, and as found in earlier studies on willow SRF [2] (figure 1A).

The stem biomass production is similar for both soil types. During the growing season, the one-year-old shoots produced about 12 10³ kg dry wood ha⁻¹, which is 65 % of the total plant biomass (wood + leaves + below ground plant parts). Despite the higher plant density, the biomass production figure is generally in accordance with

TABLE 1: Initial soil characteristics of the orthic luvisol and the orthic podzol

Soil characteristics		Loamy soil	Sandy soil
CEC (cmol$_c$ kg⁻¹) (pH 7)		13.2	10.0
Texture (%)	100-50 µm	10. 7	90.6
	50-20 µm	40.6	5.8
	20-10 µm	31.7	0.6
	10-2 µm	2.0	0.7
	< 2 µm	15.0	2.3
Total C (%)		1.0	3.7
PH (KCl)		6.9	4.6
Exchangeable cation	K	1.35	0.36
(cmol$_c$ kg⁻¹)	Ca	9.81	3.65
	Mg	1.46	0.49

other studies, reporting a biomass production of 10 to 12 10³ kg dry wood ha⁻¹ year⁻¹ (Perttu, 1991). The biomass production figure obtained with SRF is high compared with traditional forestry in long term rotations; in western Europe an average of 2.7 10³ kg dry wood ha⁻¹ year⁻¹ is produced in a pine (*Pinus sylvestris* L.) and 3.5 10³ kg dry wood ha⁻¹ year⁻¹ in a beech (*Fagus sylvatica* L.) forest [3].

The ¹³⁴Cs concentration in the wood of coppice from loamy soil decreased during the growing season, due to a dilution effect. On the sandy soil, the ¹³⁴Cs concentration increased initially. The decrease after June was not significant (Figure 1B in Bq g⁻¹). The difference in Cs evolution between a sandy and a loamy soil relates to a higher availability of radiocaesium on the sandy soil than on the loamy soil and also to the release of radiocaesium during litter decomposition in Spring. The radiocaesium concentration in previous-year-litter was about a factor 5 higher on a sandy soil (4445±768 Bq g⁻¹) than on a loamy soil (817±266 Bq g⁻¹).

The lowest radiocaesium concentration was found in the wood: 19 Bq kg⁻¹ for the loamy soil (TF: 1.9 10⁻⁶ m² kg⁻¹) and 615 Bq kg⁻¹ (TF: 61.5 10⁻⁶ m² kg⁻¹) for the sandy soil was incorporated after two growing seasons. Radiocaesium concentration was thus below the exemption limit for fuelwood in Belarus (740 Bq kg⁻¹).

The five times lower total radiocaesium uptake on the loamy soil was related with the higher radiocaesium sorption capacity and the lower radiocaesium concentration in the soil solution. For both soils, the major radiocaesium fraction was incorporated in the below-ground plant parts (roots and cuttings) (figure 2).

An important fraction of the radiocaesium recycled back to the soil during litter fall in autumn and leached off from leaves with rain: 36 % of the total amount of radiocaesium taken up by the above ground plant parts was washed off from wood and leaves for the loamy soil and 12 % from plants grown on the sandy soil.

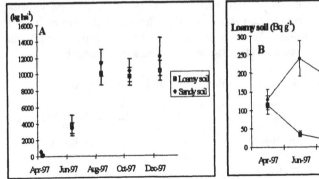
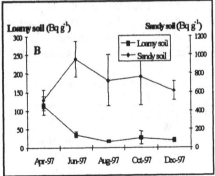

Fig. 1: Evolution of woody biomass (A) and ^{134}Cs concentration (B) during the second growing season (1997) in the willow SRF culture on a loamy and sandy soil

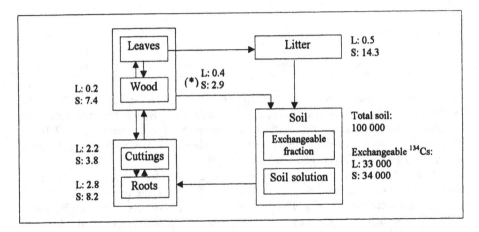

Fig. 2. Radiocaesium balance (in 10^6 Bq ha^{-1}) in a SRF culture of two-year-old plants (R2S1) on loamy (L) and sandy (S) soil. (*) Values for washing-off of caesium with rainwater were for the period August-November.

The soil compartment remains the major radiocaesium reservoir. After two years, only 0.006 % (loamy soil) and 0.034 % (sandy soil) of the soil radiocaesium was incorporated in plant material (Fig. 2).

2.2 FARMERS' SITES (SWEDEN)

To obtain more information on the effect of soil type and stand maturity on the Cs-TF to willow, coppice stands established after the Chernobyl accident were sampled in Sweden in order to be able to make predictions of the radiocaesium behaviour in the long term (25 years). At present, sites have been selected at three different locations in

N. Sweden (clay, sandy-clay, sandy soil) and soils are characterised [4] and radiocaesium and potassium content in woody biomass is determined and transfer factors calculated. The soil-to-wood TF for two clones of willow (table 2) ranged between 4×10^{-6} and 7×10^{-4} m² kg⁻¹. Variability was mainly due to differences in soil type and culture maturity, clone playing a minor role. For comparison, soil-to-wood TFs in forests are of the order of 10^{-4} - 10^{-3} m² kg⁻¹ [5]. The TF to straw, another biofuel, ranges between 10^{-3} - 10^{-2} m² kg⁻¹ [6]. From these preliminary TF results in a coppice ecosystem and the TF for other biofuels, coppice can be regarded as one of the better biofuel sources on radiological grounds.

2.3 TEST SITES (BELARUS)

Experimental sites were established in spring 1997 in Belarus on two soil types (sandy and peaty soils: analysis in progress) with 4 willow selected for high biomass production and relative frost resistance. Experimental plots were 6 by 12 m, and replicated fourfold. Plant density was 18000 cuttings per ha. From the results of this first growing season, there is no significant effect of the clone on the radiocaesium transfer. The TF is, however, significantly higher on the peaty soil (TF averaged over all clones: 2.5×10^{-3} m² kg⁻¹) than on the sandy soil (TF: 6.1×10^{-4} m² kg⁻¹) (table 2), resulting in radiocaesium concentrations of 14500 and 855 Bq kg⁻¹ wood, respectively, levels considerably above the exemption limit for fuel wood in Belarus. The TFs obtained resemble the TF in normal forest stands. Given these TF, willow wood could by safely used for fuel wood only when cultivated on sandy soil with less than about 150 kBq m⁻² and a peaty soil with less than about 30 kBq m⁻².

2.4 DISCUSSION OF RADIOCAESIUM TRANSFER DATA

Since the data obtained are for stands of different maturity, clone and biomass yield, different soil types and climates conditions, it is so far impossible to identify reliably transfer functions based on soil characteristics, stand maturity, etc. These information gaps will be coped with by additional site selection and sampling in Sweden and extended soil analysis. Notwithstanding the precautions to consider, a preliminary attempt is made to discuss, compare and point to some trends found in the TF summarised in table 2.

For the sandy soil the TF of the $R_1 S_1$ plants is about a factor 300 higher in Belarus than in Belgium, notwithstanding the about 10 years more aged contamination in Belarus, generally resulting in reduced caesium availability due to fixation on micaceous specific binding sites for radiocaesium [7, 8]. The larger fraction of bark in the slow growing willow stems in Belarus (300 kg ha⁻¹ in Belarus compared to 2000 kg ha⁻¹ in Belgium) and the about 5 times higher TF to bark compared to wood (Zabudko et al., 1995) and the extremely low K concentrations in the soil solutions of the Belarus soils may explain the results obtained. Very low potassium levels in the soil solution more than proportionally enhance radiocaesium uptake [9-11].

The TF obtained for the Swedish sandy soil are also about a factor 10 higher than in Belgium although the contamination in Sweden is much more aged. Further soils characterisation (soil solution K) is expected to explain this difference. There is also a trend of inverse relationship between the potential soil fixation capacity and the TF following peaty>sandy>loamy/clayey: 100:10:1.

TABLE 2: Transfer factors ($m^2 kg^{-1}$) for the different willow stands sampled

Country	Maturity	Clone	TF, $m^2 kg^{-1}$	Soil type
Belgium	R_1S_1	Orm	$1.9*10^{-5}$	Sand
		Orm	$8.2*10^{-6}$	Loam
	R_2S_1	Orm	$6.2*10^{-5}$	Sand
		Orm	$1.9*10^{-6}$	Loam
Belarus	R_1S_1	Orm/Rapp/Jorr/Björn	$5.8*10^{-3}$	Sand
		Orm/Rapp/Jorr/Björn	$2.5*10^{-2}$	Peat
Sweden	R_6S_2	101, 183	$4-9*10^{-6}$	Clay
	R_5S_4:101, R_6S_5: 183	101, 183	$2-3*10^{-5}$	Sand-clay
	R_3S_3	101	$7.5*10^{-4}$	Sand

3. Radio-ecological aspects: fate of radiocaesium during conversion

For estimating the transfer of radiocaesium from biofuels to ash it is important to know the ash content in the fuel and the proportion of radiocaesium escaping with the flue gasses and the condensation water. For the different biofuels studied under RECOVER, ash content is higher for straw (4-5 %) and wood (1-5 %) than for coppice (2-2.8%). Radiocaesium recoveries in the ashes depend on conversion technique and plant constructions and range between 87 and 99 % for new plants. With the TF for wood, straw and coppice in Sweden and the ash contents for the mentioned biofuels, the soil-to-ash transfer factors for willow are a factor 10-1000 lower than for the other biofuels. With the first-year-coppice TF in Belarus, which is a factor 7 to 30 higher than the TF for the Swedish trials, the soil-to-ash TFs to coppice and other biofuels become more comparable.

Fate of radiocaesium in conversion units and the chemical forms of radiocaesium in the ash will also be studied. From a literature study it was concluded that the biologically available radioacesium in wood ashes was between 30 and 80% [12]. The availability of radiocaesium in the ashes is an important parameter when discussing the fate of radiocaesium when the ashes are distributed as fertiliser or disposed off.

4. Dosimetry

Another point in the evaluation of coppice cultivation as possible remediation action for contaminated farm land is the dose received by persons dealing with the coppice culturing and the subsequent handling and combustion operations. Data are assembled on radionuclide levels in coppice (see above) and other energy crops (forestry, wheat, sugar beet, rape seed), on man power required and on the fate of the radionuclides during conversion to estimate doses to the workforce and to the public (external and internal). Since SRF is not labour-intensive, dose is expected to be low.

5. Technico-agricultural aspects

A next step in the evaluation exercise is the comparison at a technical and agricultural level of coppice with other bio-energy crops. A model was developed in order to evaluate which crop would be more appropriate in given climate and soil conditions (Belgium, Belarus, Sweden) [12]. Five annual (sugar beet winter wheat, oilseed rape, Jerusalem artichoke and sweet sorghum) and two perennial crops (miscanthus and willow) are compared on phytotechnical characteristics and requirements, on adaptability to different soil types, water regimes, climates and on production potential on energy balance. Willow is the most promising crop: adapted to various soil and climate conditions, not too fertiliser demanding and efficient to cultivate and convert from an energy point of view.

6. Energy balance and economic cost-benefit and comprehensive evaluation

The last aspects studied in the evaluation exercise are the energy balance and economical cost-benefit. Energy balance and economic cost-benefit of SRF and other bio-energy crops will be compared both for Western European (Belgium, Sweden) and Belarus conditions. Therefore, data are collected which will serve as input for RECAP, designed to model the economics of SRF for production of energy and/or heat, and for CRISP, developed under the project to model the production of oil seed rape, sugar beet and wheat and their conversion to liquid befouls.

From the radioecological, dosimetrical, technical and economic analyses, different parameters will be identified which affect the application of the short rotation forestry concept for site remediation. The ranking of these parameters will provide us with the necessary tool to validate this remediation option critically.

7. Conclusions

Some first conclusions can already be drawn from the preliminary results obtained from the radiological evaluation of short rotation forestry as an alternative land use option.

Given the low TF (also for light-textured soils) and the high biomass yields obtained under Swedish conditions coppice can be recommended for cultivation on contaminated land in Sweden on radiological and technico-agricultural grounds. In

Belarus, however, biomass production in the establishment year is a factor 2 to 8 smaller than the average production in W. Europe. A short growing season, insufficient water supply and low soil nutritive status may account for this. TFs range between 2.5 and 0.58 10^{-3} m^2 kg^{-1}. If these TF would continue to apply, coppice cultivation for energy production should thus be considered with care in the contaminated zone and be restricted to the low level contaminated sites. Appropriate fertilisation (in case K-fertilisation) may, however, lead to a higher biomass production and decrease the rather elevated through the competitive action of potassium.

Radiocaesium transfer is influenced by soil type (higher TF for light textured soils and soils with a higher organic matter content) and soil nutrition (soil solution K). Least caesium accumulates in the wood and most in the below-ground plant parts. The importance of recycling processes (litter, leaching with rainwater) was alluded upon.

On technico-agricultural grounds, SRF is a high-yielding crop, adaptable to different climate conditions and soil types and little demanding in terms of maintenance (expected dose to workers is thus low!). Its energy efficiency is higher than for other energy crops. In addition, SRF is a technically easy cultivation system and the non-food allocation of contaminated land will be psychologically more acceptable.

8. References

1. Perttu K. (1991) Enregy farming in Sweden: Biomass energy from short rotation coppice. In: Turkenburg W.C. (Ed.) *Het Perspectief van Zonne-Energie*, 3de Nationale Zonne-Energie Conferentie, 4-5 April, 1991, Leeuwenhorst, Noordwijkerhout, Holland, p 85-91.

2. [Mitchell C.P., J.B. Ford-Robertson, T. Hinckley and L. Sennerby Forsse (Eds.) (1995) Ecophysiology of short rotation forest crops. Elsevier Applied Science, London, New York.

3. Schütz P.R and G. van Tol (1990) Aanleg en beheer van bos en beplantingen. Pudoc, Wageningen.

4. Vandenhove H, Y. Thiry, A. Gommers, F. Goor, J.M. Jossart, E. Holm, T. Gäufert and J. Roed. (1998) *RECOVER: Relevancy of Short Rotation Coppice Vegetation for the Remediation of Contaminated Areas.* Mid-term report, EC-project F14-CT95-0021c). SCK•CEN, BLG 762.

5. Zabudko et al. (1995) Comprehensive radiological investigations of forests in Kaluga, Tula and Or'ol regions. IAEA technical meeting on "Clean-up criteria for forests and forestry products following a nuclear accident" Vienna, 30 October - 3 November 1995

6. Rosén, K. (1996) *The studies on the behaviour of radiocaesium in agricultural environments after the Chernobyl accident.* PhD thesis, Uppsala, Sweden.

7. Shulz, K. R. Overstreet and I. Barshad (1960) The soil chemistry of caesium 137. *Soil Science,* 89, 13-27.

8. Sawhney, B.L. (1970) Potassium and caesium ion selectivity in relation to clay mineral structure. *Clays and Clay Minerals,* 137, 47-52.

9. Shaw, G., R. Hwamanna, J. Lillywhite and J.N.B. Bell. (1992) Radiocaesium uptake and translocation in wheat with reference to the transfer factor concept and ion competition effects. *Journal of Environmental radioactivity,* 16, 167-180.

10. Smolders E., L. Kiebooms, J. Buysse. and R. Merckx. (1996) ^{137}Cs uptake in spring wheat (*Tricticum aestivum L.* cv. Tonic) at varying K-supply: 2. A potted soil experiment. *Plant and Soil,* 181, 211-220.

11. Smolders E., K. Vandenbrande, and R. Merckx. (1997) Concentrations of ^{137}Cs and K in soil solution predict the plant availability of ^{137}Cs in soils. *Environmental Science and Technology,* 31, 3432-3438.

12. Ravila, A. and E. Holm (1994) Radioactive elements in forest industry. *Science of the Total Environment* 157, 339-356.

13. Goor F. and J.M. Jossart. (1998) *Literature Review of Short Rotation Coppice: Management and Comparison with Other Energy Crops.* Deliverable to the EC-RECOVER project, FI4-CT95-0021c.

RISK ASSESSMENT MODEL AND RADIATION RESERVES AS A PERSPECTIVE OF CONTAMINATED FORESTS MANAGEMENT

E.Y.USPENSKAYA
All-Russian Research Institute of Nature Protection of GosKomEkologii of Russian Federation, 113628 Moscow, Znamenskie-Sadki, RUSSIA

A.D. POKARZHEVSKII
Laboratory of Bioindication, Institute of Ecology and Evolution of the Russian Academy of Sciences, 117071 Moscow Leninsky prospect 33, RUSSIA

Abstract

An economic use of contaminated territories, including radioactive contaminated forests represents a risk for the population and results in increased economic costs. To estimate potential environment contamination and to minimize possible ecological, economical and social consequences of such events, different models of risk assessment are applied. The complex risk assessment model, which allows us to take into account and compare different kinds of anthropogenic impacts on the environment, is discussed. Also, this approach allows us to compare effects on the different ecosystem components and determine the most reliable criteria for providing a comprehensive analysis of these effects. The way of "radiation reserves" foundation is considered to be one of the possible ways of using contaminated forest territories.

1. Introduction

The use of contaminated territories for industry, including radioactive contaminated forests represents a risk to the population and results in increased economic costs. To estimate potential environment contamination and to minimise possible ecological, economical and social consequences of such events, different models of risk assessment are applied. From a radiological protection perspective, there are fundamental *a priori* requirements for practices to be justified as offering net benefit, for them to be optimised, and to ensure that they result in compliance with dose limits for workers and members of the public. Accordingly, the implementation of radiological protection inherently relies on the prior assessment of the consequences of adopting practices.

Additional difficulties and uncertainties arise because of the fact, that during risk assessment it is necessary to take into consideration both separate and commutative influence of a variety of anthropogenic sources, like radionuclides, heavy metals, traffic

I. Linkov and W.R. Schell (eds.), Contaminated Forests, 385–394.
© *1999 Kluwer Academic Publishers. Printed in the Netherlands.*

noise and so on. Unfortunately, very often in practice, at the estimation of one kind of activity (for example, radioactive storage place) assessors consider only factors of radioactive nature, without consideration of influence of transport supplying, worker's settlements construction, and other impacts, connected with object operation.

Furthermore, we would like to present a part of the complex risk assessment model, which allows us to take into account and compare different kinds of anthropogenic impacts. This risk assessment model was developed on the basis of numerous scientific and practical works, using international experience [1]. At the end of this paper one of the possible ways of using contaminated forest territories is presented.

2. Risk Assessment Model

The main steps of the Risk Assessment Model are baseline description, determination of the space and temporary boundaries of the impacts, identification of the activities/disturbances that result in the actual interaction with the ecosystem, selection of valued ecosystem components (VEC), assessment of environmental impacts and effects, and development of remediation program.

2.1. BASELINE DESCRIPTION

Baseline description characterizes a condition of abiotic and biotic environmental compartments and resource use before activities related to the disturbance. It includes air/climate conditions, territory/soil and surface/ground water statement, description of vegetation, animals, birds, conditions, nature resource use, etc.

2.2. DETERMINATION OF THE SPACE AND TEMPORARY BOUNDARIES

Determination of the space and temporary boundaries allows us to analyze the main parameters of the impacts. A certain classification for evaluation of influence, taking into account it's most commonly characteristics is presented in Table 1.

It should be noted, that depending on the quantities of specific radionuclides, radiation exposure ranges from low (a few multiples of the natural background, an average value of the absorbed dose rate (outdoor) for the natural background is $(4.8 \pm 1.1) \times 10^{-8}$ Gy/h, [2]) to high (absorbed doses greater than about 1 Gray (Gy)). The radioactive half-lives are important in determining the duration of impacts. The chemical identity of the radionuclide influences its environmental behavior. Its radiation characteristics (i.e. alpha, beta or gamma emissions) profoundly affect the spatial dose field from any given source distribution.

2.3. IDENTIFICATION OF THE ACTIVITIES/DISTURBANCES

At the stage of identification of the activities/disturbances that result in the actual interaction with ecosystem, all kinds of impacts, connected with the given activity are determined. For example, considering discharge, it is necessary to estimate discharge to the ecosystem's various compartments.

TABLE 1. General scheme for impact classification

GEOGRAPHIC EXTENT	
Local:	Residual impact on population is measured within Local Study Area (LSA) only;
Regional:	Residual impact on population is measured within Local and Regional Study Areas (RSA);
National.:	Residual impact on population is measured beyond RSA.
DURATION	
Short-term:	Residual impact on population is not measurable beyond three years;
Mid-term:	Residual impact on population is measurable for 3 to 30 years;
Long-term:	Residual impact on population is measurable for 30 + years.
FREQUENCY	
Low:	Residual impact will result from less than 1 event per year throughout the life of the project;
Medium:	Residual impact will result from 1 to 10 events per year throughout the life of the project;
High:	Residual impact will result from greater than 10 events per year throughout the life of the project;
Continuous:	Residual impact will result from continuous events occurring throughout the life of the project.
MAGNITUDE	
Low:	Residual impact will adversely affect < 1 % of the proportion of the population residing seasonally or year-round within RSA;
Moderate:	Residual impact will adversely affect 1-10% of the proportion of the population residing seasonally or year-round within the RSA;
High:	Residual impact will adversely affect > 10% of the population residing seasonally or year-round within the RSA.

2.4. SELECTION OF VALUED ECOSYSTEM COMPONENTS

Selection of Valued Ecosystem Components (VEC) is one of the most important and difficult stages. Here we need to answer the questions:
- What objects and ecosystem parameters should be selected to estimate the anthropogenic influence in every particular case?
- How to provide the most reliable estimation of all effects on different parts of ecosystem?

At present, for conducting the risk estimation we may use a few of the main assessment parameters, like the most sensitive species and communities, generic species, endangered species and economically significant species. In some cases it is possible to select the most important from these criteria for the given Study Area. But in other cases, when all kinds of species are represented at the Study Area, all listed criteria are of importance.

The approach for the selection, represented below, allows us to compare effects on the different ecosystem components and determine the most reliable criteria for providing a comprehensive analysis of these effects. The selection of Valued

Ecosystem Components (VEC) to represent vegetation and wildlife species, groups or communities in the Investigated Area entailed four steps: the selection of evaluation criteria, the identification of vegetation species and communities in the area, the assignment of importance values, and the selection of the key indicator vegetation species or communities based on the evaluation.

Evaluation criteria for vegetation used to rate each species or community on a 3-4 point scale are shown in Table 2. These criteria may be varied according to the specific situation. Table 3 identifies the key terrestrial vegetation species or species groups in the Study Area (Demonstration example, OSLO Project Area, Canada) and the importance values assigned to each according to criteria listed in Table 2.

Similar tables are completed for key aquatic vegetation resources, vegetation communities and wildlife species (birds, mammals, etc.), that are known or expected to occur in the vicinity of the Study Area.

On the base of the conducted assessment the next VECs (terrestrial vegetation species) are selected for further evaluation in our example (Table 3): White Spruce, Aspen, Alder, Blueberry, Lichens.

TABLE 2. Evaluation criteria for selection of vegetation Valued Ecosystem Components (VEC)

Criteria	Ranking Values and Rationale
Abundance	Based on relative abundance
	3. Common
	2. Moderately abundant
	1. Uncommon
Rarity	Based on relative abundance
	3. Designated rare species, group or community
	2. Species, group or community at extreme end of range
	1. Species, group or community uncommon, but not threatened
	0. Species abundant and no concern
Ecological Importance (Sensitivity to Physical Disturbance)	Based on species' or communities' ability to recover following disturbance
	3. Unable to survive even minor changes in habitat
	2. Able to recover rapidly after minor disturbance but unable to survive extensive changes in habitat
	1. Very hardy species or communities, able to recover from a high level of disturbance
Ecological Importance (Sensitivity to Pollutants)	Based on species' or communities' ability to maintain productivity in the presence of toxic pollutants or radionuclides

	3.	Very susceptible to pollutants and least likely to recover in respite periods
	2.	Moderately susceptible
	1.	Relatively resistant to pollutant damage and highly resilient
Economic Importance (Consumptive Use)	Based on forestry, hunting and trapping	
	3.	High productivity
	2.	Moderate productivity
	1.	Low productivity
Suitability for Reclamation	Based on ability to colonize disturbed sites or to be used as seed source	
	3.	High
	2.	Moderate
	1.	Low
	0.	Inappropriate
Recreational Importance	Based on aesthetic value and political importance	
	3.	High
	2.	Moderate
	1.	Low
Diversity	Based on number and extent of species in a community	
	3.	Diverse
	2.	Moderately diverse
	1.	Simple
Availability of Information	Based on amount of information available for each species, group or community	
	3.	Abundant
	2.	Moderate
	1.	Limited
	0.	Nil

2.5. ASSESSMENT OF ENVIRONMENTAL IMPACTS AND EFFECTS

At the step of environmental impacts and effects assessment we consider each interaction with VECs separately using an Environmental Interaction Matrix (Table 4). The Interaction Matrix compares each of the activities/disturbances, determined in step 3, with the Environmental Components under potential impact. Matrix crossings may determine the different parameters (significance, probability, duration etc.). They may also define a weight of potential interaction.

TABLE 3. Rankings assigned to terrestrial vegetation species occurring within the Project area

Species	Abundance	Rarity	Sensitivity to Disturbance	Sensitivity to Pollutants	Economic Importance	Recreational Importance	Diversity	Availability of Information	Totals
Trees									
White Spruce	2	1	3	1	3	1	2	2	15
Black Spruce	3	0	2	1	1	1	1	3	12
Jackpine	1	1	2	2	1	1	2	3	13
Aspen	2	0	1	2	3	2	2	2	14
Poplar	1	0	2	2	2	1	2	2	12
Balsam Fir	1	0	3	2	1	1	1	1	10
While Birch	1	0	1	2	1	1	2	1	9
Tamarack	1	0	2	3	1	1	1	1	10
Shrubs									
Alder	3	0	1	3	2	2	2	3	16
Buffaloberry	2	0	2	2	2	2	1	1	13
Saskatoon	2	0	2	2	2	2	3	1	14
Willow	3	0	1	2	2	1	3	2	14
Blueberry	2	0	1	2	3	3	1	3	15
Herbs/Grasses									
Equisetum	3	0	2	2	1	1	0	0	9
Meadow Rue	1	0	2	2	1	1	0	0	7
Winter Green	2	0	1	2	1	1	0	0	7
Hairy Wild Rye	2	0	1	1	1	2	2	1	10
Bryoids									
Mosses	3	0	2	3	1	1	0	2	12
Lichens	3	0	3	3	1	1	0	3	14

TABLE 4. Environmental Interaction Matrix (see text for explanation)

	Terrestrial wildlife species			Aquatic wildlife Species			Rare & Endangered species	Forestry	Land Uses Consumptive (Hunting, Fishing, etc.)	Non-Consumptive
	1	2	3	1	2	3				
Activities/disturbances										
Air emission (a, b, g)										
Geosphere discharge to lake water										
Geosphere discharge to bottom of soil profile										
Geosphere discharge to well water										
Transfer from lake water to soil (irrigation)										
Transfer from air to soil (deposition)										
Topsoil removal										
Site Clearing										
Site Drainage										
Cumulative Impacts										

2.6. REMEDIATION

On the base of conducted assessment it is possible to conclude which of the ecosystem's components are most susceptible to impact, which activities/disturbances cause this impact and what are the necessary remediation steps.

3. Radiation reserves as a perspective of contaminated forests management

At present in Russia there are two basic ways of treating high-radioactive contaminated territories. The first way is quick deactivation of contaminated lands and renewal of economic activity within a short term. It is very costly at present. Remediation or decontamination of such lands from radionuclides, requires according to the different estimations (see Table 5) from more than 400.000 up to 1.000.000 $ US per hectare [3].

Table 5. Unit costs for treating contaminated soil (Figures expressed in US $/m^2)

"In-situ" treatments	
Soil washing	48-50
Bioremediation	48-50
Soil venting	5-50
"Ex-situ" treatments	
Excavation and transport of soil	15-30
Refill with clean soil	15
Disposal in sanitary landfill	100-500
Incineration of pyrolysis	100-500
Soil washing	150-200
Bioremediation	150-500
Solidification	100-150
Vitrification	Up to 250

The second way to manage sites is by temporary exclusion of contaminated territories from economic use until it is restored to the natural background due to radioactive decay. Practical radioecology in Russia and the former Soviet Union has considerable experience with such "radiation reserves" on such lands, at the nuclear testing site in Semipalatinsk, on Novaya Zemlya, and in areas of radioactive contamination in East Ural and Chernobyl.

The radioactive contaminated area in the territory of the South Ural Mountains near towns Kyshtym and Kasli (the Chelyabinsk region) was formed on September 29, 1957 after the chemical explosion of a storage tank for radioactive wastes. Near 78 PBq (2.1 MCi) of the wastes were released in the environment and precipitated mainly on the area of more than 1000 km^2. This area was named the East Ural Radioactive Track (EURT). At the moment of the explosion 66.3% of waste radiation was a portion of 144Ce + 144Pr and only 5.15% of 90Sr + 90Y. After almost 5 years the portions were accordingly 15.0 and 82.7% [4]. During two years human populations from the contaminated area were evacuated and their settlements were destroyed. The special buffer zone around the EURT was organised and protected by special police.

Agricultural lands and pastures were ploughed to decrease radioactive contamination but the level remained very high. The contamination density at the axis of the EURT in the most polluted places was from 100 to 2500 Ci of 90Sr per 1 km². At the border of the EURT the density was 2 Ci of 90Sr per 1 km².

In 1966 along the track axis the East Ural State Reserve (EUSR) was established. The total area of the EUSR was about 16 700 hectare. Forests cover 45% of the EUSR area. The share of birch forests is near 92% and the other 8% is represented by pine and aspen forests. Grasslands cover 41% of the area. The main soil type is leached chernozem, but gray forest soils and grass podzol soils are met under some forests. Ponds, lakes and rivers occupy 9% of the area. Other lands are presented by former settlements, roads and experimental plots [4].

According to a rough estimation its deactivation would require 6.680.000.000 $ US (at 1995 prices). Data from the Reserve Department (Goscomecologii RF) showed the average budget of Russian reserve for 1996 year made 635.4 million roubles (or 115,527 $ US). There fore the sum necessary for the EUSR clearing would require its functioning for 57.822 years.

The biological diversity on such protected territories, in spite of the rather high contamination, sharply exceeds biodiversity of adjacent regions, to which the data obtained in EUSR testify. Surveys of vertebrate populations in the EUSR show unusually high biodiversity when compared to surrounding areas and the rest of the Transuralian region. Practically all species recorded in both southern and central parts of the Transuralian region have been observed in the EUSR area. Forty-five species of mammals, 171 bird species, 5 reptilian species and 6 amphibian species are registered in a territory of about 300 km². These equal 82% of mammal species, 73% of bird species, 83% of reptilian species and 60% of amphibian species ever recorded for the Chelyabinsk and Sverdlovsk regions [5]. A similar situation is observed in the Chernobyl region.

4.Conclusion.

The complex risk assessment model, presented below, allows us to take into account different kinds of anthropogenic impacts on the environment, compare effects on the most important ecosystem components and determine the most reliable criteria for providing a comprehensive analysis of these effects.

The use of radiation reserves reduces the costs for the contaminated forests deactivation and also creates islands for biodiversity conservation. According to Krivolutzkii (1996) [6,7] biodiversity dynamics at radioactive contaminated areas depends on some factors. The first factor is human activity before the accident at the territory. The second one is radioactive contamination. The third one is the strong prohibition of human activity after the accident. The last one is human activity at the lands surrounding a contaminated area. We discuss vertebrate biodiversity at the EUSR taking into account these different factors of biodiversity development.

At the same time the situation with high level of biodeversity and lacking the direct evidences of radioactive contamination represents a real danger for the people of the populated contaminated territories, distant from Chernobyl. For example we may

394

consider the Bryansk and Gomel regions (about 200 km from Chernobyl), where the average dose rate is about 120 μR/h (data on May, 1998). The population of the such territories should be widely informed on the potential danger by means of mass media and informational tables, containing standards and normatives for intake of water, mushrooms and berries and other necessary information. For the most effective and ecologically safe way of management contaminated forest territories to be determined it is necessary to conduct additional economic and ecological researches.

5. References

1. United Scientific Committee on the Effects of Atomic Radiation (UNSCEAR), 1996. Effects of radiation on the environment. Forty-fifth session ofUNSCEAR Vienna, 17 to 21 June 1996.
2. Moiseev A.A., Ivanov V.I., 1990. The directory on dosimetry and radiation hygiene. Energoatomizdat. Moscow. (in Russian)
3. Carrere and A. Robrtiello, Integrated Soil and Sediment Research: a basis for proper protection,1992, "Soil clean up in Europe - feasibility and costs", Kluwer Academic Publishers.
4. Sokolov V.E. & Krivolutzkii D.A. (Eds.), 1993, Ecological after-effects of the radioactive contamination at South Ural. Nauka. Moscow, (in Russian).
5. Pokarzhevskii A.D., Kuperman, R.G., Isaev S.1., Rjabtzev, 1.A., Semenov, D.V., Uspenskaya, E.J., 1998. Do radiation reserves be the last islands of wildlife biodiversity? A case study of terrestrial vertebrate populations at the area of the East Ural radioactive track. *Biological Conservation.* (Submitted in 1997)
6. Krivolutzkii D.A.,1996. Biodiversity and ecosystems dynamic under the radioactive contamination. *General biology. Reports of Russian Academy of Science*, 347, 567-569.
7. Krivolutzkii D.A., 1996. Survival strategy of animals populations under the radioactive contamination. *General biology. Reports of Russian Academy of Science*, 349, 568-570.

COUNTERMEASURES AND RISKS ASSOCIATED WITH CONTAMINATED FORESTS

Report of the Working Group on Countermeasures and Risks[1]

B. AMIRO

Canadian Forest Service, Northern Forestry Centre, Edmonton, AB, T6H 3S5, CANADA

A. GREBEN'KOV

Institute of Power Engineering, Sosny, Minsk, 220109, BELARUS

H. VANDENHOVE

Belgian Nuclear Research Centre SCK•CEN, Boeretang 200, Mol 2400, BELGIUM

1. Introduction

This working group represented a wide range of scientific interests, encompassing the full scope of radioecological issues of contaminated forests. Our main objective was to "identify gaps in our knowledge about possible countermeasures to protect forest values, 15 to 20 years following radionuclide contamination from a Chernobyl-type accident". The focus on knowledge gaps was a key aspect because of the need to concentrate resources on key areas where decisions cannot be easily made without more knowledge. The goal of our research is to provide scientific information to help direct policy decisions, so we need to focus on areas where knowledge is limiting. Our goal was to identify a few countermeasures that have potential merit, rather than reviewing the wide range of countermeasures that have been suggested elsewhere. We also considered a broad range of forest values, not just those practices that involve forest foods or timber supply and management. The time period of 15 to 20 years after an accident was selected to concentrate on present problems posed by the Chernobyl accident; therefore, countermeasures that could be applied immediately after an accident were not

---

[1] Main working group members: B. Amiro (Canada), T. Gafvert (Sweden), M. Goldman (U.S.A.), A. Greben'kov (Belarus), A. Kupriyanchuk (Ukraine), G. Petersen (Germany), A. Rantavaara (Finland), V. Rimkevich (Belarus), V. Shutov (Russia), E. Uspenskaya (Russia), H. Vandenhove (Belgium), G. Voigt (Germany), R. Wilson (U.S.A.), V. Zaitov (Ukraine)

I. Linkov and W.R. Schell (eds.), Contaminated Forests, 395–401.
© 1999 *Kluwer Academic Publishers. Printed in the Netherlands.*

considered. The selected time period focuses on potential research that could be applied within the next three to five years and be readily implemented. Our discussions involved scientific issues to effectively utilize the expertise of about 15 scientists in the working group. These scientific issues were related to policy, economics and social issues that are linked to technical problems.

2. Presentations

The working group discussion was partially driven by presentations made by group members. The text of most of these presentations is given in this volume, of which the most important issues are presented here. Ingestion doses was a main theme of three papers, because this is the principle exposure pathway for the public. V. Shutov (Institute of Radiation Hygiene, Russia) presented data on doses from ingestion of forest foods in Russia whereas A. Kupriyanchuk and V. Zaitov (Institute of Radioecology, Ukraine) showed data from the Ukraine. In both cases, forest foods, and particularly fungi, were shown to be a main contributor to internal doses, and the exclusion of certain forest foods could reduce these doses. A possible method of education using information pamphlets explaining doses from some of the more contaminated fungi, and information on food preparation methods to reduce the ^{137}Cs content, was presented by G. Voigt (GSF Institute fur Strahlenschutz, Germany). Knowledge gaps identified in the area of forest foods included testing of the education scheme to limit fungi ingestion and validation of dose models against whole-body-count data. A. Rantavaara (STUK Radiation and Nuclear Safety Authority, Finland) presented data on the use of wood ash, and we identified the need for a comprehensive dose assessment for the practice of fertilizing forests with contaminated wood ash. V. Yoschenko (Institute of Agricultural Radiology, Ukraine) showed data on radionuclide resuspension from forest fires with the conclusion that doses from fires are quite small. E. Uspenskaya (All-Russian Research Institute of Nature Protection, Russia) outlined a scheme for ecological risk assessment of contaminated forests utilizing valued ecosystem components. H. Vandenhove (Belgian Nuclear Research Centre SCK•CEN, Belgium) outlined the current results on radiocesium cycling and biomass production in a short crop rotation coppice cultivation, and the need for improved estimates of the soil-to-plant transfer factor for a variety of situations was identified. G. Petersen (GKSS Research Centre, Germany) gave an overview on local, regional and global concerns of heavy metals, especially related to the European environment. V. Davydchuk (Institute of Geography NASU, Ukraine) illustrated maps of the wide range of landscape features in the 30-km Chernobyl zone and concluded that any selection of countermeasures must be matched to appropriate landscapes. The working group also used the information from many of the general session papers in the discussions of risks and countermeasures.

3. Dose Pathways and Countermeasures

The discussion concentrated on dose pathways and countermeasures to reduce dose through these pathways. A general plan was established to evaluate aspects of:

- the exposed individual, consisting of forestry worker, forest resident, or forest product consumer,
- exposure pathways, that included: a ground source contribution to external, inhalation (resuspended particles) and ingestion (water) doses; a tree (or wood) source contribution to external, inhalation, immersion (through fire smoke plume) and ingestion (through use of wood ash on gardens or forest soil) doses; and a food source contribution to external and ingestion doses,
- the range of doses and radionuclide concentrations encountered,
- countermeasures that could be employed,
- remediation strategies,
- the cost of a practice,
- the efficiency of a practice, and
- the application of the practice including aspects of timing and the spatial scale.

Within this framework, we discussed data bases, data inconsistencies, knowledge gaps, personal experiences and case studies. In addition to these dose considerations, it was recognized that there are policy regulations established for contaminated areas that do not strongly relate to the risk associated with the use of forest products. This issue is discussed later in this report.

Considering all dose pathways, we concluded that immersion doses from fire plumes were only of concern for forest fire fighters and we did not consider this pathway further in our analysis. Inhalation and ingestion pathways for ground sources, inhalation of tree (wood) sources, and external dose from food were estimated to be minor and need not be considered further. This left four dose pathways warranting further consideration: external doses from ground sources (mostly to workers, but also to some forest residents); external doses from tree sources to workers in specialized operations; ingestion doses to the public from tree sources related to the use of wood ash and forest organic matter on gardens; and food ingestion doses to the public.

Possible countermeasures associated with reduction of external doses from ground sources include:

- mechanical removal of the top layers of soil organic matter where most of the ^{137}Cs is still concentrated,
- plowing of the top organic layers to move contamination to deeper soil layers to decrease the external dose at the surface,
- providing physical protection for workers in the form of machinery with some shielding,
- restricting the time spent for individual forestry workers in contaminated areas,
- changing forest practices such that less time is spent in forestry operations for forest maintenance, and
- using prescribed fire to burn the top organic layers of soil so that radionuclides are solubilized and leach to lower soil layers.

We also discussed phyto-extraction but concluded that this operation cannot normally remove a sufficient amount of radionuclide to be of practical use on a wide scale; however, the potential and limitations of phyto-remediation are not sufficiently known. Of the countermeasures outlined above, we identified significant knowledge gaps in the application of mechanical removal and plowing, in the use of prescribed fire as a treatment, and in the changes in forestry practices. The other countermeasures are sufficiently well known that they can be applied without a further research investment.

Possible countermeasures associated with the reduction of external doses from trees (wood) to workers include:

- providing physical protection (shielding), both in the forest and in industrial activities,
- mechanization of forestry practices,
- removal of bark and outer parts of the tree to reduce contamination levels, and
- restriction of workers in contaminated areas or working with contaminated wood.

These countermeasures are all well known and we did not believe that there are significant knowledge gaps in their application.

Possible countermeasures associated with the reduction of ingestion doses from tree and organic matter sources used for fertilization of gardens as ash or as organic matter (compost, mulch) include:

- providing clean material, such as non-contaminated fuelwood or organic fertilizers,
- providing alternative energy sources to replace combustion of contaminated wood, and
- collection of contaminated ash followed by safe disposal.

We also discussed restricting the collection of fuelwood but this is largely impractical in rural areas unless alternative wood is supplied. The possibility of removing contaminated trees from the forest was also considered as an impractical countermeasure unless there is an additional goal for such an activity, and new trees grown would still be contaminated. Again, we did not identify any significant knowledge gaps for these countermeasures, since their application is relatively clear.

Possible countermeasures to reduce doses from food ingestion include:

- monitoring contamination of forest foods at the village level followed by exclusion of the most contaminated items from the diet,
- education programs on the risks associated with forest foods and suggestions for food preparation methods that reduce contamination levels (especially for mushrooms),
- the provision of alternative non-contaminated food supplies, and
- the removal of food from the forest, possibly as an organized activity or through purchases (or trade) with individuals who normally collect these foods.

We also discussed the restriction of public access to the forest but concluded that this is not practical, except in special cases such as the 30-km Chernobyl zone. We also believe that there are knowledge gaps in each of the four countermeasures outlined above and we should address these gaps before we can evaluate the efficiency of these measures. It is important to note that ingestion of forest foods is the most important dose pathway affecting residents of contaminated forests.

4. Practices and Other Issues

It is not always easy to divide activities into "countermeasures" and "remediation" so we consider both of these as "practices". In addition to the countermeasures outlined above, we discussed other practices that could be employed in contaminated areas, such as:

- alternative land designations, including establishment of park preserves that would contribute to national and international biodiversity goals,
- alternative land use, such as short-rotation coppice growth, and
- alternative forest products that includes wood for bioenergy production.

These practices are not necessarily linked to dose reduction, but have merit as alternative uses of contaminated forest land that provides value. In each of these cases, further research could help establish them as viable alternatives to historical forest use, provided that social and economic concerns can be satisfied.

In our discussions related to countermeasures for dose reduction, we identified the need for more case studies to validate assumptions and models. Such case studies will also help to establish the cost and efficiency of a countermeasure, which are needed before wide-scale application.

Although reduction of doses appears to be a worthy goal, there is the question of the level to which the dose should be reduced. The principle of ALARA (as low as reasonably achievable) also needs to consider social and economic values. We discussed that dose contributions from practices that result in doses exceeding 1 mSv per year could be candidates for dose reduction, since this level is the present limit established by the International Commission on Radiological Protection for members of the public for a practice. However in many cases, the doses from activities such as forest food ingestion are below this limit, and it is not clear that further dose reduction is warranted unless it can be done easily and cheaply. This is especially important to put into context when we consider that background doses are naturally higher than this, and natural variability among locations and lifestyles often cause differences in doses of greater than 1 mSv per year to individual people. Money may be better spent on resolving some more crucial economic and social issues. Despite this, we recognize that people desire to choose their own risks, and those risks associated with forests contaminated by Chernobyl are not self-imposed. Therefore, efficient countermeasures to reduce dose should still be considered, even 15 to 20 years following the accident.

We also recognized that there are policies imposed on contaminated forest lands that are not directly related to risk. For example, contamination levels for the use of wood are in place that are set at a fixed concentration (Bq/kg) which can only be related to risk if a specific end-use of the wood is considered. For a given concentration, the use of wood as fuel, furniture, or exterior building materials will all produce different doses depending on the contact with humans. It would be more economically sound if exemption limits were related to the specific end-use of forest products.

5. Research Gaps

During the working group discussions, several research gaps were identified. Of course, there are many additional gaps, but we suggest that the following require further work in the near future.

- Dose estimates from simple calculations and more complex models have many assumptions, and are usually conservative estimates of dose. However, there is evidence that these estimates are usually too high and do not fully reflect food habits and resources of people in contaminated forests. There is a need to validate these estimates against real data, such as whole body counts of forest residents and corresponding diets. Such validation could result in the relaxation of restrictions to forest food ingestion, if the actual received doses are low.
- Various schemes have been proposed to educate forest residents regarding the risks associated with ingestion of certain forest foods. However, we need to validate these schemes and to test the effectiveness of an education and information program. Relevant case studies would help address this issue.
- Full dose assessments are required for some proposed practices. As an example, the use of contaminated wood ash on gardens or for forest fertilization may increase doses, but this increase may be less than some estimates suggest. Our practical knowledge of such pathways is improving, but we still need additional work to be sure that such practices are sound.
- There is a large variability in ^{137}Cs transfer factors to many forest products in the landscape. The uncertainty associated with this range of factors makes it difficult to make decisions on the application of a practice. Work is still needed to reduce our uncertainty in transfer of ^{137}Cs and ^{90}Sr from contaminated soil to plants and other forest products.
- There are several gaps related to reduction of external doses from contaminated soil, mostly to forest workers. We need more information from pilot scale studies on the effectiveness of mechanical removal of contaminated forest floor organic material and plowing. The effectiveness of changing forestry practices, such as reduced stocking densities of new plantations or thinning existing immature stands, should also be assessed. There may be a role for prescribed fire to burn contaminated surface material, but the cycling of ^{137}Cs following such a practice is not known.

- Additional schemes to reduce ingestion doses to the public involve increased monitoring of food in villages, provisions of alternative food supplies, and purchasing or trading non-contaminated foods for contaminated forest foods that were collected by local people. In theory, these schemes appear to reduce doses, but their practical application involves many economic and social changes, and we need to know whether such schemes are effective.
- As an important part of a forest remediation strategy, controlled combustion of contaminated woody and phytomass materials resulting from some countermeasures outlined above, requires detailed evaluation with regard to economical, ecological and social factors and the assessment of potential added risk.

6. Recommendations

The working group identified a few areas where the situation could be improved, and makes the following recommendations:

- Establish further coordinated research programs to address the knowledge gaps. These programs must be structured to ensure cooperation among the affected countries.
- Gather more information on the practical costs and efficiencies of countermeasures and other practices. This requires additional case studies aimed at the broad application of a practice.
- Evaluate current policies of contamination levels in products that do not directly correspond to risk. Many potential practices are limited by these policies, and this will become especially important as the risk decreases over time.
- Consider alternative forest values such as biodiversity and preservation of natural landscapes. Perhaps some of the contaminated areas could be designated to fulfill such ecological goals and be a unique contribution to the European environment.

Discussions in the working group initiated further collaborations among scientists from NATO countries, non-NATO countries, and CIS partner countries. It is hoped that these collaborations grow so we can fully address many of the immediate issues related to risks and countermeasures in contaminated forests.

Conclusion

CONTAMINATED FORESTS: RISK IN PERSPECTIVE

After Dinner Talk

R. WILSON
Department of Physics, Harvard University
Cambridge, MA 02138, USA

There are several ways of marking the stages in a scientist's life:

- as a young student one is invited to present one's ideas to ones close associates.
- as an older student one is allowed (after extensive vetting) to present a 10 minute "contributed" paper to a society meeting
- as one's work is recognized meeting organizers invite one to give a 30 minute presentation of important work
- in due course as one becomes the world expert in a field one is invited to give a "keynote" speech at the start of a conference or a summary or "rapporteur" talk at the end.
- after that life goes downhill. Eventually, all one can do is give an after dinner talk!

That is what has happened to me at this workshop. So I enquired of my friend Professor Philip Morrison of MIT: what are the requirements for an after dinner talk? He explained them to me very clearly. It is important to have nothing to say - and to say it very slowly (especially if vodka has been served)!

This workshop has been funded by the North Atlantic Treaty Organization, NATO. Why should that be so? NATO was a military alliance to counter the USSR (which included Ukraine). Some of us think that now the cold war is over NATO should declare victory, disband, and save the money. But bureaucracies (which NATO has become) like to expand. That is not always bad. NATO always recognized the importance of good science in welding countries together. NATO science workshops were begun by my colleague, Professor Norman Ramsey when he was Science Ambassador to NATO in 1958/59. The expansion of these workshops to the countries which NATO formerly opposed may well help to ensure that we will never again become opponents.

This workshop is about transport of radionuclides through forests. But I always like to go the bottom line and ask - why are we interested? why should anyone care? I suggest the primary reason we are here is because we want to know what will happen in other situations of radioactivity release. The measurements and models are remarkably good in this respect. But I want to go back further. Over the last 25 years have been

I. Linkov and W.R. Schell (eds.), Contaminated Forests, 405–408.
© *1999 Kluwer Academic Publishers. Printed in the Netherlands.*

making comparisons of risks of life. I suggest we spend a little time today on making comparisons, and make sure that when we talk to others - politicians, newspaper reporters and so on - we make comparisons so that our auditors may gain the perspective that we have ourselves.

At the famous IAEA meeting in August 1986, a delegation of about 30 people from the USSR (not only Russians, but Armenian and Ukrainians) described to 400 assembled world experts details of the Chernobyl accident. I was present and wrote a short commentary in "Nature". Russians have always been secretive, so I noted that it may well have been the first time since Ivan the Terrible that Russians had given so much detail of a domestic accident (the USSR was of course dominated by Russia). We now know that the predictions of accumulated dose inside the USSR were somewhat overestimated. Overall in the world we anticipate a world wide population integrated dose of somewhat less than 200 million man-Rems (2 million person-Sv). This would lead on a linear dose response (but including a dose rate adjustment) to about 20,000 cancers in the world. Less reliable estimates have used a number of 300,000. Either number seems large and confuses people. But either number is a small fraction of the anticipated 500 million cancers in the world from other causes. I went further. I believe that air pollution, probably due to fine (<2 micron) particles causes chronic lung problems that lead to premature death - but delayed from the time of the pollution insult just as cancer is delayed by a latent period. My estimates are that more than 20,000 people in Russia, Ukraine and Belarus have their lives shortened by particulates every year - a catastrophe of the magnitude that happened due to Chernobyl only ONCE.

We as experts can understand this comparison. But it is vital that the citizens who have been exposed to radiation as a result of Chernobyl understand it too - and realize that the effect on public health is more limited than politicians sometimes lead them to believe.

During this meeting I have heard several participants talk about how to meet the "intervention levels". Some of us got confused and asked "what are the intervention levels?" No one asked the much more important (and politically difficult) question, "why are the intervention levels?". They are set by politicians - often without adequate technical input and sometimes contrary to professional technical advice. I illustrate this by the way in which the scientists in the UK National Radiological Protection Board (NRPB) were ignored in 1986.

In the UK "intervention levels" for destroying food are set by the Ministry of Agriculture and Fisheries, and NRPB is a part of this ministry. In 1986 NRPB was the only organization in the world that had careful thought through the proper intervention levels, to reduce exposure and save lives, but not at an expense appreciably (100 times) more than is done in other death averting measures. By coincidence the report was published about 10 days before the Chernobyl accident. It was just in time. It recommended intervention levels varying with foodstuffs of 5-10 KBq/kg. But Europe panicked. Europeans set "intervention levels" 5 to 10 times lower than the NRPB recommendation. They were encouraged by western European farmers who, in a year of good farming weather were delighted to be able to ban cheap imports from Eastern Europe. With strong scientific support Britain's Minister of Agriculture tried to head this off. But after three months (during which other EEC ministers threatened to boycott all British food exports) the Minister capitulated and changed the rules in Britain to the consternation of Welsh and Scottish farmers. (I cornered the Minister in the British Embassy in Moscow in February 1987 and he admitted that this was contrary to ALL his scientific advice). The NRPB report had pointed out that no one eats meat at the maximum level all the time. In their careful study they pointed out that an intervention level calculated on everyone eating 1 kg of the most contaminated meat a day could be safely multiplied by a factor of 30. I note that although I believe that the Minister of Agriculture was overcautious in 1986, he was lacking in caution in 1996 when he refused to recognize the risk of the "mad cow" disease - again contrary to the views of his scientific advisors.

But Sweden joined in the EEC folly. One tragic effect was on the Laplanders. Reindeer eat lichen which concentrate Cs 137. The Laplanders could not sell reindeer meat in Stockholm and (officially) were not supposed to eat it themselves.

I heard a lot at this meeting about mushrooms. I like mushrooms and eat mushrooms though perhaps not as much as those who live in Ukrainian villages. I heard that the internal radiation dose from eating mushrooms is the most important problem of living near a contaminated forest. The "intervention dose" for mushrooms is a small radiation dose - about 50 mRems/year. This leads on a (pessimistic) linear dose response relationship to an increased (lifetime) risk of 0.1% - which risk is falling with time but not much faster than the ^{137}Cs half life of 30 years. But if one takes such a calculation seriously, one should calculate the risk due to ingestion of the chemicals in the mushroom. Mushrooms contain a number of chemicals that are carcinogenic in animals, and by presumption are carcinogenic at low doses and in people also. A rough risk calculation puts the natural risk for eating the same quantity of mushrooms at 0.2% - double the present radiation risk and NOT falling with time.

There is another issue that we as professionals should also remember. When an accident is in progress how many people should one evacuate? Before Chernobyl there was a view that evacuation should be considered if the calculated time integrated dose was 25 to 75 Rems - corresponding to a cancer rate of a few percent. This has changed. I believe the present US and probably the world political view - that one should evacuate for a POSSIBLE much smaller dose of the order of 10 Rems - is wrong. There is an old WHO study that dislocation caused by evacuation causes an increase in death rate of 5%. Indeed many of the adverse effects on health noted around Chernobyl are probably NOT due to radiation but to dislocation and fear. This might well be reduced if proper facilities are made available. So I reiterate what I have said many times before. If a radiation accident is in progress and a policeman comes to tell me to evacuate because the dose might be 10 Rems I would tell him directly - "Get Lost!". Academician Romanenko, your Minister of Health in the Ukraine in 1986, told me that in June or July 1986 he went to a village and told the villagers that they had better move. He was met with loaded shot guns. I suggest that this is an example where ordinary people knew better than their leaders. It is of course particularly absurd to ask me, a 72 year old, to move because of a fear of cancer. The cancers other than leukemia have a latency period in adults of 20 years or more. It is a "no brainer" for me to decide whether to stay in my (radioactivity contaminated) home for 20 years and have even a 5% increased chance of cancer at age 92 or to face the dislocation of a move to a place I do not know.

This leads to another issue that is important for us to consider. If and when should one relocate the population back to the affected areas? The radiation dose on Kurchatov Street in Pripyat was 70 microRems per hour. This was outdoors. It would be 5 times smaller indoors. This dose is 2000 times less than the dose on the evening of April 26th 1986. The radiation dose for someone spending 2/3 of the time indoors is 200 mRem/year, and is falling with time. This dose is comparable to that in some parts of Goa (India) where there are monazite sands, or at high altitudes such as the expensive resort town of Aspen, Colorado (USA) where cosmic rays and mine tailings increase the natural background. With these facts in mind, one can ask in order:

- should one allow retirees (age 65+) to live in Pripyat and other parts of the exclusion zone? I note that some people, wiser than the bureaucrats, have taken the matter into their own hands and have moved back.
- Should one allow any persons without families to live in the exclusion zone?
- Should one allow children to be brought up in the exclusion zone?

If the answer to any of these is NO, and the exclusion zone remains uninhabited I ask the fundamental question that all persons in prosperous countries (which in spite of the recent downturn in the economy INCLUDES Ukraine) should ask:

- what right do we have to stop people from poor countries living in this place?

As an example of the implications of this question I remind you that Bangladesh, with 120 million people, is one of the poorest countries in the world. To a Bangaldeshi the chance of living in the exclusion zone (even though it is cold in the winter) might well be heaven compared to his poverty. The radiation hazard is minute compared to the hazards he faces every day. The catastrophe of Chernobyl is a Sunday School picnic compared to their catastrophe. About one quarter of all wells in Bangladesh have levels of arsenic in the well water that exceed WHO guidelines and I can testify from personal observation that 75 out of 1000 persons in a typical village have easily visible signs of chronic arsenic poisoning. Moreover the flood hazard is huge.

In your country the catastrophe of Chernobyl took place as you were engaged in a major political change from an authoritarian "paternalistic" society to a free market. You have had to learn that the government does not always tell you what to do. Less often will government pay for it. We must all remember that freedom has its duties, its responsibilities and its dangers.

If we ask a government to do something, it is we who pay for it. In this society, we, as professionals, have a duty to propose and recommend standards and guidelines that are based upon a regard for public health and not make expensive proposals based only upon imagination. The politicians and the "do gooders" in this world may not accept our standards even when we argue them well. But we would be remiss if we did not remind politicians that if they are   irrational and fail to make comparisons to aid their understanding they DECREASE rather than increase the overall public welfare.

I thank NATO for funding this workshop, the organizers for inviting me and you all for your attention.

PILOT ELICITATION OF EXPERT JUDGMENTS ON MODEL PARAMETERS AND RESEARCH NEEDS IN FOREST RADIOECOLOGY

I. LINKOV and K. VON STACKELBERG
Menzie-Cura and Associates, Inc. and Harvard Center for Risk Analysis USA.

1. Individual and Group Expert Elicitation: Design and Objectives

The high degree of uncertainty in modeling structure and input parameters used to describe radionuclide fate and transport in forest ecosystems has posed difficulties in accurately modeling radionuclide transport. Many dynamic parameters are required for modeling (such as radionuclide residence half-times in forest compartments) and these parameters are typically difficult to measure (e.g., half-lives vary over a wide range: from days to thousands of years). The NATO workshop brought together more than 70 experts in forest radioecology and ecosystem modeling encompassing a wide range of disciplines and backgrounds. It provided a unique opportunity to quantify the state of knowledge on processes and parameters used in forest ecosystem modeling.

Expert elicitation is the process of obtaining probabilistic estimates of particular parameters from experts familiar with the given topic. Expert elicitation can be used to quantify the uncertainty in specific parameters in the absence of experimental data that would provide such information. In this case, the group expert elicitation was done through discussions in working groups. Three working groups were formed during the first day of the conference:

1.   Data Collection and Measurements
2.   Processes and Modeling
3.   Risk Assessment and Remedial Policies

Each working group was Co-Chaired by both Western and Eastern Scientists. Participants were asked to nominate Group Chairs prior to the conference and the Organizing Committee then selected Group Chairs. At least one Co-Chair was selected prior to the conference. Co-Chairs developed the Working Group agendas that included oral presentations and discussions on the relevant issues. Two Rapporteurs per group were selected to record all discussions and to assist in writing the group report. Co-Chairs presented interim summaries during the conference. During the last day of the workshop, the groups reported their finding and recommendations on reducing uncertainty in current data and expanding our knowledge and understanding of the dynamics in radiologically-contaminated forests. Co-Chairs and Rapporteurs

I. Linkov and W.R. Schell (eds.), Contaminated Forests, 409–417.

were responsible for writing summaries of the group discussions for the Workshop Proceedings (see Working Group Reports in this volume).

The Individual Expert Judgment Elicitation was conducted through interviews with 8 experts. Participants were asked to nominate Experts prior to the conference. Experts were then selected by the Organizing Committee. The questionnaire was distributed to the selected experts prior to the conference together with general information on expert elicitation methodologies. During the conference, one to two-hour interviews were scheduled with each selected expert. Table 1 lists the participating experts:

TABLE 1. List of Experts

Expert	Country
R. Avila and L. Moberg[a]	Sweden
M. Belli	Italy
A. Dvornik	Belarus
A. Konoplev and A Bulgakov	Russia
A. Rantavaara	Finland
T. Riesen	Switzerland
A. Scheglov and A. Klyashtorin	Russia
Y. Thiry	Belgium

[a]Drs. Avila and Moberg provided some limited information

2. Interview Protocol

The interview protocol begins with a description of a specific forest and radionuclide deposition scenario. Forest Scenario developed for the IAEA-BIOMASS program was used [1]. The experts were asked to base all responses upon this forest ecosystem scenario. The following provides a summary of the Scenario [1]:

Source Term: Spike release and deposition of ^{137}Cs as dry aerosol. Total initial deposition at the top of the canopy is 50 kBq m^{-2}.

Deposition Date: 1st May

Topography, Climate: Deposition is to a uniform area of forest on level ground. The average annual temperature is 5.3°C. January is the coldest month (-8.5°C), and July the warmest (+19.4°C). The period of average daily temperatures above +10°C is 140-150 days. The maximum snow cover occurs from the second week of February to the first week of March and reaches 30 cm. The snow cover melts in late March, early April. The annual precipitation varies from 550 to 790 mm. About 70-75% of the total annual precipitation falls during the warm period, from April to October. Small amounts of precipitation of up to 1 mm/day contribute about 40% of all precipitation during a year.

Soil Characteristics:	The main type of soil is soddy-podzolic loamy sand formed from fluvio-glacial sand accumulation. The soils belong to the automorphic group and have a density of 1.2 g/cm^3. The main soil mineral is quartz and its content varies from 80 % to 95 % (in the 0.05 - 0.01 mm fraction). The clay content is between 0.5% and 1%. More than 95.3% of the soil consists of particles exceeding 0.01 mm (physical sand). The soils of the area are characterised by low natural fertility and unfavourable hydrophysical properties: high water permeability and low water-holding capacity, this causing rapid deep infiltration of melted snow, while considerable quantities of water are evaporated from the upper layers.
Forest Type:	The dominant species is pine (*Pinus sylvestris*) with sparse examples of birch. The rising generation includes pine (*Pinus sylvestris*) and birch (*Betula pendula*). The average age of the pine trees is 50 years. The birch trees are 40 - 50 years old. The average age of the trees at the time of contamination is 50 years. The average height of the trees is 20-25 m. The average density of wood biomass is between 120 and 160 metric tonnes per hectare.
Understorey:	The total biomass of understorey is about 1.0 kg/m^2 (10 t/ha) d.w., including small trees of the rising generation. Shrubs include rowan-tree (*Sorbus aucuparia*), alder black (*Alnus nigra*), buckthorn alder (*Frangula alnus*). The prevailing species of dwarf-shrubs are red raspberry (*Rubus idaeus*) and blackberry (*Rubus trivialis*). The main species of mushrooms are *Boletus edulis*, *Leccinum scabrum*, *Cantharellus cibarius* and *Russula* species. Grasses are rather sparse. The prevailing species are *Pteridium aquilinum* (fern), *Pyrola rotundifolia*, *Equisetum pratense*, *Calamagrostis epigeios*, *C. arundinacea*, *Deschampsia caespitosa*, *Melica nutans*, *Chamaenerion angustifolium*, *Majanthemum bifolium*. Mosses cover 90 % of the area. The prevailing species are true mosses (Bryales).

Table 2 outlines the Expert Judgment Elicitation Protocol. First, experts were questioned on conceptual models of radionuclide fate and transport in forests. These questions were structured to direct expert knowledge on specific processes in particular ways, *i.e.*, to insure a baseline level of understanding of the issues for each expert. Next, the experts were asked to quantify specific modeling parameters. We provided each expert with information from a comprehensive literature review on each parameter and asked each expert to describe a distribution of values in terms of the range and a mode. The mode corresponds to the most likely value, while the range provides the bounds on reasonably expected values. Given the constraints on time, availability, and the pilot nature of this project, these initial distributions are best characterized as triangular distributions. Finally, each expert was questioned about research priorities in forest radioecology.

412

TABLE 2. Outline of the Expert Judgment Elicitation Protocol

#	Nature of Question	Formulation
1	A series of questions designed to learn the factors and processes which you considered to be most important in modeling radionuclide fate and transport in forest ecosystems	We are interested in learning how would you conceptually model radionuclide fate and transport in forest ecosystems. We will start by asking you to think about the processes which are important for radionuclide migration and which interact to determine the answer. We will do this in two ways: 1. By asking you to give us some sense of the relative importance of key processes 2. By showing you some results from modeling and asking you to provide an assessment of how well you believe they are likely to reflect reality You responses should be based upon the forest ecosystem scenario described in the preceding section.
2	A series of questions designed to elicit subjective cumulative distributions for value of parameters used in forest ecosystem modeling	We are interested in your judgment about radionuclide residence half-times in the forest compartment within the FORESTPATH Framework. The residence half-time is the time necessary for half of the radioactivity to be removed by environmental process from one compartment to another. A paper describing the FORESTPATH model is attached. You will be asked about radionuclide half-time in the FORESTPATH compartments. We will provide summary tables from a recent literature review (Linkov, 1995). We will elicit your answer in the form of a subjective probability distribution.
3	A series of questions designed to learn expert views about research needs and appropriate priorities in the field of forest radioecology	Suppose that a 15 year research program is to be mounted that will spend $500-thousand/per year for research on forest radioecology. Assume that existing sources of support for research in forest ecology will continue, but that any current support for forest radioecology research will be subsumed with this new budget. You have 100 chips to allocate, worth $5,000 per chip. As you go through the budget exercise, please explain to us your thinking that underline the various choices you are making.

3. Results

Results are presented for each individual expert in terms of a median value and a range factor. The range factor represents the ratio of the maximum to the minimum parameter value selected by each expert. High range factors indicate significant uncertainty, while lower range factors suggest greater confidence in the estimates made by the individual experts.

Figures 1-3 compile the elicitation results for interception fraction and radionuclide residence time in trees and in the organic layers. Experts associated very little uncertainty with the interception fraction (the median value is reported within 70-89% range while the maximal range factor is about 4). The residence time in trees exhibits much higher uncertainty (range factor of up to 10), while the radionuclide retention in the organic layer compartments is highly uncertain (maximal range factor is about 20). This reflects the state of current knowledge in the field: the radionuclide interception by trees has been studied quite extensively while the processes in the Organic Layer are highly uncertain (see Working Group Reports in this volume).

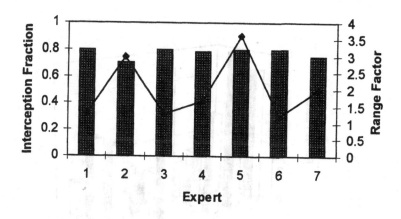

Fig. 1. Median value and Range factors (ratio of max/min) for interception fraction.

Table 3 presents the ratio of the maximum selected value for each parameter across all experts to the minimum selected value, representing an overall range factor for the minimum, mode, and maximum of the triangular distribution. For example, the ratio of the highest minimum value selected for interception fraction to the lowest minimum value selected is 2.9, as opposed to the 1.2 for the same ratio evaluated at the maximum. This indicates that the experts provided a greater range in minimum interception fraction values than for either the mode or the maximum. The bounds of the particular parameter can help to explain some difference. For example, the interception fraction can not be greater than 1, implying a bound on the ratio of one selected value to the next. By contrast, the values for long tree residence time, which do not have clear physical limits, show the same ratio of maximum to minimum across

414

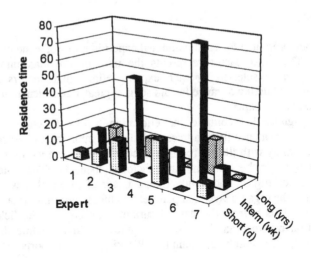

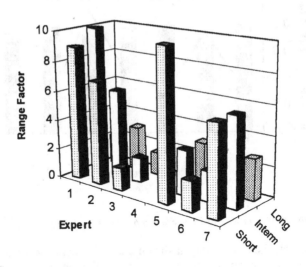

Fig. 2. Median value and Range factors (ratio of max/min) for radionuclide residence time in the Tree compartment (short, intermediate and long-term).

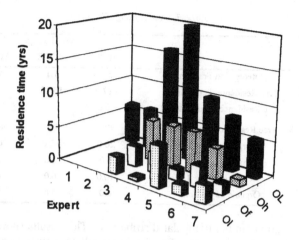

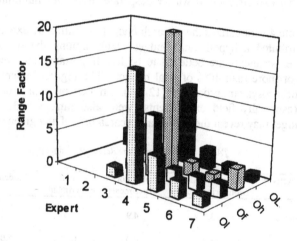

Fig. 3. Median value and Range factors (ratio of max/min) for radionuclide residence time in the Organic Layer compartments (Ol, Of, Oh and OL (total Organic Layer)).

TABLE 3. Overall Range Estimates

	Overall Range		
	MIN	MODE	MAX
Interception Fraction	2.9	1.1	1.2
Tree Residence Time (short)	112.0	133.3	200.0
Tree Residence Time (interm)	36.0	12.0	12.0
Tree Residence Time (long)	40.0	40.0	40.0
Organic Layer 1	20.0	12.0	6.3
Organic Layer 2	5.0	2.0	2.3
Organic Layer 3	10.0	6.0	6.7
Organic Layer Total	6.0	3.6	5.8

each of anchor points in the triangular distribution. This results shows consistency in uncertainty estimates by experts. For the intermediate residence time in trees, the uncertainty in the minimum value is much greater than in the maximum value, indicating that experts showed widely desperate results for minimum estimate versus maximum.

Table 4 summarizes the research budget allocation by experts. In general, the experts distributed a hypothetical budget evenly among the types of the activities (laboratory experiments are believed to be less important), but studies of ecosystem processes got more than 40% of total budget. The highest budget was allocated for modeling of ecosystem processes (12.5%). Studies of population and community dynamics (especially field experiments) were also ranked high. The allocation of research budget may reveal the particular research bias of any given expert.

TABLE 4. Allocation of the Research Budget (%)

Type of Activity Study Scale	Observational and Monitoring Studies	Laboratory experiments	Field Experiments	Modeling	Totals
Individual Species	3.5	4.9	4.5	2.4	15.3
Population & Community Dynamics	8.4	0.0	12.3	5.0	25.6
Ecosystem Processes	11.1	5.4	11.5	12.5	40.5
Regional and Larger Scale Studies	8.4	0.0	1.8	8.5	18.6
Totals	31.4	10.3	30.0	28.4	100.0

4. Conclusion

Probabilistic modeling requires specification of input parameter distributions. These distributions are difficult to obtain since experimental data are usually scarce or not available for many parameters. Expert judgment elicitation can be used to obtain parameter distributions. In this paper expert judgment were elicited for a specific forest ecosystem described in the IAEA BIOMASS program. Even though the uncertainty in many parameters are typically quite high, the expert estimates were quite consistent. The range factors reported in this study are much lower than those found in Food Chain Uncertainty Assessment [2]. This may be a reflection of the fact that the later study assessed generic model parameters, while the present study describes a specific forest ecosystem.

Formal expert elicitation studies are time and resource intensive. This study is appropriately viewed as a pilot exercise. The scientific significance of this exercise can be only viewed in conjunction with reports of the Working Groups and individual research papers published in this volume.

5. Acknowledgments

The authors would like to acknowledge enthusiasm and support of Drs. R. Avila, L. Moberga, M. Belli, A. Dvornik, A. Konoplev and A Bulgakov, A. Rantavaara, T. Riesen, A. Scheglov A. Klyashtorin, and Y. Thiry who kindly agreed to participate in this expert elicitation. The expert elicitation questionnaire was developed based on materials provided by Dr. Granger Morgan and the Department of Engineering and Public Policy, Carnegie Mellon University for elicitation of expert judgment on influence of climate change on forests. Dr K. Walker and G. Gray helped in the development of this questionnaire. This work was partially supported by a NATO Collaborative Research Grant.

6. References

1. BIOMASS Theme 3 Forest Working Group: WD1 (Preliminary Draft). IAEA Working Material Model-Model Intercomparison Study
2. Probabilistic Accident Consequence Uncertainty Analysis. Food Chain Uncertainty Assessment. 1997. NUREG/CR-6523. EUR 16771, Brussels, Luxemburg.

LIST OF PARTICIPANTS

Brian Amiro   Canadian Forest Service   e-mail: bamiro@nrcan.gc.ca
5320 122 Street, Edmonton, Alberta phone: 403-435-7217
T6H 3S5, CANADA   FAX: 403-435-7359

Gerassimos Arapis   Agricultural University of Athens   e-mail: mani@auadec.aua.ariadne-t.gr
Iera Odos 75   phone: 301-529-4465
Athens, GREECE, 11855   FAX: 301-529-4462

Andrey Arhipov   Chernobyl Scientific and Technical   e-mail: root@recover.kiev.ua
Centre ,UA 255620, Chernobyl,   phone: 380 44 93 52 201
UKRAINE   FAX: 380 44 93 52 064

Rodolfo Avila   Swedish Radiation Protection   e-mail: avila@ssi.se
Institute S 17116. Stockholm,   phone: 46 8 729 7 211
SWEDEN   FAX: 46 8 729 7 108

Maria Belli   National Environmental Protection   e-mail: belli@anpa.it
Agency (ANPA), Via Vitaliano   phone: 39 6 50072952-2924
Brancati, Roma, 00144, ITALY   FAX: 39 6 50072856-2916

Dmitriy Burmistrov   Ural Research Center for Radiation   e-mail:dima@urcrm.chel.ru
Medicine, Chelyabinsk, RUSSIA

Anatolyi Buldakov   Institute of Experimental   e-mail: typhoon@iem.obninsk.ru
meteorology Scientific, SPA   phone: 7 084 397 19 14
Typhoon, RU 249020, Obninsk,   FAX: 7 084 394 09 10
RUSSIA

Philippe Calmon   IPSN, CE Cadarache,   e-mail: philippe.calmon@ipsn.fr
IPSN/DPRE/SERE/LMODE –   phone: 33 4 42 25 71 35
13108 St Paul-lez-Durance cedex   FAX: 33 4 42 25 62 92
FRANCE

Vassili Davydchuk   Institute of Geography NASU, UA   e-mail:chronob@georg.freenet.kiev.ua
252034, Kiev, UKRAINE   phone: 380 44 224 14 51
FAX: 380 44 224 32 30

Janos Dombovari   Irrigation Research Institute 5540   FAX: 3666-311-178
SZARVAS Szabadsag u.2,
HUNGARY

Alexander Dvornik   Forest Institute of NAS Belarus,BY   e-mail: dvornik@fi.gomel.by
246654, Gomel, BELARUS   phone: 375 232 53 83 41
FAX: 375 232 53 53 89

419

Serguei Fesenko — Russian Institute of Agricultural Radiology, RU 249020, Obninsk, RUSSIA — e-mail: acr@storm.iasnet.com — phone: 7 084 39 248 02 — FAX: 7 095 255 22 25

Torbjorn Gafvert — Dept. of Radiation Physics, University Hospital, Lund, SWEDEN 222 85 — e-mail: Torbjorn.Gafvert@radfys.lu.se — phone: 046-173122 — FAX: 046-127249

Eugenyi Garger — Institute of Radioecology UAAS,UA 252033, Kiev, UKRAINE — e-mail: — phone: 38 044 227 43 60 — FAX: 38 044 220 93 46

Marvin Goldman — University of California, Dept. Surgical and Radiological Sciences University of California-Davis CA 95616, USA — e-mail: mgoldman@ucdavis.edu — phone: (530)752-1341 — FAX: (530)752-7107

François Goor — Catholic University of Louvain Place Croix du Sud 2 bte 15 Louvain-la-Neuve 1348 BELGIUM — e-mail: goor@ecop.ucl.ac.be — phone: 32 10 473818 — FAX: 32 10 473455

Alexander Greben'kov — Institute of Power Engineering 2 Zhodinskaya Str. Minsk 220141, BELARUS — e-mail: greb@sosny.bas-net.BY — phone: 375172639201 — FAX: 375172642315

Vladislav Golikov — Institute of Radiation Hygiene, 197101 St-Petersburg, Russia — e-mail: ira@protection.spb.sk — phone: 7812 233 48 43 — Fax: 7 812 232 04 54

Valeryi Giriy — Institute of Radioecology UAAS,UA 252033, Kiev, UKRAINE — e-mail: — phone: 38 044 227 43 60 — FAX: 38 044 220 93 46

Kathryn Higley — Oregon State University Dept. of Nuclear Engineering 100 Radiation Center, Corvallis, OR 97331-5902, USA — e-mail: higleyk@ccmail.orst.edu — phone: 541-737-0675 — FAX: 541-737-0480

Yuri Ivanov — Institute of Agricultural Radiology, UA 255205,Kiev, UKRAINE — e-mail: root@inrad.kiev.ua — phone: 380 44 266 75 31 — FAX: 380 44 266 71 75 — phone: 380 44 228 78 58

Nikolay Kaletnik — Ministry of Forest Industry of Ukraine, RY 252001, Kiev, UKRAINE

Mikhail Kanevsky — Institute of Nuclear Safety, RU 113191, Moscow, RUSSIA — e-mail: mkanev@ibrae.msk.su — phone: 7 095 334 08 02 — FAX: 7 095 230 20 65

Valerii Kashparov	Institute of Agricultural Radiology , UA 255205,Kiev, UKRAINE	e-mail: root@inrad.kiev.ua phone: 380 44 266 75 31 FAX: 380 44 266 71 75
Peter Kiefer	GSF Institute of Radiation Rrotection section Riskanalysis, Ingolstädter Landstr. 1, Neuherberg, D-85764, GERMANY	e-mail: pkiefer@gsf.de phone: 49 89 3187 2789 FAX: 49 89 3187 3363
Eckehard Klemt	Weingarten University, Postfach 1261, GERMANY, D-88241	e-mail: klemt@fbp.fh-weingarten.de phone: 751 501 578 FAX: 751 49240
Alexey Klyashtorin	Soil Science Departament Moscow State University,RU 119899,Moscow, RUSSIA	l:klia@kliash.soils.msu.su hone: 095 939 25 08 FAX: 095 939 09 89
Alexey Konoplev	Institute of Experimental meteorology Scientific,SPA Typhoon,RU 249020,Obninsk, RUSSIA	e-mail: konop@hotmail.com phone: 7 084 397 18 96 FAX: 7 084 394 09 10
Vladimir Krasnov	Polesskya Agro-forest-ameliorative scientific research station,UA 262004, Zhitomir, UKRAINE	e-mail: station@wrs.zhitomir.ua phone: 38 0412 26 86 38 FAX: 38 0412 26 86 28
Nikolay Kuchma	Chernobyl Scientific and Technical Centre ,UA 255620, Chernobyl, UKRAINE	e-mail: root@recover.kiev.ua phone: 380 44 93 52 201 FAX: 380 44 93 52 064
Yuri Kutlakhmedov	Institute of Cell Biology and Genetic Engineering NASU, UA 252022 , Kiev, UKRAINE	e-mail:e-mail@booz.freenet.kiev.ua phone: 380 44 263 61 67 FAX: 380 44 220 93 46
Vlada Kutlakhmedova-Vyshnyakova	Institute of Cell Biology and Genetic Engineering NASU, UA 252022 , Kiev, UKRAINE	e-mail:e-mail@booz.freenet.kiev.ua phone: 380 44 227 43 60 FAX: 380 44 220 93 46
Igor Linkov	Menzie-Cura and Associates, Inc, One Courthouse Ln. Suite 2, Chelmsford, MA 02138, USA	e-mail: linkov@huhepl.harvard.edu phone: 1- 978-453-4300 ext 15 FAX: 1 –978-453-7260
Sibylle Nalezinski	Federal Office for Radiation ProtectionDepartment of Radioecology (S 3.4), Oberschleissheim 85 762 GERMANY	e-mail: snalezinski@bfs.de phone: 0049-89-31603-286 FAX: 0049-89-31603-111

Francisco Neves — Évora University, Department of Mathematics, Rua das Alcaçarias, 9 Évora, PORTUGAL 7000 — e-mail: francisco_neves@hotmail.com phone: 06625063 FAX: 066743303

Alexandr Orlov — Polesskya Agro-forest-ameliorative scientific research station,UA 262004, Zhitomir, UKRAINE — e-mail: station@wrs.zhitomir.ua phone: 38 0412 26 86 38 FAX: 38 0412 26 86 28

Ahmet Ozcimen — Molecular Oncology Hardal Sk. Somtas B2 Blok, D:3 Istanbul, TURKEY — e-mail: ata@unimedya.net.tr phone: 90 212 6946754 FAX: 90 212 6946754

Alexandr V. Panfilov — Radioecology Division, Ministry of Forestry, RUSSIA

Gerhard Petersen — GKSS Research Centre, Institute of Hydrophysics, Max-Planck-StrasseGeesthacht D-21502, GERMANY — e-mail: petersen@gkss.de phone: 49-4152-871847 FAX: 49-4152-871888

Oleg Povetko — Oregon State University, Department of Nuclear Engineering 131 NW 4th Street #284 Corvallis , OR 97330, USA: — e-mail: povetkoo@ucs.orst.edu phone: (541) 752-0462 FAX: (541) 737-7070

Aino Rantavaara — STUK-Radiation and Nuclear Safety Authority, Laippatie 4 (PO Box 14) FIN-00880, FINLAND — e-mail: aino.rantavaara@stuk.fi phone: 358-9-75988436 FAX: 358-9-75988498

Barbara Rafferty — Radiological Protection Inst Of Ireland 3 Clonskeagh Square Dublin 14 IRELAND

Thomas Riesen — Paul Scherrer Institute, OSUA / 105 Villigen 5232, SWITZERLAND — e-mail: thomas.riesen@psi.ch phone: 41 56 310-2341 FAX: 41 56 310-2309

Vitaliy Rimkevich — Institute of Power Engineering 2 Zhodinskaya Str. Minsk 220141, BELARUS — e-mail: greb@sosny.bas-net.BY phone: 375172639201 FAX: 375172642315

William R. Schell — Dept. of Environmental and Occupational Health, University of Pittsburgh, Pittsburgh, PA 15261, USA — e-mail: wschell@vms.cis.pitt.edu phone: 412-687-3105 FAX: 412-624-3040

Aleksey Shcheglov — Soil Science Departament Moscow State University,RU 119899,Moscow, RUSSIA — l:klia@kliash.soils.msu.su : 7 095 93922 11 FAX: 7 095 939 09 89

Azat M. Shagiakhmetov — Intitute of Problem of Apply Ecology and Natural Resouces Use, 10/1, 8-March Str., Ufa, RUSSIA — e-mail: : ipendy@diaspro.com phone:28-87-75FAX: 28-87-90

Vladimir Shutov — Institute of Radiation Hygiene, 197101, St. Petersburg, RUSSIA — e-mail: ira@protection.spb.su phone: 7 812 232 73 46 FAX: 7 812 232 04 54

Boris Sorochinsky — Institute of Cell Biology and genetic engineering of NASU,UA 252022, Kiev, UKRAINE — e-mail: bvs@phyto.kiev.ua phone: 38 044 266 71 04 FAX: 38 044 252 17 86

Eiliv Steinnes — Norwegian University of Science and Technology, Department of ChemistryN-7034 Trondheim, NORWAY — e-mail: eilivs@alfa.avh.unit.no phone: 47-73596237 FAX: 47-73596940

Eileen Seymour — Department of Experimental Physics, University College Dublin, Belfield, Dublin 4, IRELAND — e-mail: eseymour@ollamh.ucd.ie phone: 353-1-7062220 FAX: 353-1-2837275

Yves Thiry — Belgian Nuclear Research Centre SCK-CEN, Boeretang, 200 Mol 2400, BELGIUM — e-mail: ythiry@sckcen.be phone: 32 14 33 52 82 FAX: 32 14 32 03 72

Elena Uspenskaya — All-Russian Research Institute Of Nature Protection,Ru 113628, Moscow, RUSSIA — e-mail: rinpo@glasnet.ru phone: 7 095 423 13 01 FAX: 7 095 423 23 22

Hildegarde Vandenhove — Belgian Nuclear Research Centre SCK-CEN, Boeretang 200 Mol 2400, BELGIUM — e-mail: hvandenh@sckcen.be phone: 32-14-335280 FAX:32-14-320372

Ansie Venter — QuantiSci; 12 Acacia House, Henley-on-Thames, Oxfordshire RG9 1AT, UK — e-mail: aventer@quantisci.co.uk phone: 44 1491 410 474 FAX: 44 1491 576 916

Gabriele Voigt — GSf-Institut f]r Strahlenschutz Risikoanalyse, Ingolstddter Lands. 1 D-85764 Neuherberg, GERMANY — phone: 49 89 3187 4005 FAX: 49 89 3187 3363

Richard Wilson — Department of Physics, Harvard UniversityCambridge, MA 02138, USA — e-mail: wilson@huhepl.harvard.edu phone: 1- 617 495 3387 FAX: 1 - 617 495 0416

Boris Yakushev   Institute of Experimental Botany   FAX: 375 17 68 48 53
Bel. Academy of Sciences
BELARUS

Ivan Yaskovets   Institute of Radioecology UAAS,UA phone: 38 044 227 43 60
252033, Kiev, UKRAINE   FAX: 38 044 220 93 46

Satoshi Yoshida National Institute of Radiological e-mail: s_yoshid@nirs.go.jp
Sciences, 3609 Isozaki, Hitachinaka- phone: 81-29-265-7139
shi 311-1202, JAPAN   FAX: 81-29-265-9883

Vladimir Zaitov Institute of Radioecology UAAS,UA phone: 38 044 227 43 60
252033, Kiev, UKRAINE   FAX: 38 044 220 93 46

Tatiana   Forest Institute of NAS Belarus,BY e-mail: dvornik@fi.gomel.by
Zhuchenko   246654, Gomel, BELARUS   phone: 375 232 53 83 41
FAX: 375 232 53 53 89

Jeran Zvonka   Josef Stefan Institute   e-mail: zvonkajeran@js-si
Lsubljana, Jahova 39,   phone: 38 661 1885281
SLOVENIA

AUTHOR INDEX

SUBJECT INDEX